CLIMATIC DESIGN

CLIMATIC

ENERGY-EFFICIENT BUILDING

DESIGN

Donald Watson, FAIA, and Kenneth Labs

PRINCIPLES AND PRACTICES

McGRAW-HILL BOOK COMPANY *New York St. Louis
San Francisco Auckland Bogotá Hamburg Johannesburg London
Madrid Mexico Montreal New Delhi Panama Paris
São Paulo Singapore Sydney Tokyo Toronto*

Library of Congress Cataloging in Publication Data

Watson, Donald, date.
  Climatic design.

  Bibliography: p.
  Includes index.
  1. Building — Energy conservation. 2. Solar energy —
Passive systems. 3. Architecture and climate.
I. Labs, Kenneth. II. Title.
TJ163.5.B84W38 1983   697.9   82-21681
ISBN 0-07-068478-2

1234567890 HAL/HAL 89876543

ISBN 0-07-068478-2

Graphic Credits

photos by the architects, except as noted below:

Robert Perron: cover, pages 1, 8, 9, 12 and 13
Hedrich-Blessing: page 6
Bob Super: page 17
Karl H. Riek: page 18, left
Susan Benson: page 18, right
Cervin Robinson: page 251

cover design: Abner Graboff
graphics and layout: Sandra Olenik and Susan Simler

Contents

Acknowledgements

In preparing this book, the authors reviewed the more than fifty-year record of research in building bioclimatology and architecture. Among the contributions to which any contemporary research in this field is indebted is the pioneering work of Victor and Aladar Olgyay, who gave to the field the term "bioclimatic design", James Marston Fitch, the prime-mover of widespread architectural interest in climatic design in the late 1940s, and Baruch Givoni whose building bioclimatic chart is the basis of the statistical analysis of climate data in Part III of this book and whose contributions will be evident to the reader throughout the book. Among our colleagues and mentors, the authors are also indebted to Arthur Bowen, Jeff Cook, Ralph Johnson, Murray Milne, Bill Shurcliff, and John Yellott, alongside many others whose comments, critiques and encouragement cannot be properly acknowledged except in the way they helped improve the work. Computer analysis for the Part III weather data was executed by Robert Frew, Keith Harrington, and Carl Williams of ARGA Associates, New Haven. Both authors have been assisted in carrying out the research in this book by support from the Rockefeller Foundation, the U.S. Department of Housing and Urban Development, the National Association of Home Builders Research Foundation, the National Endowment for the Arts, and the U.S. Department of Energy's Solar Cities and Towns Program.

CLIMATIC DESIGN

"Dog-trot" house plan, indigenous to the U.S. southeast, provided natural cooling by cross-ventilation, shaded porch and outdoor hallway.

"San Francisco", a restored plantation manor near New Orleans, LA. Its natural cooling elements included a ventilated roof, cross-ventilation through a central hallway, and shaded porticos.

Introduction

Introduction

Present throughout the entire history of building and architecture is the response the designer makes to climate. Even in so-called "primitive" architecture, climatic design finds subtle and sophisticated expression, whether in the wind-protected setting and south orientation of a Swiss alpine farmhouse or in the plan of the traditional courtyard house, proportioned to trap and "pool" cool night air in hot-arid locales. Such "climatic design" intelligence is evidenced in the colonial architecture of early United States settlers, from the Salt Box of the Massachusetts colonies to the raised veranda designs of Southern plantations. In these indigenous examples and the architectural styles that derived from them, knowledge of climate served as the basis of human livability and, ultimately, the aesthetic expression of its design.

This book describes the basis of climatic design—also referred to as building bioclimatology—its scientific principles and its practical application. This is not the first text in the topic. Part IV provides a complete bibliography of references in climatic design upon which this text builds. This book brings together comprehensive data about all climatic impacts that ought to be understood in designing a building to suit its local climate. Recent emphasis in energy-efficient building design has been upon solar heating, and rightly so, in view of the cost of heating fuel. But to design a building only from the standpoint of winter heating is incomplete, and possibly counter-productive, if the sun is not excluded from the building by proper shading during summer months. Many climatic design techniques, such as using insulation or earth-sheltering can reduce both heating and cooling energy costs. Natural ventilation can provide comfort in summer, when used with the other cooling techniques . . . costly air conditioning is often required in a building simply because its window location or lack of shading makes it a solar oven in summer. Even during the months when it is comfortable outside, an improperly designed building can be uncomfortable due to the designer's lack of understanding of climatic design principles and practices.

This book does not provide a simple solution or formula for climatic building design. This is always a matter of choice, not only in terms of the art of building design, but also in terms of bioclimatology: in all but the most severe climates, there are different options to achieve mutually satisfactory results in human comfort and energy efficiency. The information in this book does not replace the designer's judgment to be selective and to be artful. It does provide the designer with factural data about climate (Part III shows how to accurately document climatic design criteria). It includes an extensive illustrated catalog of climate design techniques (these comprise Part II). For the student of building bioclimatology and architecture, it provides a technical discussion of the theory and principles of climate design (in Part I).

This combination of theoretical principles and practical applications may appear overly exhaustive. It is doubtful that a reader needs to master all parts to become a knowledgeable practitioner of climatic design. Instead, readers can select from this book information needed for the particular climatic locales where they are designing and building. If the scope of the book appears ambitious, it is in response to a manifest desire expressed by designers and builders for a mastery of all aspects of climatic design.

The topic of climatic design is also of general interest beyond the building profession. Central to understanding and appreciating architecture of any epoch and locale is how a building design fits its particular climate. How a building takes advantage of sun, breeze, vegetation, and creates a unique microclimate is one of the subtle but enduring measures of the designer's skill.

Climatic design is the one approach by which to reduce the energy cost of a building comprehensively: the building design is the first "line of defense" against the stress of outside climate. In all climates, buildings built according to climatic design principles reduce the need for mechanical heating and cooling by using "natural energy" available from the climate at the building site. The long-term energy cost savings that result make climatic

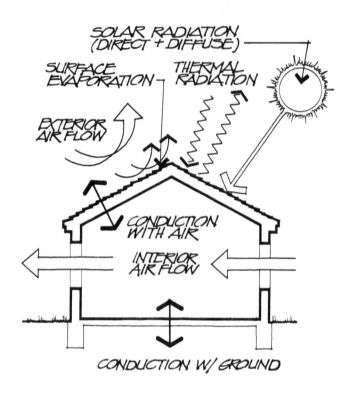

Fig. A. Paths of Heating Energy Exchange at the Building Microclimate

design techniques the best financial investment for any building owner. Many are "no cost" techniques, requiring only climatic design knowledge. Other techniques are easily incorporated into conventional construction, as illustrated in Part II.

Designing buildings to take advantage of natural energy flows has other benefits: buildings are made more comfortable by climatic design practice. Rather than imposing "brute force" tasks upon mechanical heating and cooling systems, the building itself can do the work, quietly, without fans and other machines, and without adding to the peak demands of central power systems. And because we build buildings not only to survive the rigors of climate but to grace our lives and to give the places we live an expression of health and beauty, it is by means of climatic design techniques that a building responds to the natural environment, whether through a window, a skylight, a greenhouse, a covered porch or a protected courtyard. These establish the settings, the places of balance, between ourselves and the outside world.

The entire topic of climatic design can be understood quite simply. The physical comfort we feel in building is the result of the heating energy balance between ourselves and the surrounding space. Heating energy is exchanged between ourselves and the physical surfaces and material of our buildings in seemingly complex, interactive ways, but these are described by four principles of the physics of heat flow: conduction, convection, radiation and evaporation. Each of these and the physical basis of climatic design principles are detailed in Part I. As an introduction, it can be seen that sun, wind, precipitation and the resulting temperatures in the air and stored in the ground create sources of natural heating (and cooling) energy (*Fig. A*).

In winter (or the "underheated season" when heating energy is desired), the objectives of climatic design are to *resist loss* of heat from the building interior and to *promote gain* of solar heat, such as directly through south-facing windows. In summer (or the "overheated season" when cooling is needed), these objectives are reversed, to *resist gain* of solar heat, such as through sun-shading, and to *promote loss* of heat from the building interior. To achieve these objectives, there are nine practical climatic design principles (*Fig. B*). Once a designer understands the local climate (from analysis of weather data tabulated in Part III), the set of climatic design principles appropriate to that climate can be elaborated and climatic design choices compared.

The term "climatic design practice" used in this book refers to specific building techniques in design or construction which serve to reduce heating or cooling costs and which use

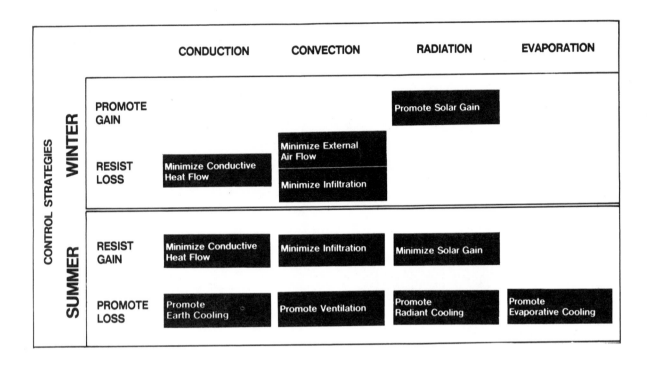

Fig. B. Summary of the Principles and Strategies of Climatic Design

natural energy flows to create human comfort conditions in buildings. Familiar examples of such practices include insulation, windows for winter solar gain, ventilating roof monitors and so on. Fifty distinct practices or techniques are catalogued in Part II. They are numbered from 1 to 50 for easy reference, beginning with those that should be considered in site planning, then in building massing and planning, and finally in the detailed design of the building envelope and openings. In other words, the order follows the logical sequence of the design decision process. Each technique is also tabulated in terms of the climatic design principle that it follows. The Part II catalog can thus be read in several ways, *consecutively* as a way of reviewing each practical technique in the order it might be considered in the design sequence or *selectively* to review those techniques that fulfill a particular principle. In the latter case, all of the practical ways to promote solar gain, for example, can be reviewed as a set of choices from site planning to building opening details.

There is a third way to group the climatic design practices. Each of the 50 techniques is a way of fulfilling one of the following "generic" concepts of climatic design:

1. Wind breaks — winter
2. Plants and water — summer
3. Indoor/outdoor rooms — winter and summer
4. Earth sheltering — winter and summer
5. Solar walls and windows — winter
6. Thermal envelope — winter
7. Sun shading — summer
8. Natural ventilation — summer

The index at the beginning of Part II tabulates the 50 practices as sets of techniques that relate uniquely to these generic concepts through which climatic design can be undertaken. What this indicates is that the topic of climatic design can be understood from many perspectives, from that of theoretical design principles described in Part I and from design decisions and practices catalogued in Part II as techniques and design concepts. All of these are ways of structuring our knowledge and understanding of climatic design, each with its own advantages to the learner. In the end, each of these will contribute, if only in part, in making knowledge of building climatology useful to designers and builders so that they can embark upon the art of climatic design with understanding and confidence.

✱ All plans in the portfolio section are reproduced at the same scale, 1' = 1/16". North is at the top of the page unless otherwise noted.

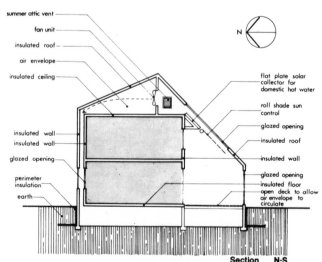

summer attic vent
fan unit
insulated roof
air envelope
insulated ceiling

N

flat plate solar collector for domestic hot water
roll shade sun control
glazed opening
insulated roof
insulated wall
glazed opening
insulated floor
open deck to allow air envelope to circulate

insulated wall
insulated wall
glazed opening
perimeter insulation
earth

Section   N-S

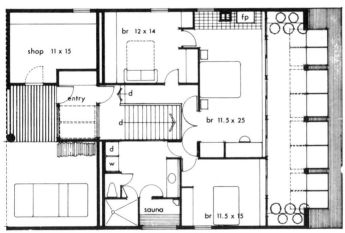

2

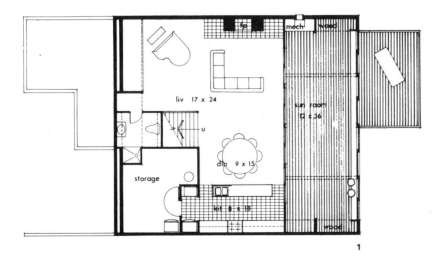

1

ONTARIO, CANADA

House design is based upon the double-shell or "envelope" principles, whereby solar gain from sunspace is circulated around the entire perimeter. Additional northern climatic design features include high levels of insulation; compact and wind-protected massing; predominant openings facing south.

John Hix, Architect
Toronto, Canada

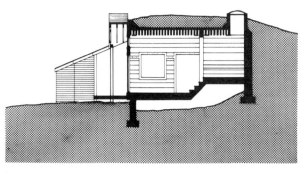

NORTHERN VERMONT

A simple, compact earth-sheltered house plan heated by wood stoves and solar gain from south-facing windows.

Mark Simon, Architect
Moore Grover Harper, PC
Essex, CT

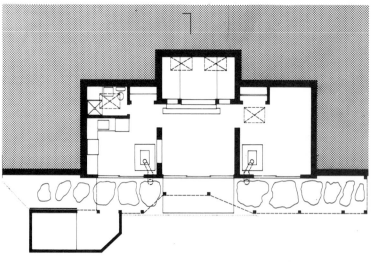

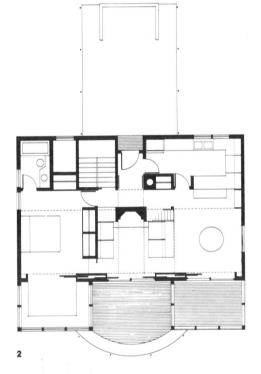

2

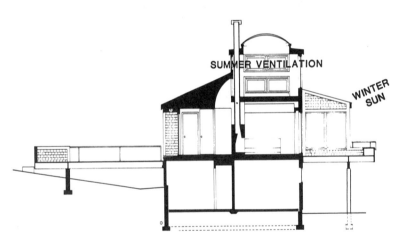

SUMMER VENTILATION

WINTER SUN

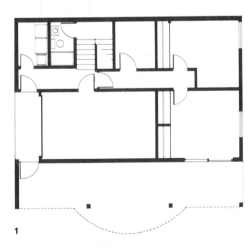

1

BLOCK ISLAND, RI

Utilizing design style traditional to Block Island, climatic design features include south-facing windows and sunroom for winter heating and shaded deck and screen porch for summer. Upper level belvedere is also used to trap winter solar heat and to create vertical-flow ventilation in summer as well as to provide 360° views of the surrounding terrain.

Donald Watson, FAIA
Branford, CT

9

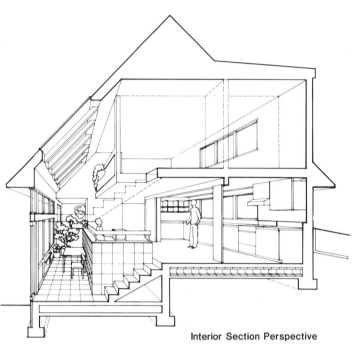

Interior Section Perspective

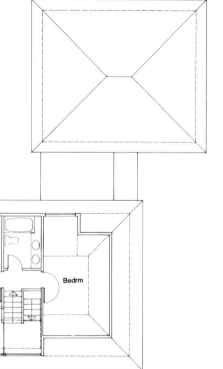

Bedrm   Bedrm

2

SOUTHWESTERN VERMONT

A builders model home that combines south-facing skylight and windows for winter solar gain; thermal storage via interior masonry wall and fireplace, plus underfloor concrete-block "plenum"; high levels of insulation, insulating night-shades, and earth-berming on north and west sides. Shading fabric is installed over skylight in summer.

Donald Watson, FAIA
Branford, CT

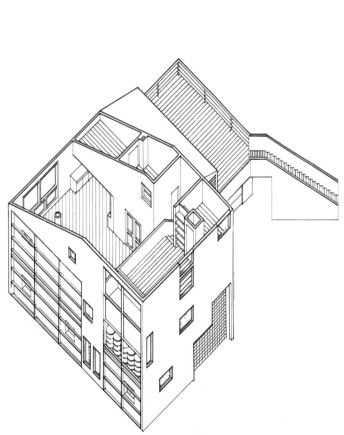

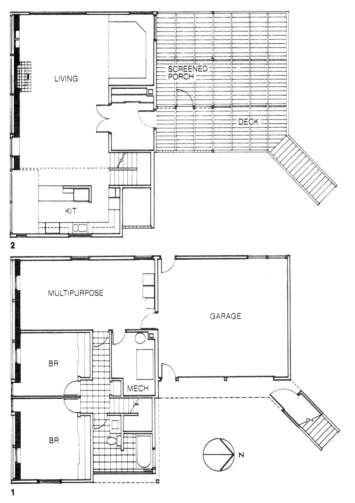

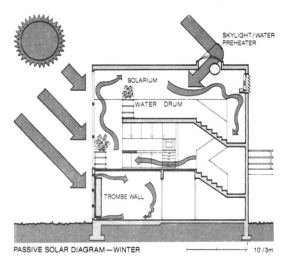

PASSIVE SOLAR DIAGRAM—WINTER   10'/3m

NEW JERSEY

A builders model home that integrates a number of passive solar features, including south-facing windows, and Trombe masonry wall and water barrels for heat storage. A screened porch is situated over the garage on the northwest side for summer use.

Kelbaugh & Lee Architects
Princeton, NJ

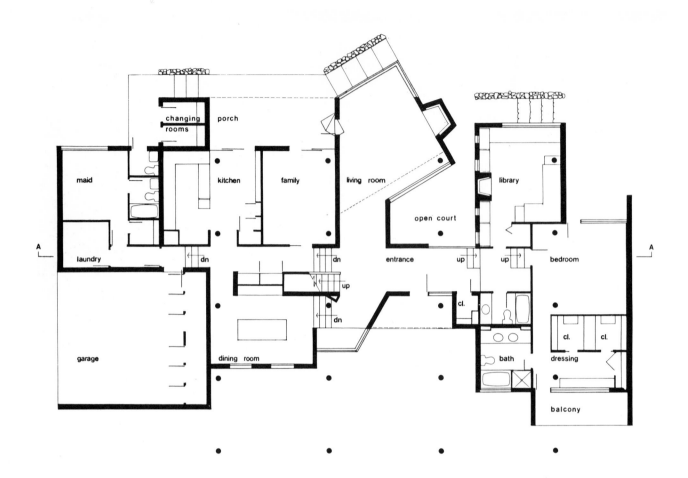

maid

changing rooms

porch

kitchen

family

living room

library

open court

A

laundry

dn

dn

entrance

up

up

bedroom

up

cl.

dn

cl.

cl.

garage

dining room

bath

dressing

balcony

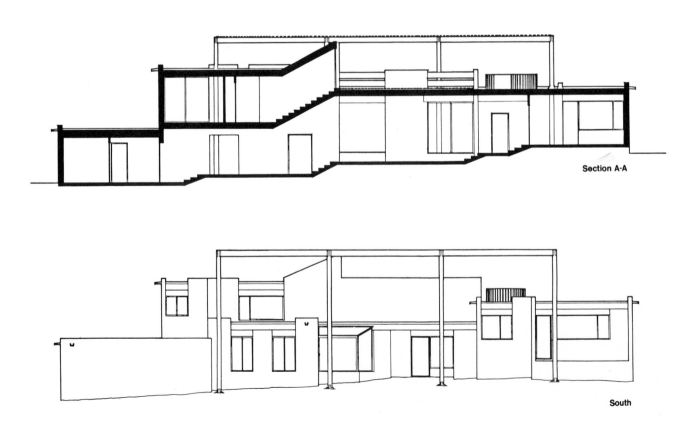

Section A-A

South

ARIZONA

A house design based upon traditional climatic design responses of the U.S. southwest, including heavy masonry construction for thermal "flywheel" effect; courtyard and verandas shaded by trellis or "ramada"; orientation to diurnal breezes that flow up and down nearby mountain slopes.

Judith Chafee, Architect
Tucson, AZ

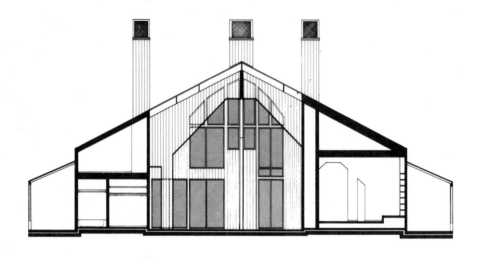

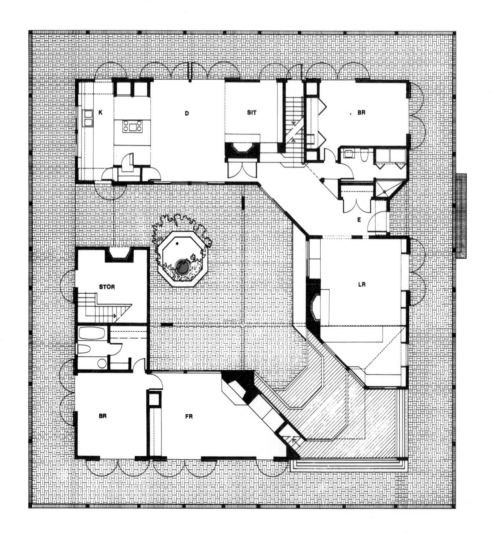

16

CENTRAL CALIFORNIA

House is designed around shaded courtyard-atrium and also features verandas and louvered window-walls for shading and cross-ventilation.

William Turnbull, Architect
MLTW/Turnbull Associates
San Francisco, CA

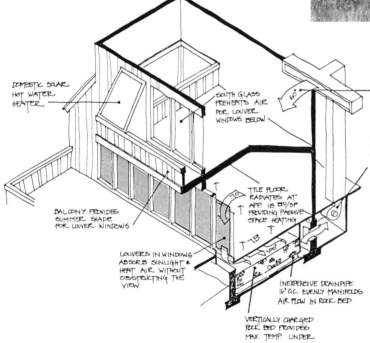

Labels on cutaway diagram:

DOMESTIC SOLAR HOT WATER HEATER

SOUTH GLASS PREHEATS AIR FOR LOUVER WINDOWS BELOW

COOL ROCK BED EXHAUST PREVENTS OVERHEATING IN LOFT

SMALL FAN DRAWS AIR THROUGH WINDOWS & CHARGES ROCK BED ON SUNNY DAYS

TILE FLOOR RADIATES AT APP. 18 BTU/SF PROVIDING PASSIVE SPACE HEATING

BALCONY PROVIDES SUMMER SHADE FOR LOUVER WINDOWS

LOUVERS IN WINDOWS ABSORB SUNLIGHT & HEAT AIR WITHOUT OBSTRUCTING THE VIEW

INEXPENSIVE DRAINPIPE 16" O.C. EVENLY MANIFOLDS AIR FLOW IN ROCK BED

VERTICALLY CHARGED ROCK BED PROVIDES MAX. TEMP. UNDER FULL RADIANT FLOOR AREA

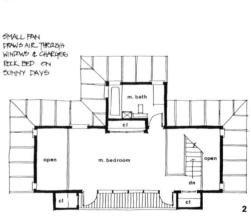

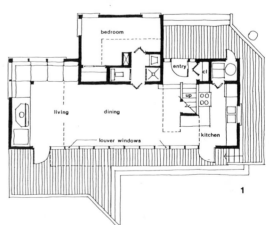

NORTHERN CALIFORNIA

Climatic design features include shaded patio and balconies for summer and "passive-active" winter solar heating system: heat gain from louvered windows is ducted to underfloor rock-bed plenum, which radiates collected heat to floor above.

Peter Calthorpe, Architect
Van der Ryn, Calthorpe & Partners
Sausalito, CA

Housing at Sea Ranch, CA, designed by Esherick, Momsey, Dodge & Davis. Ecological site planning and climatic design elements include sunspace, sod roofs, and wind-diverting fence design to protect garden-courtyards.

Part I: Principles

1. Heat and People

To design energy-efficient buildings, it is essential to understand the principles of heat exchange. Energy flows in and out of buildings – through walls, roofs, windows, floors, and other routes of heat exchange. Why this occurs is simple enough for our purposes: lost heat energy seeks its own "lowest level", that is, it always migrates from spaces or materials of higher temperatures to those of lower temperatures. The "how" is not so simply stated; the purpose of this book is to identify the hows – so that designers can build in controls on heat flow in order to minimize the amount of purchased energy required to maintain comfortable conditions in houses and similar small buildings. In Part I, the principles of climate-based energy flows in buildings are described in detail. It provides a summary of the physics of building climatology, which is the basis of the practices and designs described in Part II.

We think of energy through our experience of movement, sound, light, and the feel of heat, but it is also stored in the materials of our surroundings. In building engineering in the United States, heat energy is measured in British thermal units or Btu. Btu is defined in terms of heat storage: one Btu is the amount of heat which is absorbed by a pound of water (1.92 cups) in raising its temperature by one degree on the Fahrenheit scale. Since it is defined in terms of other units (mass and temperature), the Btu is called a *derived* unit of measure. An ordinary match gives off about 1 Btu of heat energy. A simple calculation shows that a little less than 170 Btu are required to bring a pint of tap water at 50F to the boiling point, and 1010 Btu are required to boil it away. In smaller units, 1 Btu will raise 1 tablespoon (Tbl) of water slightly less than 31F, so 5 Btu will take 1 Tbl of water at 58F to the boiling point, and another 32 Btu will boil it away.

Other units of heat energy can be converted into Btu and into one another. The calorie (cal), or more properly, the gram-calorie (or "small" calorie), was once used by chemists and biologists, but it has given way to the joule in the *Système International d'Unités* (SI). Like other units of heat measure, the calorie is "derived", defined as the amount of heat necessary to raise one gram of water (1 milliliter, or about 1/5 teaspoon) by one degree on the Celsius scale. The calorie is thus a much smaller unit of heat than the Btu; it takes 252 calories to make 1 Btu.

The kilogram-calorie (kcal or Cal, or "large" calorie) is the unit of food energy used by dieticians. However, unless a discussion is about nutrition or weight loss, one might assume that when a calorie is mentioned, it is the gram-calorie that is meant. In subsequent discussions here, food calories will be represented by the scientific notation kcal. One kcal describes the amount of heat required to raise 1 kilogram of water (1 liter) by 1°C; this is about four times as much energy as a Btu (see *Table 1d* for conversion factors).

Another useful unit of energy is the watt-hour (Wh), and its 1,000 watt-hour equivalent, the *kilowatt-hour* (kWh). In the English system, watt-hours are used to measure electrical energy consumption. Because electrical energy eventually always degrades into heat energy, the watt-hour conversion is essential in determining internal heat gains generated in buildings by lighting and appliances. One watt-hour equals 3.41 Btu.

Table 1a lists the energy content of fuels used in different kinds of heating systems and of other familiar substances. These are expressed in unusually small quantities (one tablespoon) to aid in visualizing amounts of energy. The body, for example, is a kind of furnace which is fueled by food, and the energy content of things we eat can be expressed in the same units as heating fuels; some are given in *Table 1b*.

Power and Heat Flux.

The expenditure of energy over time – or the energy use or production rate – is described as *power*. In the English system, heat energy consumption or production has the dimensions of Btu per hour, or Btuh. Electrical power is expressed in watts (W) or kilowatts (kW). The relationship between energy and power is simply defined by time: energy = power × time; or, power = energy/time. Power conversion factors are given in *Table 1d*.

Furnaces, hot water heaters, air conditioners, and appliances are rated according to their power demands and power output. The body can also be rated according to its heat output, and its rate of heat production depends on its activity level. Bodily heat production is usually expressed by physiologists as heat produced in an hour through a unit area of skin surface, rather than for the body as a whole. The rate of energy transfer across an area of surface is described as *heat flux.* In building engineering, heat flux through the building skin has dimensions of Btuh/ft^2 or W/m^2; physiologists, however, measure bodily heat flux in kcal/m^2 (hr), or in Met units, where 1 Met unit equals 50 kcal/m^2 (hr), the approximate rate of heat production of a sedentary person. Total bodily heat production depends on body surface area. Big people produce more heat than little people. An estimate of body surface area can be made by the formula,

$$A = 0.202(wt^{0.425})(ht^{0.725})$$

where area (A) is in m^2, weight (wt) in kilograms, and height (ht) in meters. An average size man (145 lb., 5'10'') has a body surface area of about 1.82 m^2 or 20 ft^2.

A short selection of human heat production rates is given in *Table 1c,* both in terms of heat flux (Met units) and total heat power output for an average-size man. Our sleeping person produces about the same amount of heat as a table lamp. One-half of an electric blanket for a full-size bed draws about 90W, and produces a heat flux of about 20 Btuh/ft^2, both of which closely approximate the output of the body.

In order to heat oneself throughout a day, a sedentry adult of average size requires about 2200 kcal or 8700 Btu, the energy content of 14 Hershey bars, 4 cans of baked beans, 1 cup of No. 2 fuel oil, or 2560 Wh of electricity. It is not far-fetched to think of heating the body with electricity. Professor R. V. Pound, a physicist at Harvard University, has speculated on the energy-conserving potential of warming the body with electrically-produced microwave energy (with a wave-length of about 1 cm) in lieu of space heating. Prof. Pound [1980] writes,

> Human thermal comfort is believed to require maintenance of skin temperature at about 34°C, which most persons achieve in a sedentary state in surroundings at about 22°C with normal indoor clothing. The 60W generated in the metabolic processes evidently supports a 12°C rise of the skin above the surroundings in dissipating the metabolic heat, through all the mechanisms of bodily heat loss. If 60 additional watts of energy were externally supplied to the surface of the body, it seems reasonable to suppose that the desired skin temperature could be maintained in surroundings 24°C below skin temperature, or 10°C. In this way a 10°C room temperature should be rendered as comfortable as one at 22°C without radiation.

Prof. Pound speculates that the body could be kept comfortable at an air temperature of 50F if the room were provided with a microwave generator of 60W output, and the room surfaces made reflective of microwave energy so that the body would be the primary sink for the 60W of power. A similar concept, utilizing infrared instead of microwave radiant energy, was investigated in the early 1950s by Dr. Clarence A. Mills, Professor of Experimental Medicine at the University of Cincinnati. Mills had two difficulties with his approach, one being deficiencies with the mechanical (radiating) devices he used at the time. The other was the problem of finding aesthetically acceptable infrared-reflective interior surface finish materials. Although aluminum foil works well and proved the worth of the idea in the laboratory, wallpapering every room in the house with Reynolds Wrap does not accord with most popular notions of the good life, and so the concept of reflective radiant conditioning has largely been forgotten. More recently, David Lee Smith [1979], Professor of Architecture at the University of Cincinnati, and anthropologist Edward T. Hall [1979] have both urged reinvestigation of the concept. Practical and experimental (for the do-it-yourselfer) applications might be found in bath and dressing rooms, where quick heating is desirable, and where aluminum-faced walls might be accepted.

It is useful to imagine building heat fluxes in terms of units of energy that can be easily grasped. With a 50F temperature difference between indoors and out, for example, a north-facing 4 by 6 foot double-pane window loses 720 Btu every hour; one can visualize this as 1⅓ tablespoons of heating oil lost through the window every hour. On the other hand, if it were a south-facing wall in Topeka, over 3 cups' worth of oil could be gained from the same area (24 ft^2) every day in January. In a similar manner, the solar equivalent of one barrel of oil splashes over every square meter of ground every year in Boise, Idaho. And in a less fantastic sense, one can think of the geothermal heat flux (the heat produced at the center of the earth), which by the time it reaches a cornfield in Illinois, has been diffused over such a large area that its meager 160 Btu/(ft^2) year is insufficient even to melt a layer of ice ¼ inch thick.

While some of these energy conversions are facetious, we will see in sections that follow how energy conversions in the natural world — like the amount of heat absorbed in the evaporation of an inch of water — are responsible for the temperature balance of the environment; and how, given an understanding of the processes and magnitude of such conversions, we can use them to control the interior environment of our buildings.

Fuel	lb./ft^3	Btu/lb.	Btu/gal.	Btu/0.5 oz^a	Btu/Tbl.
Toluene	55.31	18,501	136,794	578	534
Benzene	55.18	18,184	134,134	568	524
Isopropanol[b]	49	14,220	93,146	444	364
Ethanol[c]	49.48	12,962	85,732	405	335
Methanol[d]	49.60	9,755	64,681	305	253
Acetone	49	13,212	86,536	413	338
Gasoline	46.08	20,260	124,800	633	487
Naphtha	48.62	20,020	130,100	626	508
Kerosene	51.46	19,750	135,800	617	530
No. 1 Fuel Oil[e]	50.87	19,800	134,647	619	526
No. 2 Fuel Oil[f]	53.86	19,496	140,371	609	548
Carbon		14,093		440	
Charcoal (Air Dry)		12,850		393	
Anthracite Coal[g]		12,700		397	
Bituminous Coal[g]		11,000—14,000		344—438	
Subbitum Coal[g]		8,500— 9,000		266—281	
Wood, Oven Dry		8,300— 9,150		260—286	
Coffee Grounds[h]		10,058		314	
Newspaper[h]		7,883		246	
Wheat[h]		7,532		235	
Paraffin	56	18,612		582	
Napthalene[i]	71.48	17,303		541	
Butane[j]		21,180	3200[k]	662	
Propane[j]		21,560	2500[k]	674	
Acetylene[j]		21,502	1574[k]	672	
Methane		23,875	1070[k]	746	
Natural Gas[j]			1025[k]		

a 0.5 ounce avdp., approximately the weight of 1 Tbl. water, or 6 dimes.
b Isopropyl, or rubbing alcohol
c Ethyl, or grain alcohol
d Methyl, or wood alcohol
e Used in vaporizing pot burners
f Domestic heating oil
g Typical "as received" values
h Dry basis
i Moth balls
j Typical commercial values
k Btu/ft^3 gas

Table 1b — Energy Value of Selected Foods, 1 Tablespoon (½ fl. oz.) Serving

Food	kcal	Wh	Btu
Wesson Oil	120	140	476
Skippy Peanut Butter	101	117	401
Dairy Butter	100	116	397
Honey	61	71	242
Grandma's Molasses	60	70	238
Strawberry Jam	53	62	210
Cream Cheese	50	60	206
Sugar	46	53	183
Caviar	42	49	167
Hershey's Syrup	37	43	147
Rum, 80 Proof[a]	33	38	131
Guacamole Dip	33	38	131
Graham Cracker, 2½" square	30	35	119
Sour Cream	28	33	111
Lean Ground Beef	25	29	100
Campbell's Pork and Beans	19	22	74
Cottage Cheese (4%)	15	17	60
Stewed Rhubarb	14	16	56
Whole Milk (3.5%)	9	11	37
Mashed Potatoes	8.5	10	34
Coca-Cola	6	7	24
Miller Lite	4	5	16

a Applies to all 80 proof liquors; see Ethanol, *Table 1a.*

Table 1c — Human Energy Consumption — Production Rates

Activity	Met Units[a]	Energy rate for average size man[b]		
		kcal/hr.	Btuh	Watts
Sleeping	0.7	64	253	74
Reclining	0.8	73	289	85
Sitting, sedentary	1.0	91	361	106
Drafting / Standing, relaxed	1.2	109	433	127
Typing / Eating	1.3	118	469	138
Walking, 2 mph	2.0	182	722	212
3 mph	2.6	237	939	276
4 mph	3.8	346	1372	403
Sawing by hand	4.4	400	1588	466

a One Met Unit = 50 kcal/hr(m^2) = 18.4 Btuh/ft^2 = 58.2 W/m^2
b 145 lbs., 5'-10"

Energy

Btu	X	252	=	cal	X	0.003968			
		0.252	=	kcal	X	3.968	=	Btu	
		0.2929	=	Wh	X	3.414			
		1.0551	=	KJ[a]	X	0.9478			

Power, Thermal Transmission (Energy/Time)

Btuh	X	0.2929	=	W	X	3.414			
		0.252	=	kcal/hr	X	3.968	=	Btuh	
		0.07	=	cal/sec	X	14.286			
		8.33×10^{-5}	=	ton[b]	X	12,000			
ton[b]	X	3.5168	=	kWh	X	0.2843	=	ton	

Heat Flux [Energy/(time x area)]

Btuh/ft^2	X	3.152	=	W/m^2	X	0.3172			
Btu/ft^2 (hr)		0.2929	=	W/ft^2	X	3.414			
		2.712	=	kcal/hr(m^2)	X	0.3687	=	Btuh/ft^2	
		0.00452	=	ly[c]/min	X	221.2		Btu/ft^2 (hr)	
		0.2712	=	ly[c]/hr	X	3.687			
		0.0543	=	Met units	X	18.4			

Miscellaneous

ft	X	0.3048	=	m	X	3.28084	=	ft
ft^2	X	0.092903	=	m^2	X	10.7639	=	ft^2
lb	X	0.453592	=	kg	X	2.20462	=	lb
(F - 32)	X	5/9 = °C			(°C X 9/5)	+ 32	=	F

a 1 joule = 1 watt/sec

b ton of refrigeration, the cooling effect produced when 1 ton (2000 lb.) of ice at 32F melts to water at 32F in 24 hours

c ly = langley = 1 cal/cm^2; commonly used in meteorology as a measure of solar heat flux

2. Comfort and Indoor Climate

The purpose of climatic design is to maintain, or to minimize the energy cost of maintaining, thermal comfort conditions within building interiors. Maintenance of thermal comfort is a problem of heat balance between the body and its surroundings.

The body exchanges heat with its environment through four processes: 1) conduction (contact); 2) conduction-convection (air movement); 3) evaporation-convection of skin moisture; 4) radiation (solar and thermal). The body itself generates heat, the amount of which varies according to level of activity. The metabolic rate of heat production plus environmental heat sources and sinks govern the heat budget of the body. Factors governing the rate of heat exchange and, consequently, the sense of comfort, comprise (corresponding respectively to the foregoing four processes): a) thermal resistance of clothing and temperature of surfaces in contact with the body; b) thermal resistance of clothing, air temperature, and speed of air movement; c) (water) vapor pressure of the air; d) temperature of surrounding surfaces (and area of the body exposed).

Comfort Conditions and Standards.

For building design and engineering purposes, human *thermal comfort* can be defined as the state of mind which expresses satisfaction with the thermal environment. Many researchers prefer the term "thermal neutrality" to "thermal comfort," to emphasize the qualification that the subject feels neither too hot nor too cold, nor feels any local discomfort due to asymmetric radiation, drafts, cold floors, non-uniform clothing, and so forth.

Numerous tests have been conducted to determine what sets of conditions are judged most comfortable by the agreement of test volunteers. Investigators working at Kansas State University found that the most comfortable condition for subjects wearing light clothing suitable for office wear (0.4-0.6 clo) corresponded to a dry-bulb temperature of 79F at 50 percent relative humidity, for air velocity less than 35 feet per minute [ASHRAE *Fundamentals* 1981]. A comfort zone enveloping the range of satisfaction of 80 percent of the subjects can be drawn on the psychrometric chart (*Figure 2a*). Other responses of the KSU test subjects have been analyzed and reduced to a series of equations describing "cold" to "hot" conditions. These are given in *Table 2a*, and are also plotted on *Figure 2a*. For all tests, subjects were sedentary (activity level = 1 Met).

In addition to its influence upon the rate of skin moisture evaporation, vapor pressure has been described in terms of its effect upon one's psycho-physiological state. Le Roy [Landsberg 1972] has ascribed the designations, "healthy," "soothing," "depressing," and "debilitating," to vapor pressure ranges between 12 and 16, 16 and 21.2, 21.2 and 26.4, and in excess of 26.4 millibars, respectively. These are plotted on *Figure 2b*. Discomfort limits of 5 and either 17 or 18 millimeters of mercury have been cited by other researchers. A related scale of climatic sensation developed by Brazol has been reported by Landsberg. It is based on the total heat content of the air (enthalpy), and therefore uses wet-bulb temperatures to characterize comfort conditions. This is an incomplete conceptualization as far as bodily heat balance goes, but is no less distorted than judging heat or cold stress by dry-bulb temperatures alone. Brazol lists wet-bulb temperatures in excess of 101F as "lethal heat" (*Figure 2c*).

Comfort ranges and limits have been identified and published as standards for building engineering and for prescribing safe job conditions for workers. The Occupational Safety and Health Administration (OSHA), for example, has established recommended limits for different levels of work activity. These are stated for two different air speed ranges (*Table 2b*), according to a simplified heat-stress index known as Wet-Bulb Globe Temperature (WBGT). The WBGT is a function of wet-bulb and globe temperatures, the latter of which is a combined measure of air and radiant temperatures (for indoor environments),

$$WBGT = 0.7\ WBT + 0.2\ GT + 0.1\ DBT$$

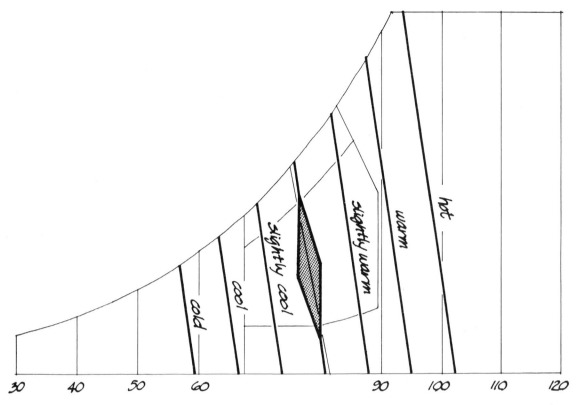

Figure 2a. Characterization of thermal sensations according to studies conducted for ASHRAE at Kansas State University. The shaded zone indicates conditions described as comfortable for 80% of test subjects at an activity level of 1 Met and with 0.5 clo of clothing.

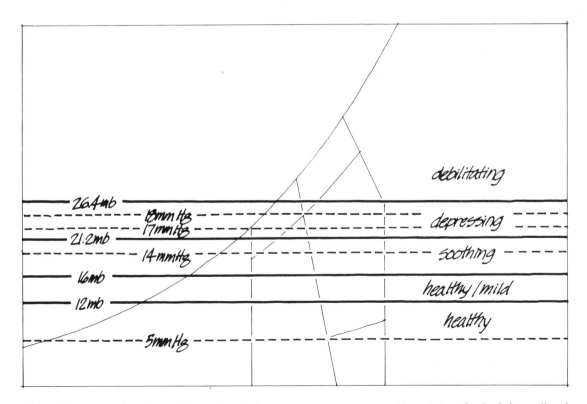

Figure 2b. Characterization of the effect of vapor pressure upon the state of mind described by LeRoy (graphed from data cited in [Landsberg, 1972]).

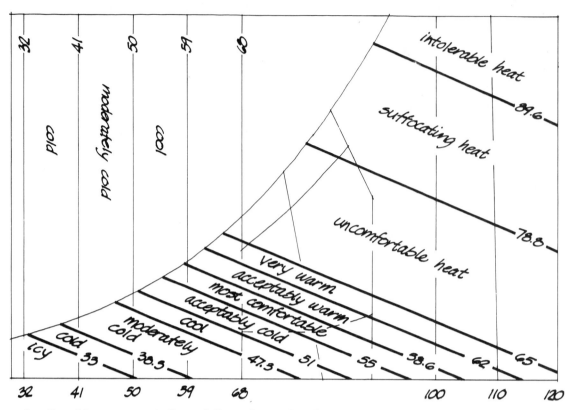

Figure 2c. Graphic representation of Brazol's scale of climatic sensation based on wet-bulb temperatures (graphed from data cited in [Landsberg, 1972]).

If the mean radiant temperature is assumed equal to the dry-bulb temperature, then GT = DBT, and the OSHA limits can easily be plotted on the psychrometric chart (*Figure 2d*). Although the OSHA limits are not perfect indicators of physiological stress, they give some additional human meaning to the psychrometric chart. Also depicted on *Figure 2d* is 95ET*, which has been described as the "Danger Line for Heat Stroke."

The most familiar set of comfort conditions is that described in ASHRAE Comfort Standard 55-74. It is based on the responses of sedentary adults (activity level 1.0-1.2 Met) wearing light office clothing (0.5-0.7 clo). The ASHRAE 55-74 comfort zone extends from 72F to 78F on the most recently developed Effective Temperature (ET*) scale, and is bounded by vapor pressures of 5 and 14 mm Hg (*Figure 2d*). It assumes air movement rates of less than 45 fpm. Like the KSU studies, the ASHRAE 55-74 comfort zone represents limits within which 80 percent of subjects, tested at the J. B. Pierce Foundation at Yale University, expressed satisfaction with their thermal environment under carefully controlled conditions. The 2 degree difference between the optimum of 76ET* in the Pierce Foundation study and 78ET* in the KSU experiments is accounted for by slight differences in test conditions; lighter clothing is characteristic of the KSU conditions (0.5 vs 0.6 clo), and the Pierce Foundation tests involved slightly greater work activity rates (1.1 vs 1.0 Met units). With these reconciled, results of the independent studies are virtually identical.

Effect of Clothing.

Aside from increasing metabolic heat production or ingesting hot or cold foods or liquids, the only practical opportunity that the individual has to alter his or her comfort condition is to add or subtract clothing, or seek a different environment (which is not applicable to the present discussion). According to a rule of thumb cited by Goldman [1978], air temperature departures from the optimum for 0.6 clo resistance can be offset by 1F for each 0.1 clo deviation from the standard for sedentary individuals (light office work, 100-200 kcal/hour), and by 2F for each 0.1 clo deviation at higher work levels. For moderate work activity levels, therefore, comfort can be achieved at 68-70F simply by increasing clothing. Fanger [1972] has published a series of widely utilized charts which allow the comfort standards to be corrected for other conditions. According to Fanger's charts, comfort is attained at 68F by increasing clo value to 0.9 to 1.4, when air velocity is 20-30 fpm or less.

The acceptability of 68F as a comfort standard has been widely discussed. Gonzalez [1977] has verified its acceptability in experiments in which he found that 80 percent of test subjects

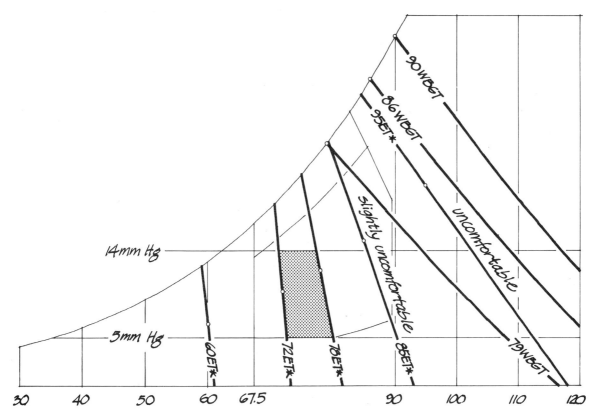

Figure 2d. ASHRAE 55-74 comfort zone (shaded) depicted in relation to thermal sensation characterizations determined by workers at the J.B. Pierce Foundation, and to OSHA work stress limits (see Table 2b). The index of 95ET* has been cited as a danger limit for heat stroke.

expressed thermal satisfaction with 68F temperature with clothing levels of 0.9 to 1.2 clo—somewhat less than predicted by the Fanger charts. (A clo value of 1.0 corresponds to heavy slacks, a light sweater, blouse and jacket for women, and heavy trousers, sweater, shirt and jacket for men.) Practical limits on how low air temperatures can go with compensation by increased clothing are set by the temperature of fingers and other relatively exposed parts of the body, such as the ankles. While a loss of dexterity may occur when air temperature falls below 65F, no degradation is expected at 68F. The conclusions of Gonzalez and others is that thermal comfort can be achieved at 68F without extraordinary or burdensome increases in the amount of clothing worn.

Effect of Air Movement.

Air movement influences bodily heat balance and, hence, thermal comfort by: 1) affecting the rate of conductive-convective heat transfer between the skin and the air; and by 2) affecting the rate of bodily cooling through evaporation of skin moisture. The former is governed by air dry bulb temperature. Increasing air speed increases the rate of heat transfer, but the direction of heat flow depends upon whether the temperature of the air is greater or less than skin temperature (about 90-95F). Heat is removed from the body by conduction-convection when air temperature is less than 90F, but heat is added to the body by conduction-convection when air temperature approaches and exceeds skin temperature. The rate of evaporation is governed by both air speed and vapor pressure. Increasing air speed always increases evaporative cooling effect, although at high vapor pressures, the overall effect may be small.

Within certain conditions, the effect of increased air movement is to extend the upper limit of the comfort zone to higher temperatures. A recent paper by Arens *et al.* [1980] illustrates these extended limits for a variety of different air speeds on the psychrometric chart (*Figure 2e and 2f*). Because the rate of heat exchange caused by increasing air movement is coupled to vapor pressure as well as to dry-bulb temperature, the lines of equal comfort are skewed, favoring higher temperatures at low vapor pressure, and lower temperature at high vapor pressure. Note also that increase in air speed is able to offset a higher temperature increase at low vapor pressure than at high vapor pressure.

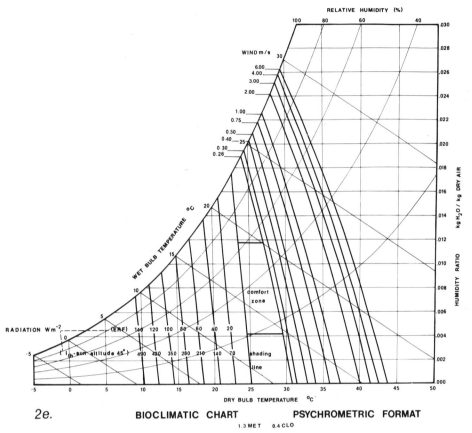

2e.   **BIOCLIMATIC CHART   PSYCHROMETRIC FORMAT**

1.3 MET   0.4 CLO

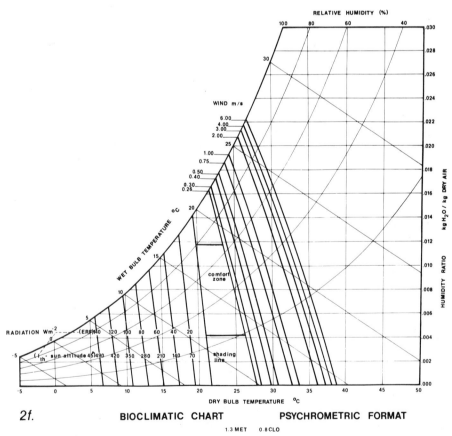

2f.   **BIOCLIMATIC CHART   PSYCHROMETRIC FORMAT**

1.3 MET   0.8 CLO

Figure 2e. 2f. Bioclimatic Charts for normal summer (0.4 clo) and winter (0.8 clo) dress based on a thermal comfort model developed at the J.B. Pierce Foundation.

Effect of Mean Radiant Temperature (MRT).

Heat is exchanged between the body and the surrounding environment by thermal radiation. The temperature of surrounding surfaces is assumed to equal air temperature in most comfort standards, such as the relationships depicted in *Figures 2a and 2e and 2f*. Under actual conditions, the temperature of room surfaces vary, and may significantly differ from air temperature. This is especially true of the warm interior face of Trombe Walls (sun-heated masonry) and radiant-heating floors and ceilings, and the cool faces of window glass surfaces and subgrade walls.

Although the rate of radiant heat exchange between the body and its surroundings is dependent upon surface temperature differences between these, several factors must be considered in analysis of the heat balance. Among these are air temperature, air speed, and clothing level. Fanger's charts show that for sedentary adults at typical summer clo levels of 0.5, and at relatively still air conditions (velocity less than 20 fpm, or 0.1 m/s), for an increase in air temperature from 78 to 88F, the mean radiant temperature (MRT) must drop from 78 to 68F in order to maintain a sensation of equal comfort. At low air speeds of 40 fpm (0.2 m/s), a 10F increase in air temperature from 80 to 90F must be offset by a decrease in MRT of 14F, from 80 to 66F. Under nearly still air conditions, therefore, a decrease of 1 to 1.4F in MRT is required to offset every 1F increase in air temperature. At the unique air temperature of 87F (according to Fanger's charts), an MRT of 69F is necessary to maintain comfort conditions, regardless of air speed, when it is less than 1.5 m/s.

Mean radiant temperature is defined as "the uniform surface temperature of an imaginary black enclosure with which man exchanges the same heat by radiation as in the actual environment." In other words, MRT is the average of all room surfaces, weighted according to emissivity (which is nearly constant for most building materials). The radiant heating or cooling ability of any surface, therefore, must be evaluated in the context of its area in proportion to the area and temperature of other surfaces in the room. The angle of exposure of surface to body and the orientation of exposed parts of the body must also be considered.

It is noteworthy that, whereas with a still air temperature of 90F, an MRT of 67F is required to achieve comfort (at 0.5 clo), an equal sensation of comfort with 90F air temperature is achieved with a ventilation air speed of 20 fpm (1 m/s) at 0.4 clo. The latter may be read directly from *Figure 2e*, while the former can be deduced from Fanger's charts.

Table 2a		KSU Thermal Sensation Equations
Exposure duration (hr)	Sex	Regression Equations ($T_a = °C$; $P_v = mm\ Hg$)
1	Male	$Y = 0.220 T_a + 0.031 P_v - 5.673$
	Female	$Y = 0.272 T_a + 0.033 P_v - 7.245$
	Combined	$Y = 0.245 T_a + 0.033 P_v - 6.475$
2	Male	$Y = 0.221 T_a + 0.036 P_v - 6.024$
	Female	$Y = 0.283 T_a + 0.028 P_v - 7.694$
	Combined	$Y = 0.252 T_a + 0.032 P_v - 6.859$
3	Male	$Y = 0.212 T_a + 0.039 P_v - 5.949$
	Female	$Y = 0.275 T_a + 0.034 P_v - 8.622$
	Combined	$Y = 0.243 T_a + 0.037 P_v - 6.802$

Note: Applies to young sedentary adults with clo value of 0.5; MRT = DBT; air velocities less than 0.2 m/sec. Y values will range from −3 to +3, where −3 is cold; −2 is cool; −1 is slightly cool; 0 is comfortable; +1 is slightly warm; +2 is warm; +3 is hot.
Source: ASHRAE *Fundamentals* [1981] p. 8.19.

Workload	WGBT in degrees F	
	Low air velocity 300 fpm	High air velocity 300 fpm
Light (level 2) (<200 kcal/hour)	86	90
Moderate (level 3) (201-300 kcal/hour)	82	87
Heavy (level 4) (> 300 kcal/hour)	79	84

*WBGT = 0.7 WBT + 0.3 GT, for indoor use (assumes DBT = GT)

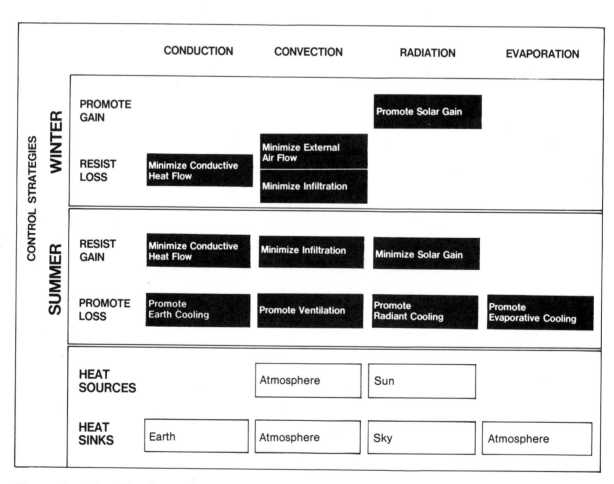

Figure 3a. Identification of hypothetical and practicable strategies of climate control.

3. Strategies of Climate Control

The building envelope is a device through which heat exchange between the interior and exterior environments is controlled. The building envelope intercedes with the external climate, creating a new interior microclimate zone. The fundamental control options consist of 1) admitting or 2) excluding heat gain from external energy sources, and 3) containing or 4) rejecting heat energy present in the interior. Most of the manifestations of these controls are *static*, or fixed in place; examples include insulation placed in wall and ceiling cavities, and the area and orientation of glazing in the building shell. Some of the most important controls, however, are *dynamic;* these include operable window sash, movable window insulation, and a variety of adjustable sun-shading devices.

Application of the four heat flow control options to the three mechanisms of heat transfer plus the adiabatic phase change process (ordinarily coupled with convection) produces a set of sixteen hypothetical strategies of climate control (*Figure 3a*). Not all of these actually avail themselves to use: the sun, for example, is the only passive source of heat energy (except in rare areas of accessible geothermal energy), while the usefulness of the different potential heat sinks depends on locale and meteorological conditions. As a result, only eight of the sixteen strategies can effectively be exploited in design, as indicated in *Figure 3a*, although there are numerous ways of executing each.

The eight practicable strategies can be separated into those appropriate to underheated ("winter") conditions and those appropriate to overheated ("summer") conditions. In predominantly overheated or underheated climates, preference can usually be given to the more appropriate set, while the remaining strategies are used, if at all necessary, in a secondary way. In temperate regions, however, such as in most of the United States, both sets of strategies ("summer" and "winter") are applicable for significant fractions of the year. Failure to address both heating and cooling needs—or using one to the exclusion of the other—will result in sacrifices in overall annual comfort and energy performance.

Climatic Analysis.

The appropriateness of a building design strategy of climate control under any set of ambient temperature and humidity conditions is determined by the analysis of weather data and the requirements for human comfort. The concept of relating coincident temperature and humidity conditions to the needs for climate control in building design was first given a well-defined structure by Victor and Aladar Olgyay in the early 1950s [Olgyay and Olgyay 1953]. The Olgyay brothers' *Bioclimatic Chart* is a temperature-humidity diagram used to display the "comfort needs" of a sedentary person. The Bioclimatic Chart of the Olgyays has been updated by Arens et al. [1980], and is reproduced here in *Figure 3b*. It should be noted that this is exactly the same diagram as in *Figures 2e and 2f*, except that the latter are presented in the standard psychrometric chart format.

An important extension of the Olgyays' work was made by Baruch Givoni [1976; Milne and Givoni 1979], who determined limits of effectiveness of different building practices in meeting bioclimatic comfort needs which previously were only identified. The limits for well-developed executions of each climate control strategy identified in *Figure 3a* can be plotted on the psychrometric chart, producing a new diagram which Givoni termed the *Building Bioclimatic Chart.* The effective limits of each strategy creates within them an effectiveness zone which can be visualized as an extension of the comfort zone (*Figures 3c-3f*). The Building Bioclimatic Chart indicates that whenever ambient outdoor temperature and humidity conditions fall within the designated limits of a control strategy, then the interior of a building designed to effectively execute that strategy will remain comfortable. The psychrometric limits of these strategies and the building characteristics necessary to effectively execute them are summarized in the following discussions. The principles and physics of the control strategies are elaborated in succeeding sections.

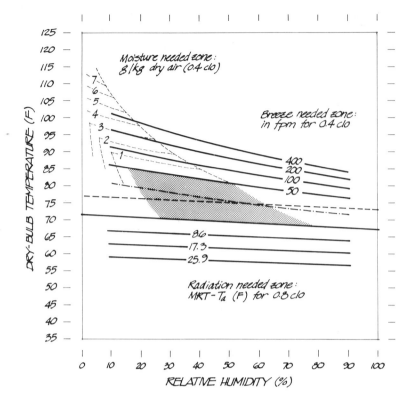

Figure 3b. Updated Bioclimatic Chart in the format originally used by Olgyay and Olgyay, based on a comfort model developed at the J.B. Pierce Foundation.

Promote Solar Gain, Minimize Conduction, Minimize Infiltration

The need for passive solar heating—to the extent practical within a region—and the heat conservation strategies of minimizing conduction and infiltration heat losses is indicated whenever outdoor air temperature falls below the lower comfort zone limit of 68F. Houses with even small heat storage capability can easily maintain comfortable conditions internally without special design attention so long as daily average outdoor air temperature remains above 65 or 68F (interior temperatures are normally raised by 5F or more by the heat generated by people, appliances, cooking, and other activities such as showering and laundering).

Milne and Givoni [1979] suggest that the heating demand for a small 1000 ft² building with 200 ft² of south-facing glass can theoretically be satisfied for one day without auxiliary heat under the solar and air temperature conditions used in *Table 3a.* This house is assumed to be well insulated, and the values given are based on an overall solar heating efficiency of 0.33 for collection, storage, and distribution of the solar heat (See *Figure 3a*).

The *balance point* of a house is the lowest outdoor air temperature at which the interior remains within comfort limits without either a net gain or loss of heat under a specified solar contribution. The values given in *Table 3a* show that the balance point of a house varies with the solar climate; it also varies with insulation level, air tightness, and other factors. A useful simplified method for determining the balance point of a house has been proposed by G. Hans [1981]. He offers a means of examining the successfulness of solar house design within the framework of bioclimatic analysis.

Minimize Solar Gain

The need for solar gain controls is indicated whenever outdoor air temperatures exceed the lower limits of the comfort zone (68F), since the comfort conditions adopted here are defined with the stipulation that mean radiant temperature and air temperature are equal. Shading is the only strategy necessary when outdoor air temperatures fall in the range 68F to 78ET*, and it is also required — although it is no longer sufficient — for climate control when outdoor air temperatures exceed 78ET*.

Promote Ventilation

The ability of air movement to produce comfort cooling of the body has been discussed in a previous section. The limits of effectiveness of ventilation in producing body cooling in building interiors are based on the assumption that indoor air temperature and vapor pressure are identical indoors and out, and that the mean radiant temperature of the building interior (i.e., the

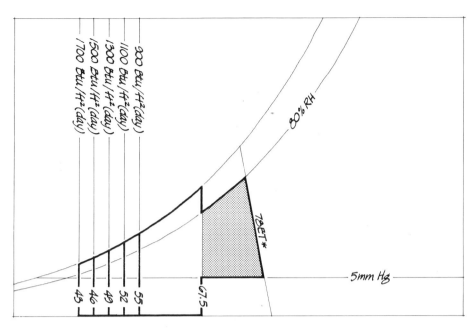

Figure 3c. Example of zones of solar heating effectiveness for a small dwelling described by Milne and Givoni (1979).

average temperature of all interior surfaces "seen" by the body) is approximately the same as that of the air. Both assumptions are sufficiently valid if the interior is well ventilated, and if the building exterior is of a light color and the shell well insulated, so as to restrict excess solar heat gain to a negligible level.

The upper limit of the ventilation effectiveness zone is determined by the greatest wind speed that will not cause annoyance. Givoni [1976] sets this speed as 1.5 m/sec, which results in a limit loosely approximated by the air density of 0.0704 pcf (14.2 ft^3/lb specific volume) when vapor pressure exceeds 17 mm Hg. High rates of air movement become less desirable as the moisture content of the air decreases, so a dry-bulb temperature of 89F is accepted as a limit for vapor pressures less than 17 mm Hg.

Promote Radiant Cooling/Thermal Mass

The thermal mass approach to temperature control is limited by the assumptions that 1) the exterior shell is massive enough so that it damps out daily temperature fluctuations, causing the interior surfaces to assume something near the daily average outdoor temperature, and 2) the building is closed during the daytime to minimize intrusion of heat. The second assumption necessitates an upper limit of vapor pressure of 17 mm Hg, the maximum humidity at which one can feel comfortable in the absence of air movement. The upper temperature limit is conveniently (although not exactly) satisfied by the air density of 0.0707 pcf (14.15 ft^3/lb specific volume).

The dry-bulb temperature limit is much greater under arid conditions because 1) the body is comfortable at higher temperatures at low humidities (as seen in the slope of the 78ET* line), and 2) diurnal temperature range increases as absolute humidity decreases, so for any given daily average temperature, a dry climate will exhibit a higher daily maximum, balanced by an equally lower daily minimum, temperature than may be found in a humid climate. Diurnal range in air temperature is associated with vapor pressure according to the empirical relation, after Givoni [1976],

$$\text{Diurnal range (F)} = 47F - \frac{1.5F}{mm\ Hg}\ P_V \text{ in mm Hg}$$

where P_V is the average local vapor pressure. A diurnal range of about 22F is expected in climates where P_V = 17 mm Hg(corresponding to a dew point temperature of 67F), and a range of 40F is expected in climates where P_V= 5 mm Hg (dew point temperature of 35F). Adding the corresponding half-ranges of these figures (to produce the expected daily maxima) to the respective 78ET* comfort zone limits of 76.5F and 80F at 17 and 5 mm Hg yields temperature limits of the mass effectiveness strategy of 87.5F at 17 mm Hg and 100F at 5 mm Hg.

A building structure designed to satisfy the daily temperature averaging function must be thermally massive and must have a highly reflective outer surface (whitewash, for example), or be well shaded so that no significant solar excess heat is propagated through the interior. The effectiveness of this strategy is enhanced—and its dry-bulb temperature limits extended—by ventilating the interior at nighttime so that both interior and exterior surfaces are exposed to the daily minimum temperatures of night air, while the building is closed during the day so that only the exterior is exposed to the daily maximum temperature. This nocturnal ventilation practice and the preference for high mass interiors is elaborated upon by Milne and Givoni [1979] and Danby [1973].

Promote Evaporative (Space) Cooling

The evaporative cooling process referred to here applies to direct evaporation of water into air drawn from the out-of-doors as it is admitted to the interior space. This process can occur by several methods, including spraying a mist into the intake air stream or blowing the air through a wetted mat. The change of phase from liquid to water vapor takes place by absorption of sensible heat from the air; because no heat is gained or lost from the water-air system, the process is one of constant enthalpy (that is, the sum of sensible and latent heat remains constant, the former merely being converted into the latter), and the wet-bulb temperature of the air remains the same. One limit of the evaporative cooling strategy, therefore, is the maximum wet-bulb temperature acceptable for comfort. This is selected as 71.5F, coincident with the uppermost limit of the comfort zone.

The dry-bulb boundary, in Milne and Givoni's words,

> is defined only by the cooling capacity of the volume of air that can be comfortably moved through the interior of the building. This also assumes that sufficient amounts of water are available. For all practical purposes, 25F temperature reductions are about the limit of what can be achieved at reasonable indoor air velocities.

The limit, accordingly, is taken as 105F (80 + 25F).

A number of other evaporative cooling techniques can be used, some of which employ a heat exchanger so that the moisture content of the incoming air is not increased. Some of these devices are described in references listed in the bibliography. The regional range of evaporative cooler effectiveness has been mapped by Lee [1978] among others. Greater effectiveness in evaporative cooling can be obtained in very arid regions by nighttime operation. The cooled air is often forced through rock bins or pebble beds for storage rather than circulating it at night through the space itself.

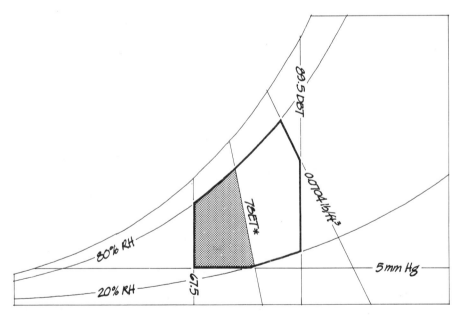

Figure 3d. Psychrometric limits of the capability of ventilation to produce comfort conditions at acceptable indoor air speeds.

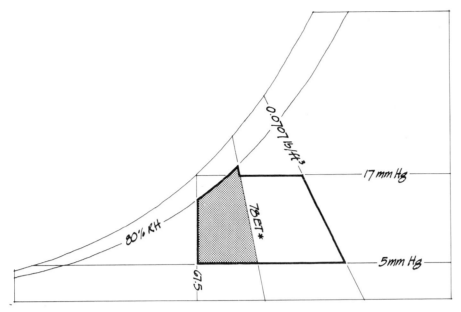

Figure 3e. Psychrometric limits of the capability of a high thermal mass building shell to maintain comfortable indoor conditions.

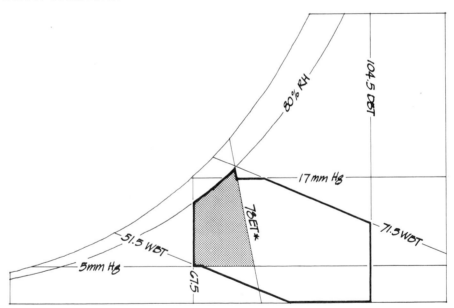

Figure 3f. Psychrometric limits of the capability of direct evaporative cooling to produce comfortable indoor conditions.

Table 3a	Balance Point Temperature for a Solar House[a]
Daily Minimum (Balance Point) Temperature (F)	Insolation on South-Facing Vertical Surface
43	1700 Btu/(ft^2) day
46	1500
49	1300
52	1100[b]
55	900[b]

a House described in text; adapted from Milne and Givoni [1979]

b Inferred by linear extrapolation

4. Promote Solar Gain

The sun is the only significant natural source of energy available for passive heating of buildings. The intensity of solar energy received upon earth varies with latitude and local sky clearness, however, so the feasibility of solar heating depends on the relationship between solar energy received and the winter temperatures which determine heating load.

Terminology.

Solar energy is intercepted by the earth's surface, where it is measured as heat flux, in terms of energy per unit area and time. *Insolation (incident solar radiation), solar irradiation,* and *solar intensity* are all synonymous with solar heat flux. In the English system, it has the dimensions of Btu/ft^2(hr) or Btu/ft^2(day), and is usually represented by the capital letter "I". Solar flux can also be expressed in watts/m^2, or langleys per second, minute, hour, or day, where one langley is defined as 1 cal/cm^2.

Solar radiation arriving at a building surface consists of *direct, diffuse,* and often *reflected* components. Direct—or beam—radiation arrives in parallel rays, directly penetrating the earth's atmosphere. Diffuse radiation results from the scattering of sunlight among dust particles, water droplets (clouds), and aerosols in the atmosphere. The building surface may also receive appreciable amounts of solar energy reflected from materials on the ground and from roofs, walls, and other surrounding surfaces.

Even on the clearest days, some diffuse radiation is received, although it is not significant in comparison to the magnitude of the direct beam flux. As a matter of reference, the amount of diffuse radiation received is often described as a ratio of its contribution to that of the maximum direct radiation possible at a given place and time. As is evident from *Table 4a,* the diffuse component supplies the majority of available solar heat on overcast days.

Irradiation Measurement and Expression.

Sometimes the direct and diffuse components of solar heat flux are measured and reported separately, but most often insolation is expressed as the sum of the two. This is usually given as either the maximum possible value for a time and place under a clear sky, or as an average of long term observations or estimates. Traditionally, insolation has been reported as solar intensity upon a horizontal surface without any reflected contribution, or as direct beam irradiation, that is, normal (perpendicular) to the solar beam. With increasing interest in passive solar applications, insolation data for vertical, south-facing surfaces have become available in a number of published sources [Balcomb 1980a; Kusuda and Ishii 1977; ASHRAE *Fundamentals* 1981].

The intensity of solar radiation received on any surface can be computed by the *cosine law* if the direct beam intensity and the angle of incidence θ of the solar beam for the surface are known:

$$I_s = I_n \cos\theta$$

where I_s = intensity of radiation falling on surface
$\quad\quad$ I_n = intensity of radiation normal to the solar beam
$\quad\quad$ θ = angle between the solar beam and a line normal to the surface

The angle of incidence θ between the solar beam and a line normal to a vertical surface (wall) is found by the spherical cosine equation, which states,

$$\cos\theta = \cos\beta\cos(\phi-\psi)$$

where β = solar altitude
      φ = solar azimuth angle, measured clockwise in degrees from north
      ψ = wall orientation (wall azimuth) angle, measured clockwise in degrees from north

Once $\cos \theta$ is obtained from the spherical cosine equation, it can be inserted into the cosine law equation to compute incident radiation. It is not necessary to know the value of θ itself for the substitution, although this is found by the relation, $\theta = \arccos(\cos\beta\cos(\phi-\psi))$. Because solar intensity upon a surface is at maximum when the solar beam is normal to it ($\theta = 0°$), solar heating value decreases (being diluted over a larger area) as the incidence angle increases; the wall falls in shade at the moment θ exceeds 90°.

Solar Geometry.

Although a detailed discussion of solar geometry is beyond the scope of this book, a few more definitions will be provided. *Solar altitude* (β) is defined as the vertical angle between the horizon and the line of sight to the sun. It is measured in degrees. *Solar azimuth* (ϕ) or *solar bearing,* is the horizontal angle between the projection on the ground of the line of sight to the sun and the north-south axis. Azimuth may be measured from the axis facing either north or south; in sun-angle calculations it is usually expressed as deviation in degrees from the south unless specified otherwise (as in the spherical cosine equation preceding). When measured from the north, as in navigational charts, values increase in a clockwise direction.

Surface or wall azimuth (ψ) refers to the horizontal angle made by a line drawn normal to a wall and the north-south axis. Like solar azimuth, wall azimuth may be measured from either north or south, although deviation from south is most commonly used in the United States. Wall azimuth is simply an angular expression of orientation (see *Table 4g*).

The *surface solar azimuth* (γ) is the horizontal angle made by a line drawn normal to a wall and the projection on the ground of the line of sight to the sun. If solar and wall azimuths are taken with a north origin, then the surface solar azimuth is given by the difference ($\phi-\psi$). If solar and wall azimuths are taken with a south origin, then γ is found according to the procedures given in *Table 4b*.

The *profile* or *shadow line angle* (Ω) is defined as the angular distance between a horizontal plane and a plane tilted about a horizontal axis in the plane of a vertical window until it intersects the sun. In terms of the solar altitude β and the surface solar azimuth γ, the profile angle Ω can be calculated by the relation,

$$\tan \Omega = \tan \beta / \cos \gamma$$

The profile angle can be described as the projection of the altitude angle of the sun upon the cross-sectional picture plane of the interior. The profile angle and the solar altitude angle are equal only at the moment when the projection of the line of sight to the sun on the ground is perpendicular to the window plane; at all other times, the profile angle is greater than the altitude. The profile angle is important in designing exterior sunshades and in determining the depth of solar penetration into a room when viewed in cross-section.

Solar Time.

The time we keep our clocks by rarely coincides with solar time, so it is often necessary to convert from the local standard time to *Apparent Solar Time* (AST). Two corrections are necessary to make this conversion, one for the "Equation of Time," and the other to accommodate for the time of rotation of the earth with respect to the sun between the standard time meridian of the local time zone and the longitude of the site in question.

The earth is divided into 360 longitudinal segments or meridians of longitude. The sun may be thought of as passing over these at the rate of 360° longitude per 24 hours, or 1 meridian every 4 minutes. Standard time for each time zone is referenced to every fifteenth meridian (15 x 4 min = 1 hr); in the United States, these comprise: 75° for Eastern ST, 90° for Central ST, 105° for Mountain ST, 120° for Pacific ST, and 150° for Alaska-Hawaii ST.

Apparent Solar Time is related to the local standard time and place by the correspondence,

$$AST = LST - ET - 4(LON-LSM)$$

where: ET = "Equation of Time," values from *Table 4c*
        LST = local standard time
        LON = longitude of site, degrees
        LSM = local standard time meridian, degrees
           4 = minutes of time for 1° rotation of earth

For example, we can calculate the AST for 12:00 noon EST on January 21 in Gambier, Ohio, which lies near the 82nd meridian:

              AST = 12:00 - 11.4 (82° - 75°)
              AST = 12:00 - 39.4 minutes = 11:20.6 a.m.

On January 21, solar time lags behind standard time by 39.4 minutes in Gambier. Consequently, solar noon occurs at 12:39.4 p.m. One must also not neglect to accommodate for Daylight Savings Time (DST) during summer. In Gambier, for instance, on July 21, 12:00 noon EDST is really 11:00 a.m. EST, so solar time is found:

$$AST = 11:00 - 6.2 - 4(82° - 75°)$$
$$AST = 11:00 - 34.2 \text{ minutes} = 10:25.8 \text{ a.m.}$$

This indicates a lag of 34.2 minutes behind EST. In terms of Eastern Daylight ST, therefore, solar noon occurs in Gambier at 1:34.2 p.m.

Greenhouse Effect.

All solar heating systems exploit the *greenhouse effect,* a phenonemon easily appreciated by anyone who has entered a car that has been in the sun with closed windows. The greenhouse effect describes the temperature elevation of an enclosed space with a glazed aperture exposed to the sun; the increase in internal temperature is produced by the glazing's 1) transmission of inward solar heat gain and 2) suppression of outward convective heat loss. The simplicity of this effect is commonly confused by the additional fact that glass is 80 to 90 percent transparent to shortwave (solar) radiation, but is opaque to longwave (infrared) radiation. This selectiveness of transmittance at different wavelengths often is erroneously thought to be primarily responsible for the net heat gain of greenhouses, solar collectors, and other glazed enclosures. While ordinary window glass does not transmit longwave radiation, it does not appreciably reflect it, either; thermal radiation received by a glazing sheet is absorbed and converted into sensible heat which is quickly lost to the out-of-doors by conduction-convection, and radiation emitted by the exterior surface of the glass. The warm interior thermally "sees" the window at a temperature close to that of the outside, even though it doesn't "see through" the glass.

The relative insignificance of the longwave opacity of glass was demonstrated experimentally as early as 1909 by the British physicist R. W. Wood. He compared identical glazed enclosures, one covered with a plate of glass and the other with a plate of rock salt (which is transparent to longwave radiation). He found [Wood 1909], "scarcely a difference of one degree (Celsius) between the temperatures of the two enclosures." More recent analytical refutations of the "mousetrap theory" of the greenhouse effect have been published by J. A. Businger [1963] and R. Lee, [1973, 1974. See also Silverstein 1976]. In summarizing his own analyses, Lee concludes [1974],

> Greenhouse glass does not trap radiant energy, but it does trap air. The trapped air is relatively motionless and inefficient (compared with the normal atmosphere) in removing energy from the surface, so the temperature of the surface must be greater in order to dissipate the same radiant energy flux. The secret of the greenhouse is that it permits a relatively normal radiant energy exchange, while trapping a small volume of air near the surface.

The longwave transmission of glazing materials seems to be important in space conditioning applications in overheated arid regions, where greenhouses are used for creating a humidified growing environment. In order to prevent excessive heating of the interior, it is desirable to promote as much radiant heat loss through the structure as possible. Thermally transparent glazings such as polyethylene film and other plastics are often favored for this use [Sayigh 1981].

Solar Heating Systems.

In addition to the glazed aperture or collector, solar heating systems contain two or three components, 1) an absorbing surface which converts the solar radiation into thermal energy, 2) the space to be heated, and 3) an optional medium for heat storage. In *passive* solar heating systems, heat energy is transported from one component to the other through naturally-driven mechanisms of heat transfer. Heat energy is transported within *active* systems principally by the power of pumps or fans. Systems in which heat transfer occurs primarily by natural processes, but which may be assisted by mechanical devices are called *hybrid* systems. This book deals with passive means of climate control, as these fall within the province of the architect and builder. Active system design falls within the discipline of mechanical engineers and mechanical systems contractors; they will not, consequently, be discussed in this book [See Watson 1977]. Hybrid systems and techniques will be described where applicable.

Passive system types are distinguished from one another by the physical relationships between apertures, absorbers, thermal storage media, and conditioned spaces. Thirty different conceptual passive solar systems can be identified by considering the range of possible component interrelationships, as diagrammed in *Table 4d.*

Table 4a — Diffuse Solar Radiation Received on a Horizontal Surface

Sky condition	Ratio of Actual Direct to Maximum Direct Radiation	Ratio of Diffuse to Maximum Direct Radiation
Clear	1.00	0.12
Clear, Slightly Hazy	0.80	0.25
Hazy	0.60	0.35
Overcast	0.40	0.55

Source: Milne [1981]

Table 4b — Azimuth Relationships

Surface azimuth (ψ), in degrees

Orientation	N	NE	E	SE	S	SW	W	NW	N
South origin[a]	180	135	90	45	0	45	90	135	180
North origin[b]	0	45	90	135	180	225	270	315	360

Surface solar azimuth (γ), in degrees

For surfaces facing east of south,

$$\text{morning:} \quad \gamma = \phi - \psi$$
$$\text{afternoon:} \quad \gamma = \phi + \psi$$

For surfaces facing west of south,

$$\text{morning:} \quad \gamma = \phi + \psi$$
$$\text{afternoon:} \quad \gamma = \phi - \psi$$

a convention commonly used in the U.S.
b practice preferred by the International Solar Energy Society and used in navigation

Table 4c

Values From Equation of Time, Minutes and Seconds

	Day of Month							
	1	**5**	**9**	**13**	**17**	**21**	**25**	**29**
January	+3'14"	5'6"	6'50"	8'27"	9'54"	11'10"	12'14"	13'5"
February	+13'34"	14'2"	14'17"	14'20"	14'10"	13'50"	13'19"	
March	+12'38"	11'48"	10'51"	9'49"	8'42"	7'32"	6'20"	5'7"
April	+4'12"	3'1"	1'52"	+0'47"	−0'13"	1'6"	1'53"	2'33"
May	−2'50"	3'17"	3'35"	3'44"	3'44"	3'34"	3'16"	2'51"
June	−2'27"	1'49"	1'6"	−0'18"	+0'33"	1'25"	2'17"	3'7"
July	+3'31"	4'16"	4'56"	5'30"	5'57"	6'15"	6'24"	6'23"
August	+6'17"	5'59"	5'33"	4'57"	4'12"	3'19"	2'18"	1'10"
September	+0'15"	−1'2"	2'22"	3'45"	5'10"	6'35"	8'0"	9'22"
October	−10'1"	11'17"	12'27"	13'30"	14'25"	15'10"	15'46"	16'10"
November	−16'21"	16'23"	16'12"	15'47"	15'10"	14'18"	13'15"	11'59"
December	−11'16"	9'43"	8'1"	6'12"	4'17"	2'19"	−0'20"	+1'39"

Source: *Smithsonian Meteorological Tables,* 6th Revised Edition, pp. 445-446

Table 4d

Conceptual Passive Solar System Relationships

Aperture Options		Absorber: Space Relationship		Absorber: Storage Relationship		Storage: Space Relationship
wall roof		direct indirect isolated		none integral remote		(none) integral remote
2	X	3	X	[1 + (2	X	2)]

5. Minimize Conductive Heat Flow

The process of heat transfer through solid building materials is described as *thermal conduction*, in which heat energy is transmitted from particle to particle within the material. The rate of heat transfer through a material or composite such as a building assembly is termed the *thermal transmission*. Thermal transmission is actually an expression of thermal power or flow rate, and it has the dimensions of Btu/hr or Btuh. Whereas the upper case "Q" usually designates total heat energy, thermal transmission is denoted by the lower case "q." Heat flux, introduced in an earlier discussion, can be defined as thermal transmission across a unit area, and it can be thought of as the energy density flow rate. Interrelationships between these expressions are clarified below.

> Heat Q (Btu)
> Thermal transmission $q = Q/\text{time}$ (Btuh)
> Heat flux $= q/A$ (Btu/ft^2 (hr) or Btuh/ft^2)

Conductance (C).
When a temperature difference occurs between opposite faces of a wall or roof section, heat will flow through the assembly from the warm to the cold side, in an attempt to equalize the thermal imbalance. As long as the faces are maintained at two different and constant temperatures, the heat flow rate between them will also remain constant; this is described as a *steady state* condition. The magnitude of the heat flow rate is governed by a property of the material or assembly section known as its *conductance*. Conductance is defined as the time rate of heat flow through a unit surface area (1 ft^2) of a body from one of its faces to the other, for a unit temperature difference (1F) between them. Conductance is denoted by the upper case letter "C", and in the English system it has dimensions of Btu/hr(ft^2)F. In symbolic terms, conductance is defined by the relation:

$$\text{heat transmission } q = C(T_h - T_c)A$$

$$\text{conductance } C = \frac{q}{(T_h - T_c)A} \qquad \frac{\text{Btuh}}{\text{F(ft}^2)}$$

where T_h is the temperature of the warmer face (F)
T_c is the temperature of the cooler face (F)
A is the surface area (ft^2)

The conductance of a homogeneous material is found simply by dividing its conductivity by its thickness. The conductance of an assembly such as a wall or roof section is found by summing the reciprocals of the conductances (i.e., the *resistances*) of each layer. Conductances of most building materials are published by their manufacturers and in handbooks of building thermal engineering.

Conductivity (k).
The *conductivity* of a material is its conductance for a standardized unit thickness. In building engineering this is normally 1 inch, although for some materials such as soil and snow a unit thickness of 1 foot is used. Conductivity is defined as the time rate of heat flow through a unit area and unit thickness of a homogeneous material under steady state conditions, when a unit temperature difference is maintained between its faces. It is usually represented by the lower case letter "k" (sometimes being called the "k *factor*"), and it has the English system dimensions of Btu/ft^2 (hr)F/in for building materials and Btu/ft (hr)F for earth materials. Because conductivity is stated for a standard thickness, it allows quick comparison between different materials of the ease with which they transmit heat.

In symbolic terms, conductivity k is defined by the relation:

$$\text{heat transmission } q = \frac{k}{x}(T_h - T_c)A$$

$$\text{conductivity } k = \frac{qx}{(T_h - T_c)A} \qquad \frac{\text{Btuh(in)}}{\text{F(ft}^2)}$$

where "x" is the thickness of the material, in inches. From these expressions, it is clear that C and k/x are interchangeable, such that

$$\text{conductance (C)} = \text{conductivity (k)} / \text{thickness (x)}$$

Thermal Resistance R, R-value.

Designers and builders are more inclined to talk about the *thermal resistance* of a building component or assembly than its conductance or the conductivity of its parts. Thermal resistance, designated by the capital letter "R", is simply the reciprocal of thermal conductance. It is defined by the relation:

$$\text{heat transmission (q)} = \frac{(T_h - T_c)A}{R}$$

$$\text{resistance } R = \frac{(T_h - T_c)A}{q} \qquad \frac{ft^2(hr)F}{Btu}$$

The preference of building professionals for the R-value designation probably has to do with the fact that its units are usually whole numbers instead of fractions, and that they are more easily remembered and manipulated than conductances. The dimensions of R are always, in the English system, $ft^2(hr)F/Btu$. Since these dimensions are awkward, and since they are standardized in the construction industry, they are usually dropped and we speak of the resistance of a building material or section as, for example, "R15" or or "R33". R-values are also used in the metric world, and these have the dimensions of $m^2 (°C) / W$. Because neither system commonly states the dimensions, one must note carefully when reading foreign literature and building standards which system is being used. To convert metric R-values into English system units, multiply by 5.678 (see *Table 5a*).

Resistances are additive, so that the total resistance of a building section is the sum of resistances of each of its component layers. Conductances are not additive; in order to find the conductance of a building section, one must first find its total resistance. For a wall section of four layers, one writes,

$$\text{wall } R = R_1 + R_2 + R_3 + R_4$$

$$\text{wall } C = \frac{1}{\text{wall R}} = \frac{1}{R_1 + R_2 + R_3 + R_4}$$

U-value.

The units discussed so far have been defined in terms of heat transmission from one *surface* through a material to the other *surface*. Surface-to-surface calculations are suitable for stating material properties and thermal standards for construction, but in practice we usually do not know the actual surface temperatures. Instead, we assume indoor and outdoor design air temperatures; the effect of the transition between air and surface temperature must, therefore, be incorporated in the heat flux equation. This is done by assuming a conductance value for the layer of air adjacent to outside and inside surfaces. These conductances are known as the *surface* or *film conductances*, and their values are sometimes referred to simply as the *film coefficients*. They are usually represented by the symbols h_o and h_i for the outdoor and indoor coefficients, respectively. Values of h_o vary with wind speed, and h_i is related to surface orientation. (These are dicussed in the section "Promote Ventilation").

The heat transmission equation given previously can now be rewritten for the air-to-air heat exchange:

$$\text{heat transmission } q = \frac{(T_i - T_o)A}{1/h_i + R + 1/h_o} \qquad \text{(Btuh)}$$

where T_i = indoor air temperature
T_o = outdoor air temperature
h_i = indoor film conductance
h_o = outdoor film conductance

The reciprocal of the sum of all resistances, including that of the surface film coefficients, is called the *(overall) coefficient of heat transfer*, the *overall conductance coefficient*, the *thermal transmittance*, or simply the *U-value*:

$$\text{thermal transmittance (U)} = \frac{1}{1/h_i + R + 1/h_o} \qquad \frac{Btuh}{F(ft^2)}$$

$$\text{thermal transmission (q)} = U(A)\Delta T$$

where ΔT is a shorthand expression for the temperature difference between indoor and outdoor air, $(T_i - T_o)$.

Conductive Heat Flow Controls.

The heat transmission equation has just three variables of major significance, namely, 1) surface area, 2) thermal resistance of the building envelope, and 3) temperature differential. Within each of these variables are numerous opportunities — or substrategies — for reducing conductive heat losses and gains. Reducing surface area, for example, reduces heat flow; substrategies related to this approach include minimizing the surface-to-volume ratio (making the building more compact) and reducing the relative area of highly conductive envelope elements such as windows and doors. The design approach of multiple-family construction also falls in this category, by minimizing the surface area-to-unit relationship.

As in most systems, a multiplicity of control options are possible in the design of the building. What may be inefficient in terms of excessive surface-to-volume ratio can be compensated for by increasing the thermal resistance of the shell itself. Often this is desirable, since elongated or sprawling plans may be desirable to increase the capture of breezes or to take best advantage of a view or of other conditions of the site.

Decreases in temperature differential can be achieved through selection of favorable building sites, by building underground, and by creating outdoor "sun pockets" adjacent to the structure.

Table 5a **Thermal Conduction Conversion Factors**

Conductance C
Transmittance U, Coefficient of heat transfer U
Surface (film) coefficients h_i, h_o

		5.678	=	W/m^2 (°C)	X	0.1761		
Btu/ft^2 (hr) F	X	0.2929	=	W/ft^2 (°C)	X	3.414	=	Btu/ft^2 (hr) F
		1.356×10^{-4}	=	cal/cm^2 (sec) °C	X	7373.5		
		4.882	=	$kcal/m^2$ (hr) °C	X	0.2048		

Conductivity k

		0.1441	=	W/m (°C)	X	6.938		
		3.445×10^{-4}	=	cal/cm (sec) °C	X	2903		
Btu/ft^2 (hr) F/in	X	0.124	=	kcal/m (hr) °C	X	8.064	=	Btu/ft^2 (hr) F/in
Btu (in)/ft^2 (hr) F		1.240	=	cal/cm (hr) °C	X	0.8064		Btu (in)/ft^2 (hr) F
		0.0833	=	Btu/ft (hr) F	X	12.0		

Resistance R (RSI, metric system)

ft^2 (hr) F/Btu	X	0.1761	=	m^2 (°C)/W	X	5.6783	=	ft^2 (hr) F/Btu

6. Minimize Infiltration

Infiltration refers to the entry of cold air through joints, cracks, and faulty seals in construction, and around doors and windows. Infiltration — and the accompanying *exfiltration* of heated air — is considered the largest and potentially most intractable cause of heat loss in a residence, once practical insulation measures have been taken. Some writers discuss infiltration and exfiltration under the designation *air leakage*, although a technical distinction is made between these: infiltration is always balanced by an equal amount of exfiltration, whereas air leakage is the sum of all parallel air flows through cracks or other openings into or out of a building, regardless of flow direction (as when a building is pressured by its mechanical system).

In the same way that noise can be regarded as unwanted sound, infiltration can be characterized as unwanted ventilation. Infiltration is driven by the same two forces of wind pressure and thermal buoyancy as is ventilation, and its rate is calculated in much the same manner. While ventilation of occupied spaces is always necessary, we seek to control this by making the building as airtight as possible and providing air change through devices in which it can be regulated. This approach allows further gains in energy efficiency when heat recovery systems are installed.

Sources and Effects of Infiltration.

Pathways for infiltration are provided through cracks and joints in the building envelope, through leaky window sash and faulty dampers in other penetrations through the building shell, and through the "pumping action" of the occupants' opening and closing of doors. *Table 6a* lists the most common sources and frequency of infiltration sites found in retrofit studies in New Jersey. Collins [1981] cites similar studies of new homes in Dallas carried out by Texas Power and Light Company, in which major sources of leakage were found at the sole plate (25%), wall outlets (20%), windows, including the frame (12%), and ductwork (14%). Collins' own retrofit studies in the Denver area revealed the frequency of air leakage locations listed in *Table 6b*.

Some ingenious low cost methods have been developed for locating infiltration sites in existing houses. Perhaps the simplest of these uses a high-pitched sound source indoors (such as a vacuum cleaner or noisy hair dryer) while the investigator on the outside searches for noise transmission through potential leakage sites with a listening device. Suitable instruments include a mechanic's stethoscope, plastic headsets of the type used on airlines, and rubber hoses from auto parts stores. On windy days, a burning incense stick carried around indoors helps reveal leakage currents in and out of walls and around windows and doors.

While infiltration is primarily associated with heat loss, there is a further liability: direct leakage of moisture-laden indoor air out through cracks into cold wall cavities and attics often results in condensation on building material surfaces. As houses are built more and more air- and vapor-tight, indoor humidity levels rise so that where leaks do occur, a tremendous amount of moisture can be carried through them; condensation problems in contemporary construction are now attributed mostly to moisture transfer via air leakage rather than by migration (permeation) through building materials. The consequences of condensation include destruction of surface finishes, reduction in the thermal resistance of insulation, and structural damage due to rot and the expansive force of freezing condensate. Air leakage also results in the entry of dirt (which can sometimes help to identify leakage sites), odors, and drafts.

Infiltration Calculations.

Two hand methods of calculating infiltration rates are commonly used. The *air change method* gives gross estimates of likely rates at which the entire indoor air volume of a house is replaced, or "turns over." It is based on past experience with average houses under average weather conditions, and it is not, therefore, a suitable tool for precise prediction of infiltration loads in new, relatively airtight construction. The measure of the amount of time that it takes for the volume of air in a house to turn over is useful, however, so we usually speak of the airtightness of a house in terms of air changes per hour. Houses having turnover rates in excess of 2/3 air change per hour might be viewed as "leaky" according to the current trends in construction; in well-built houses, rates of less than 0.2 air change per hour have been measured in the field by tracer gas techniques.

The *crack method* predicts infiltration rate by relating it to the wind pressure difference between inside and outside and the cross sectional area of open cracks and joints. Although the crack method is more analytical and flexible in use than the air change method, its accuracy depends on uncertain estimates of wind pressure and the effective length and width of cracks. The crack method is commonly applied to evaluating window performance.

The most revealing aspect of the crack method is that it relates infiltration rate linearly to wind pressure, which is a function of the square of the wind velocity:

$$P = 0.0026V^2$$

where P = static pressure applied by wind to vertical surface normal to wind direction (psf)
V = wind speed (mph)

One can see from this relationship that the energy-saving benefit of shielding the building from winds increases with increasing wind speed.

Infiltration heat load.

Calculation of the heat load caused by infiltration is simple if the air change rate and volume of the building are known. The calculation is made in two parts. The first part determines "q_s" the amount of (sensible) heat required to raise absolutely dry air at outdoor temperature to indoor temperature. If the indoor air must continually be humidified to compensate for moisture lost in exfiltration, then the second part of the calculation is made to determine "q_e" the amount of (latent) heat required to evaporate the necessary amount of water into the air.

The first component of the calculation is written:

$$q_s = 0.240V\rho_o(T_i - T_o)$$

where q_s = heat (sensible) required to raise air from temperature T_o to T_i (Btu/hr)
V = volume of air entering building (ft^3/hr)
ρ_o = density of air at temperature T_o (lb/ft^3)
T_i = temperature of indoor air (F)
T_o = temperature of outdoor air (F)
0.240 = specific heat of dry air (Btu/lb F)

Inserting a representative value ρ_o = 0.080 lb/ft^3 for the density of dry air (at a T_o of 35F):

$$q_s = 0.019V(T_i - T_o)$$

The second component, for the latent heat load q_e, is written,

$$q_e = V\rho_i(W_i - W_o)\mathscr{L}$$

where q_e = heat (latent) required to evaporate water into the air to maintain indoor (absolute) humidity W (Btu/hr)
W_i = humidity ratio of indoor air (lb/lb)
W_o = humidity ratio of outdoor air (lb/lb)
$\mathscr{L}$ = latent heat of vaporization of water at indoor temperature T_i (Btu/lb)
ρ_i = density of air at temperature T_i (lb/ft^3)

The latent heat of vaporization $\mathscr{L}$ of water at 70F is about 1054 Btu/lb, and substituting in a representative value of ρ_i = 0.074 lb/ft^3, we can rewrite,

$$q_e = 78V(W_i - W_o)$$

The total infiltration heat load then is the sum ($q_s + q_e$)

Infiltration Control Strategy.

Controls on infiltration can be implemented at all stages of the design process, beginning with careful siting and ending with protective landscaping. The building itself can be shaped and oriented (including underground placement) to minimize its exposure to winds. Air tightness is attended to by detailing and care in construction of the building shell, and in selection of window type and quality, as well as in securely-closing dampers for other ventilating devices.

Table 6a   **Sources of Infiltration**

Location	% Contribution*	Responsibility
Envelope, wall: sill plate-foundation joint, electrical outlets, plumbing penetrations, top plate-attic joint, gaps in sheathing	18-50 (35 typ)	architect's detailing, builder's workmanship
Envelope, ceiling: leaks into attic via light fixtures, plumbing and electrical penetrations, attic access	3-30 (18 typ)	architect's detailing, builder's workmanship
Doors and windows: generic type rather than manufacturer, most important	6-22 (15 typ)	architect's design, specifier's selection
Exhausts and vents: fireplaces (indoor air supply, faulty dampers)	0-30 (12 typ)	architect's design; hardware, accessories
Exhausts and vents: vents in conditioned spaces (usu. lacking dampers)	2-12 (5 typ)	architect's design, specifier's selection
Heating system: location of and leaks in ductwork, absence of outdoor combustion intake	3-28 (15 typ)	engineer's design, installer's workmanship
Door openings and closings	significant* (varies)	occupants' habits; architect's plan and space relationships

*Values from ASHRAE *Fundamentals* [1981] ; does not account for door operations.

Table 6b   **Frequency of Air Leakage Locations**

Path or location of leakage	% of houses treated
Bottom of drywall	100
Window fit, including sill	86
Plumbing fixtures, inside and outside walls	79
Electric fixtures, including medicine cabinet	76
Bathroom vent	59
Outside door fit	55
Access to attic space	52
Basement door fit	48
Fireplace fit	45
Stair steps and risers over unheated space	45
Garage door fit	38
Clothes dryer vent	34
Garage-house connection	31
Fireplace damper	28
Heating ducts	24
Bathtub fit	24
Kitchen fan vent	24
Closet door trim	17
In-wall air conditioner	17
Sill plate	17
Door to unheated storage	14
Door bell	14
Smoke alarm	14
Crawl space opening	14
Baseboard heater	14
Crawl space vent	14
Shower stall fit	14
Closet door runners	10
Kitchen cabinets, behind or on top	10
Philips control box	10
Sewer pipe penetration	7
Wood paneling on studs or furring	7
Intercom	7
Cellar floor drain	7
Toilet paper holder	7
Construction discontinuities	7
Telephone cord	7
Abandoned furnace flue	3
Soil pipe to basement	3
Bathroom cabinets, behind	3
Door latch	3
Skylight	3
Masonry seems porous	3
False ceiling beam	3
Stove damper	3

*Data from Collins [1981] for sites where retrofit treatment was found necessary for 29 houses in the Denver area.

7. Minimize Solar Gain

When solar energy strikes an exposed material, some of the energy is absorbed, some is reflected, and a portion may be transmitted through it. The properties of a material that govern the disposition of energy it receives are known as its *absorptance* (α), *reflectance* (ρ), and *transmittance* (τ). They are expressed as decimal fractions or percentages, and the sum of the three must alway equal one (see *Table 7a* for definitions):

$$\alpha + \rho + \tau = 1$$

The values of α, ρ and τ depend on the angle of incidence the solar beam makes with respect to the building surface. When comparing the solar-optical properties of different materials, it is important to make sure the values are given for identical test conditions.

Glazing materials are characterized by their ability to transmit some useful amount of light. Regardless of their apparent transparency, all glazing materials absorb and reflect some amount of radiation, although it may be small in comparison to the amount transmitted. Ordinary 1/8 inch double-strength sheet glass (DSA) is commonly used as a reference material in describing properties of glazing; it has 0.86 transmittance, 0.08 reflectance, and 0.06 absorptance at an incidence angle normal to the surface. DSA glass has an accepted reference U-value of 1.04 Btuh/ft^2 ($h_o = 4.0$, for a 7.5 mph wind) for summer, and 1.10 for winter ($h_o = 6.0$, for a 15 mph wind).

The absorption of solar energy by an outdoor building surface raises its temperature above that of the air by some degree which depends on the surface color and radiation intensity, and the countervailing rate at which its temperature is reduced by the wind and radiant loss to cooler surroundings. The effect of the solar heating of the wall can be represented as a "fictitious" temperature of outdoor air which would produce the same rate of heat entry into the surface as occurs due to the actual combination of incident solar radiation, radiant exchange with the sky and other surroundings, and convective heat exchange with outdoor air. This fictitious temperature is known as the *sol-air temperature* (SAT):

$$\text{SAT} = T_o + \frac{I\alpha}{h_o} - \frac{\varepsilon\Delta IR}{h_o}$$

where SAT = sol-air temperature (F)
     T_o = outdoor air temperature (F)
     I = incident solar heat flux (Btuh/ft^2)
     α = surface absorptance (dimensionless)
     h_o = coefficient of heat transfer by longwave radiation and convection (Btu/hr(ft^2)F)
     ε = hemispherical emittance of surface (dimensionless)
    ΔIR = difference between longwave radiant heat flux incident on surface from sky and surroundings and the radiant heat flux emitted by an ideal radiator (black body) at outdoor temperature (Btu/ft^2)

The temperature of the environment surrounding a vertical wall is usually so close to its own that ΔIR can be assumed to equal zero, so that the last term can be disregarded. For the case of vertical surfaces, therefore,

$$\text{SAT}_{\text{vert}} = T_o + \frac{I\alpha}{h_o}$$

Sol-air temperatures for roofs are discussed in the section on radiant cooling.

The *sol-air excess temperature* (SAXT) is simply that part of the sol-air temperature that is attributable to solar irradiation alone, that is, the elevation above the outdoor air temperature caused by solar absorption and accompanying convective losses. For vertical surfaces,

$$\text{SAXT}_{\text{vert}} = (\text{SAT}_{\text{vert}} - T_o) = \frac{I\alpha}{h_o}$$

The *solar excess heat flow rate* (SXQ) is the rate of heat flow per unit area delivered through an opaque wall, in excess of that driven by outdoor air temperature, as a result of the solar heat absorption of the wall. It is given by the relation (in Btuh/ft^2):

$$\text{SXQ}_{\text{vert}} = \frac{I\alpha U}{h_o}$$

The *solar gain factor* (SGF) is the ratio of heat flow through an opaque or translucent building assembly that results from solar absorption at the surface (i.e., the SXQ) to the intensity of radiation received at the surface. It represents the fraction of incident solar radiation that is actually delivered to the interior. The solar gain factor is defined by the relation,

$$\text{SGF} = \frac{\text{SXQ}}{I} = \frac{\alpha U}{h_o} \quad \text{(dimensionless)}$$

It can be seen from this relation that the solar gain factor is a constant, dimensionless property of the building assembly. Koenigsberger *et al.* [1974] recommend that the SGF should not exceed 0.04 in warm humid regions, or 0.03 in the hot, dry season of composite climates, when ventilation is reduced.

Fenestration Factors.

The term *solar heat gain* refers only to the heat entering the interior through a glazing material exposed to the sun due to its transmission of solar radiation and due to the glazing's inward liberation of heat absorbed within it. Solar heat gain (SHG) may be expressed symbolically, for single glazing (in Btuh/ft^2),

$$\text{SHG} = I\tau + \frac{I\alpha U}{h_o} = I(\tau + \frac{\alpha U}{h_o})$$

The terms collected in parentheses are together known as the *solar heat gain coefficient*, designated F (dimensionless). The determination of F is more complex for multiple glazings, and will not be discussed here [see ASHRAE *Fundamentals* 1981].

In order to simplify solar heating- and cooling-load calculations, ASHRAE has computed hourly and total daily solar heat gains for various window orientations and locations in the United States, using standard reference values for double-strength sheet glass. The heat fluxes found by substituting the properties of DSA glass into the solar heat gain equation are called *solar heat gain factors* (SHGF). Because reflectance τ and absorption α vary as the angle of incidence θ changes, the value of F is not constant for latitude, but changes with solar position according to the time of year and hour of the day. In symbolic notation (Btuh/ft^2),

$$\text{SHGF} = IF_{\text{DSA}} = I\left(\tau_{\text{DSA}} + \frac{\alpha_{\text{DSA}} U}{h_o}\right)$$

The *shading coefficient* (SC) is used to describe the transmittance of glazing materials, using DSA glass as a common reference. The shading coefficient is the dimensionless ratio of F for any fenestration material or fenestration accessory (blinds, drapes, louvers, etc.) to the F of DSA glass under the same conditions:

$$\text{SC} = \frac{F \text{ of fenestration material}}{F \text{ of DSA glass}}$$

Under most circumstances, manufacturers' published values of SC are computed using a transmittance value for an angle of incidence of zero (light beam normal to the glazing). For this condition, F for DSA glass is found to be 0.87, and so for other materials,

$$\text{SC} = \frac{F \text{ of fenestration}}{0.87} = (1.15)F \text{ of fenestration}$$

Obviously the SC of DSA glass is 1.00. The solar heat gain through any glazing material can be computed if its SC and the SHGF for the time, place, and orientation are known:

$$SHG = SC \times SHGF$$

Solar Control Strategy.

Solar gain controls can be grouped into three basic categories, 1) interception, 2) reflection, and 3) selection of fenestration area and orientation. Interception techniques range from trees and roof overhangs to lightweight ventilated shading panels attached to walls and roofs. For the most part, these are whole-building and window shading devices; they are usually the most effective and desirable of controls, provided they do not interfere with acceptance of solar gain when it is desired. In northern regions where building shells are normally well-insulated and therefore able to resist inward conduction of heat absorbed at the surface (solar excess heat flow), shading devices are often the only mechanisms needed for solar control.

In southern regions where building shells are not well insulated and where air temperatures frequently exceed comfort limits, reflection of solar heat required at the skin of the building is an important complement to shading. In all regions, the area and orientation of fenestration is of major significance, particularly for windows which cannot be shaded easily. The shape and orientation of the building shell itself is important in overheated areas, but its significance diminishes as the thermal resistance of the shell increases.

The amount of heat transmitted as sunlight to the interior can be reduced by intercepting this within the glazing system itself, by using tinted or heat-absorbing glass, or glazings with water-filled cavities (see [Weichman 1981; Sayigh 1981]). The problem with this tactic is that, unless ventilated internally, the temperature of the glazing itself is significantly raised in the process, and much of this heat is released to the interior through thermal radiation and convection from the interior surface of the glazing.

Table 7a   Definitions of Material Properties

Absorptance is defined as the ratio of the amount of radiant energy a particular surface absorbs to the total amount of radiant energy incident upon it; *absorptivity* refers to the ability of a material, independent of its geometry or surface condition, to absorb radiant energy.

Emittance is defined as the ratio of the amount of radiant energy released (emitted) by a particular surface at a specified wavelength and temperature to the emittance of an ideal "blackbody" at the same wavelength and temperature; *emissivity* refers to the ability of a material, independent of its geometry or surface condition, to emit radiant energy.

Reflectance is defined as the ratio of the amount of radiant energy reflected by a particular surface to the total amount of radiant energy incident upon it; *reflectivity* refers to the ability of a material, independent of its geometry or surface condition, to reflect radiant energy.

Transmittance is defined as the ratio of the amount of radiant energy transmitted through a specified thickness of a substance at a specified wavelength to the total amount of radiant energy incident upon its surface; *transmissivity* refers to the ability of a substance, independent of its geometry or surface condition, to transmit radiant energy.

8. Promote Ventilation

Ventilation comes from the Latin word, *ventus,* and means the movement of air. Ventilation is defined by the air-conditioning industry as the process of supplying or removing air by natural or mechanical means to or from a space, usually through air exchange with the out-of-doors. Ventilation serves three ends in the environmental control of buildings: it is used 1) to satisfy the fresh air requirements of the occupants ("health ventilation"), 2) to increase the rate of evaporative and sensible heat loss from the body ("comfort ventilation"), and 3) to cool the building interior by exchanging warm indoor air for cooler outdoor air ("structural ventilation").

Convection comes from the Latin expression meaning "to carry," and in physics and engineering it is used to describe the transfer of heat by the movement of a liquid or gas. Convection by air cannot occur in the absence of air movement, but air movement can occur without the transfer of heat. Convection can also occur within a closed system such as a room or building and in no way does it imply an exchange of heat or air with the out-of-doors.

Modes of Ventilation and Convection.

Air is made to move by differences in density and differences in pressure. When a mass of air is heated, as in a fireplace, it expands and, becoming less dense, rises. Conversely, a cold mass of air, as in a draft falling along a cold window, seeks its lowest level, thereby displacing surrounding warmer air upward. Under such conditions, we say that the air is driven by *thermal force, thermal buoyancy, or buoyant draft.* When thermal force discharges air from a building, the action is referred to as the *chimney effect,* or *stack effect.*

Convective heat transfer by thermal force is described as *free convection* or *natural convection* when it occurs in an open system, as in warm air lifting off a parking lot, or the narrow plume of heated air rising from a cigarette, made visible by its own smoke. When thermally-driven convection occurs in a closed circuit, it is described in architectural parlance as *thermosiphoning,* or *gravity circulation.* Thermosiphoning systems are common in passive solar design, while "gravity flow" describes generations of conventional steam and hot water domestic heating circulation systems.

When convective heat transfer is propelled by pressure differences, it is described as *forced convection.* Forced convection may be induced by pumps, fans, or blowers, or acting upon the exterior of buildings, by the wind (although wind is purely natural convection occuring on a regional scale). Other ventilation terms describe the origin of the force; thus we have ventilation by *wind force,* or *dynamic draft*—more commonly called *cross ventilation,* and *fan-forced* or *power* or *mechanical ventilation.*

Natural ventilation applies to air flow which is driven by pressures or thermal forces created by or converted from meteorological events. This includes both cross-ventilation and ventilation by stack effect. "Natural" convection, in the language of heat transfer engineering, refers only to heat transported by air motion induced by thermal force, whereas "natural ventilation" may be driven by either wind or thermal force.

Rate of Heat Transport by Convection.

The amount of heat able to be carried by the air is a simple function of its heat capacity and rate of flow (unit volume per unit time), and the temperature difference between the incoming and outgoing air. The convective heat transmission rate for air exchanged between indoors and out of doors can be written (Btu/hr),

$$q_{conv} = CFM\,(60)\,\rho c\,(T_o - T_i)$$

where CFM = air flow rate (ft^3/min)
      60 = minutes per hour conversion
      ρ = air density (lb/ft^3)
      c = specific heat of air (Btu/(lb)F)
      T_o = outdoor air temperature (F)
      T_i = indoor air temperature (F)

The heat capacity (ρc) of air depends primarily on its temperature, but it also depends on its moisture content. Perfectly dry air has a specific heat c of 0.24 Btu/(lb)F, while water vapor has a specific heat of 0.45 Btu/(lb)F. The water content of the air determines its specific heat, which may be computed by the relation:

$$c = (0.24 + 0.45\,W)$$

where c = specific heat of moist air (Btu/(lb)F)
        W = humidity ratio (lb/lb)

The humidity ratio W is a measure of the moisture content or absolute humidity of the air, and it can be estimated from *Table 8a* if the ambient dew point temperature T_d is known.

Air density ρ also depends on temperature, and it is easily found on the psychrometric chart if other factors are known. The density of air corresponding to conditions for which ventilation is effective for comfort (body) cooling ranges between 0.071 and 0.073 lb/ft^3. A typical value is 0.072 lb/ft^3, which represents a range of dry-bulb temperatures extending from roughly 75F at saturation to 91F for dry air. In the case of structural cooling, we are concerned with air temperatures within or below the limits of the comfort zone, and this means that the air is denser. A typical value (ASHRAE "standard" air) is 0.075, which represents air at 60F at saturation and 69F dry.

For warm, humid conditions of T_d = 70F and ρ = 0.072 lb/ft^3, the expression for q_{conv} simplifies to,

$$q_{conv} = 1.07 CFM(T_o - T_i) \quad [\text{warm, humid}]$$

For the cooler, drier conditions of T_d = 50F and ρ = 0.075 lb/ft^3, the equation reduces to,

$$q_{conv} = 1.10 CFM(T_o - T_i) \quad [\text{cool, dry}]$$

It is clear from the small differences resulting from the varied conditions exemplified here that very exact values of ρc are not warranted. In fact, we can state the grosser rule of thumb that *the cooling rate in Btu/hr is roughly equal to the air flow rate in CFM times the temperature difference F*.

Stack Effect.

The stack effect is most easily understood by considering the interior of a building as a column of air which connects two sets of openings at a vertical distance z apart. We assume the interior to have an average temperature of T_i and density of ρ_i. Surrounding the building is a pool of outdoor air which, over its height between the building's openings, has a lower average temperature T_o, and consequently a greater density ρ_o. The indoor column of air suspended between the openings exerts a unit pressure of $\rho_i(z)$ on the horizontal plane passing through the center of the lower opening, while the corresponding height of outdoor air exerts a unit pressure of $\rho_o(z)$. Because the heavier outdoor air has access to the interior, it will displace the lighter indoor air, which escapes through the upper opening. The driving pressure difference ΔP_{stack} is called the *stack effect pressure* or the *buoyant draft*; it is simply the difference between the unit forces exerted by the columns of indoor and outdoor air over the height z:

$$\Delta P_{stack} = (\rho_o - \rho_i)z$$

where ΔP_{stack} = driving pressure difference (psf)
         z = vertical distance between opening centers (ft)
        ρ_o = density of outdoor air (pcf)
        ρ_i = density of indoor air (pcf)

Under summer ventilating conditions, the difference in density between inside and outside is practically a function of absolute temperature alone. The density difference $(\rho_o - \rho_i)$, therefore, can be written

$$(\rho_o - \rho_i) = \rho_i \left[\frac{\mathring{T}_i}{\mathring{T}_o} \right] - \rho_i \left[\frac{\mathring{T}_o}{\mathring{T}_o} \right] = \rho_i \left[\frac{\mathring{T}_i - \mathring{T}_o}{\mathring{T}_o} \right]$$

where $\mathring{T}_i$ = absolute temperature of indoor air (°R)
        $\mathring{T}_o$ = absolute temperature of outdoor air (°R)

The important factor here is the *difference* in density between indoor and outdoor air, and not the absolute values themselves. As a result, we can substitute in a representative value for ρ_i (which for typical conditions when ventilation is desirable is about 0.071, or 1/14 pcf):

$$\Delta P_{stack} = z \left[\frac{\mathring{T}_i - \mathring{T}_o}{14\mathring{T}_o} \right]$$

This expression reveals 1) that stack effect increases linearly with the height z between openings, and 2) that stack effect increases linearly with the difference in temperature between inside and outside air. Inserting values of z = 10 ft and 85F and 75F for T_i and T_o (545°R and 535°R, respectively) yields a pressure difference of 0.013 psf; this, as will be seen in the next section, is very small compared to the pressure differences created by breezes.

For summer ventilating conditions when T_o and T_i are close to 80F, the rate of air flow resulting from stack effect through an interior where there is no significant resistance to air flow is given by the expression,

$$CFM = KA \sqrt{z(T_i - T_o) / T_i}$$

where CFM = rate of air flow (cfm)
A = free area of inlets or outlets, assumed equal (ft^2)
z = height from inlets to outlets (ft)
T_i = average temperature of indoor air over the height z (F)
T_o = temperature of outdoor air (F)
K = constant of proportionality, dimensionless value of 9.4 to 7.2

In this formulation, the area of inlets and outlets is assumed equal. The constant of proportionality of 9.4 includes a value of 65 percent to allow for effectiveness of openings; if conditions are not favorable, opening effectiveness should be reduced to 50 percent, giving a K of 7.2. The greatest flow rate per unit area of openings is obtained when outlets and inlets are equal in size; however, increasing either opening size can increase the flow rate up to a maximum of a little less than 40 percent more than when both are equal. Adjustment factors to be applied to the CFM expression are given in *Table 8b*. Ultimately, the air flow rate is always governed by the size of the smaller opening.

The form of the expression for CFM assumes that T_i is greater than T_o, so that air movement in the stack is upward. If the air in the interior is cooler than the out of doors, the direction of flow reverses, and the value of T_o should be substituted for T_i in the denominator of the quantity under the radical sign.

Cross Ventilation.

The pressure that a breeze exerts on a building surface is a function of wind speed and the angle at which it strikes the surface. The distribution of pressures over the surface depend on its proportions and surface area. If the surface is a wall that faces into the wind, then there occurs some location on the wall where the full kinetic energy of the wind is transformed into pressure; this is the maximum pressure that the wind can exert, and it is termed the *stagnation pressure*. The point on the wall where the stagnation pressure occurs is found where the wind stream separates to slip around the structure. Its magnitude is found by Bernoulli's equation,

$$\Delta P_{stag} = \rho_o V^2 / 2G \quad \text{[in English units]}$$

where P_{stag} = stagnation pressure of wind (psf)
ρ_o = density of outdoor air (pcf)
V = wind velocity (fps)
G = gravitational conversion constant (32.2 slug-ft/lb-sec^2)

This appears to be a nonhomogeneous equation in English units, but in fact it is internally consistent. One "pound" is a unit of force which represents a 1 slug mass acted upon by gravity; 1 lb = 1 slug-ft/sec^2. The equation actually reads,

$$\frac{lb}{ft^2} = \frac{slug}{ft^3} \left[\frac{ft^2}{sec^2} \right] \frac{lb\text{-}sec^2}{32.2\ slug\text{-}ft}$$

Under typical summer ventilating conditions we can again assume an air density ρ_o of 0.071 or 1/14 pcf, so we can rewrite

$$P_{stag} = \frac{.071(FPS)^2}{64.4} = \frac{(FPS)^2}{902}$$

For winter conditions, a more representative density would be 0.08 pcf, which corresponds to dry-bulb temperatures between 33 and 36F. Wind speed in mph is converted into fps by multiplying mph by a factor of 1.467.

Prediction of wind pressure distribution is complex and is usually studied in wind tunnels. Here, only the stagnation pressure will be considered, realizing that the actual pressure applied by wind over the face of a building will almost always be less (coefficients which correct the stagnation pressure for individual building: wind relationships are discussed by Akins, Aynsley, Sachs, and others in the literature of wind engineering). It is, however, important for the designer to note that wind speed at ground level is zero, and that velocity increases exponentially with height. Wind speed measurements are made at a standard height of 30 feet in the U.S.; wind speeds at lower and higher levels can be estimated by a power equation, if the surface condition is known:

$$V_x = V_r \left(\frac{x}{r} \right)^K$$

where V_x = mean wind speed at height x (any units)
       V_r = known mean wind speed at reference r (same units as V_x)
       K = exponent for best fitting curve (dimensionless), with values of:
           0.16 for flat open country
           0.28 woodland forest
           0.40 urban area

The power equation cannot be expected to estimate accurately the velocities within the boundary layer itself (forest, or example). For correcting standard weather station readings of heights less than 30 feet, therefore, it should be considered valid only over open terrain, utilizing K = 0.16. If measured wind speed at 30 feet is 3 mph, therefore, we compute the speed at 10 feet to be,

$$V_{10} = 3 \text{ mph} \left(\frac{10}{30} \right)^{0.16} = 2.5 \text{ mph}$$

It can be seen that this wind speed produces a stagnation pressure exceeding the buoyant draft created by a temperature difference of 10F over a one-story height, which was found earlier to be 0.013 psf:

$$2.5 \text{ mph} \times 1.467 = 3.7 \text{ fps}$$

$$P_{stag} = \frac{(3.7)^2}{902} = 0.015 \text{ psf}$$

Because the stagnation pressure is a function of the square of the wind speed, the driving potential for cross ventilation increases rapidly with small increases in wind velocity. From this we conclude that, while the stack effect is useful under windless conditions, even very low wind speeds offer much greater ventilating potential.

The rate and volume of flow through building interiors due to cross ventilation is not well researched. ASHRAE *1981 Fundamentals* (Chapter 22) indicates that the quantity of air flow forced through a ventilation inlet opening by the wind can be computed by the simple formula,

$$(CFM) = KAV$$

where CFM = volume of air flow (ft^3/min)
       A = free area of inlet opening (ft^2)
       V = wind velocity (mph)
       K = effectiveness for openings, taken as 0.5 to 0.6 for perpendicular winds and
           0.25 to 0.35 for diagonal winds

This relationship oversimplifies the problem, as it does not account for the air-damming action of the wall. More correctly, air flow through a building opening should be modeled as an *orifice flow,* analogous to the way in which the rate and volume of water rushing out of a punctured barrel can be computed as a function of water pressure (level) behind the opening. By such a method, the velocity of a wind current entering a window is given by an expression of the general form,

$$V_o = KA(\Delta P)^m$$

where V_o = air flow rate through opening
       K = discharge coefficient
     ΔP = total pressure difference driving the flow
       m = air exponent, usually 0.5, but ranging between 0. and 1.0

The physics of estimating air flow rates and volumes need not be known by the building designer, although mechanisms of increasing flow rates are very important. Such practices are discussed in Part II of this book. Wind tunnel studies to devise calculation procedures and to explore the effectiveness of different design alternatives are discussed in an excellent paper by Aynsley [1980b]. Other more qualitative discussions of particular note are found in Givoni [1976] and Van Straaten [1967].

An observer's index of wind speeds known as the *Beaufort Scale* is presented in *Table 8e* for general reference.

Conduction-Convection; Convection Coefficients.

Often the most critical factor in a heat transfer calculation is the rate of energy exchange between a solid surface and the air bounding it. This heat exchange occurs between the solid surface and the air by conduction, although once the air is heated, it is convected away. The rate of conductive heating of the air depends on how long the air remains in contact with the surface; the shorter the duration of time, the faster the overall rate of transfer. Air rapidly moving over a surface gives off or takes on much more heat than a layer of still air that clings to the surface; we know this by experience, and it is the basis for the often-cited "wind chill index," based on the rate of heat loss of a one liter cylinder of water exposed to different wind speeds. The wind chill index is given by the relation,

$$WCI = (10.45 + 10\sqrt{V} - V)(33 - T_a)$$

where WCI = wind chill index (kcal/(m^2)hr)
      V = wind velocity (m/sec)
      T_a = air temperature (°C)
The *equivalent wind chill temperature* T_{WCI} is computed (in °C),

$$T_{WCI} = -0.04544 \, (WCI) + 33$$

Because the rate of conductive heating of the air depends on the rate of air movement, the overall process is referred to as *conduction-convection*, and the factor describing the numerical rate of heat transfer is known as the *coefficient of convective heat transfer* h_c. The convection coefficient is expressed as a conductance, in Btu/ft^2(hr)F. When radiant heat exchange effect between the surface and surroundings is lumped together with the convection coefficient, the resulting parameter is termed the (combined) *surface or film conductance* h (see discussion on conduction, elsewhere). Convection coefficients have been determined empirically for a wide variety of both free and forced conditions.

In natural (free) convection, the rate of transfer between a surface and the air depends on the orientation of the surface and the direction of the heat flow. Heat is lost more rapidly from a warm horizontal surface to the air above it than from warm air to a cold surface beneath it; similarly, heat is exchanged more quickly between warm air and a cold horizontal surface overhead than vice-versa. Each condition is represented by a different coefficient, the most useful of which in building heat transfer calculations are given in *Table 8c*.

Under conditions of forced convection, where the air blows across a surface, surface orientation makes a difference only at low rates of flow. An often accepted value for the forced convection coefficient is $h_c = (1 + 0.3V)$, for wind velocity V in mph and h_c in Btu/ft^2(hr)F. The rate of transfer is increased by surface roughness and is also related to the length of the surface. In practice, generalized values like those given in *Table 8d* are used.

Ventilation Strategy.

Two primary strategies can be used for ventilating buildings for comfort control, and the suitability of each is related to the (absolute) humidity of the regional climate. We can describe these as "continual venting" and "nighttime venting."

Comfort is attainable by natural means in very humid overheated regions only with a constant movement of air across the skin. In such regions, diurnal temperature range is small, owing to the suppression of thermal radiation to the sky by a humid atmosphere; the greatest thermal advantage in design is obtained with use of lightweight building shells that cool off quickly at night. Daytime temperature control is maintained by ventilating as effectively as possible, both for dissipation of solar heat absorbed by the building shell as well as for body cooling. In the extreme case, the best structure is no structure at all, except for a canopy to provide shade. In practice, elevated structures, roof ventilators and wind-catching verandas are found as both traditional and contemporary elements of humid-region ventilation design. Field tests of ventilated building shells are discussed by Shaaban [1981].

In arid regions, conditions are usually so dry and exceed comfort zone limits so greatly during the daytime that ventilation is undesirable from both the standpoint of bodily water balance as well as that of thermal comfort. The clarity of the atmosphere allows temperatures to fall deeply at night, so buildings constructed with massive walls and roofs can maintain relatively moderate temperatures throughout the daytime. The ventilating strategy for such structures is to vent the interior at night, and to close the building during the day, when the heat of the air would only extract the "coolth" stored in the structure. (The term "coolth storage" first used by Yellott refers to the heat storage capacity of a material as a function of its temperature and thermal response). Such diurnal ventilation is usually accomplished via small building openings and a variety of devices which aid in precooling incoming air. Traditional mechanisms include evaporative screens and pools, wind towers, and ventilating tunnels. Modern practices also exploit evaporative cooling, driven by fans [Cook 1979].

In the case of either daytime or nighttime ventilation, the cooling sink is the air, and the index of the cooling potential is the dry-bulb air temperature. The suitability of either strategy depends, in essence, on whether air movement at acceptable velocities is capable of maintaining comfort during typical daytime conditions and, if not, if nighttime temperatures are low enough to counter-balance the overheatedness of the day. Important opportunities for promoting natural ventilation occur at every level of design, all of which can be utilized by the designer.

Table 8a Dew Point Temperature — Humidity Ratio Relationships

Dew Point T_d (F)	Humidity Ratio W	Dew Point T_d	Humidity Ratio W
35	0.00428	60	0.0111
40	0.00522	65	0.0133
45	0.00633	70	0.0158
50	0.00766	75	0.0188
55	0.00923	80	0.0223

Table 8b Increase in Flow Caused by Excess of One Opening Size Over the Other

Ratio of Outlet to Inlet		Increase in Percent
1:1	1:1	0
1:1.5	1.5:1	17.5
1:2	2:1	26
1:2.5	2.5:1	31
1:3	3:1	34
1:3.5	3.5:1	36
1:4	4:1	37
1:6	6:1	38

Source: Adapted from ASHRAE *1981 Fundamentals,* p. 22.7.

Table 8c Coefficients for Free Convention h_c (Btu/ft² (hr) F)

	Coefficient of Free Convection only, h_c	Combined coefficient h_i[a]		
		$\varepsilon = 0.9$	$\varepsilon = 0.2$	$\varepsilon = 0.05$
vertical surface, horizontal heat flow	$0.19\,(T_s - T_a)^{1/3}$	1.46	0.74	0.59
horizontal surface: warm surface facing up cool surface facing down	$0.22\,(T_s - T_a)^{1/3}$	1.63	0.91	0.76
horizontal surface: warm surface facing down cool surface facing up	$0.12\left[\dfrac{T_s - T_a}{L}\right]^{1/4}$	1.08	0.37	0.22
sloping surface: warm surface facing up at angle $30° < \beta < 90°$	$0.19\,([T_s - T_a]\sin\beta)^{1/3}$			
45° sloping surface warm side facing up		1.60	0.88	0.73
45° sloping surface warm side facing down		1.32	0.60	0.45

a Combined coefficient assumes both convective and radiative heat transfer for $T_s = 70F$ and surface-air temperature difference of 10F
Source: Compiled from ASHRAE *Fundamentals* [1981] and other sources.

Table 8d   **Coefficients for Forced Convection h_c (Btu/ft^2 (hr)F)**

	Coefficient of Forced Convection h_c	Combined coefficient h_o[a]
Any orientation[b]	(1.0 + 0.3 MPH)	
Any orientation[c]	(1.0 + 0.22 FPS)	
Horizontal roof[d]	(0.49 + 0.24 MPH)	
Vertical surface indoors, $V <$ 16 FPS[e]	(0.99 + 0.21 FPS)	
Vertical surface indoors, $16 < V <$ 100 FPS[e]	$0.5 \, (FPS)^{0.8}$	
Horizontal surface of length L(ft), $VL >$ 15[f]	$0.54 \left[\dfrac{(FPS)^4}{L} \right]^{0.2}$	
Winter outdoor design coefficient (V = 15 MPH)[g]		6.0
Summer outdoor design coefficient (V = 7.5 MPH)[g]		4.0

a includes ΔIR effects; $\varepsilon = 0.9$
b attributed to Hottel and Woertz, *Trans. ASME,* Vol. 64, No. 2, 1942
c attributed to Inst. Htg. & Vent'g. Engrs. (U.K.)
d Clark, Loxsom, Shutt & Faultersack [1981]
e ASHRAE *Fundamentals* [1981], p. 2.15
f Kreider and Kreith [1977], p. 255
g ASHRAE *Fundamentals* [1981], p. 23.12

Beaufort No. and description		Effects	Velocity at 30 ft. height		
			mph	fps	m/sec
0	Calm	Smoke rises No perceptible movement	< 1	< 0.7	0–0.2
1	Light air	Smoke drift shows light air movement; Wind vanes don't move; Tree leaves barely move	1 – 3	0.7 – 2.3	0.3 – 1.5
2	Light breeze	Wind felt on face; Leaves rustle; Newspaper reading becomes difficult; Weather vane shows direction	4 – 7	2.4 – 5.0	1.6 – 3.3
3	Gentle breeze	Wind extends light flag; Hair is disturbed; Clothing flags; Leaves and small twigs in contant motion	8 – 12	5.1 – 8.5	3.4 – 5.4
4	Moderate breeze	Small branches moved; Hair disarranged; Dust and loose paper are raised	13 – 18	8.6 – 12.5	5.5 – 7.9
5	Fresh breeze	Force of wind felt on body; Small trees in leaf begin to sway; Limit of agreeable wind	19 – 24	12.6 – 16.7	8.0 – 10.7
6	Strong breeze	Umbrellas hard to use; Difficulty in walking; Large branches in motion; Whistling telegraph wires	25 – 31	16.8 – 21.5	10.8 – 13.8
7	Near gale	Whole trees in motion; Strong inconvenience when walking against wind	32 – 38	21.6 – 26	13.9 – 17.1

9. Promote Radiant Cooling

Thermal radiation is defined as the transfer of heat energy through space by electromagnetic radiation. Radiant heat waves pass from one object to another by "line of sight" without warming the air in between. Unlike conduction and convection, radiation is impeded, rather than conveyed, by a medium interposed between the regions of heat exchange. We experience this firsthand when someone has stepped between us and blocked our "view" of a nearby fire or as we've felt the chilling effect of a cloud blocking the sun.

Thermal Radiation Spectrum.

All bodies having temperatures greater than absolute zero (−459.7F, or 0° on the Rankine (°R) temperature scale) emit thermal radiation. The wavelength at which the most radiant energy is emitted is related to absolute temperature, and can be found by *Wien's Displacement Law,*

$$\lambda_{max} = \frac{5216}{\overset{\circ}{T}}$$

where
λ_{max} = wavelength at which greatest amount of energy is emitted (μm)
$\overset{\circ}{T}$ = absolute temperature of radiator (°R)
5216 = a constant having the dimensions (μm°R)

Bodies radiate over a range of wavelengths. Wien's Law gives the wavelength at which the greatest emission occurs; about 25 percent of the energy is radiated at wavelengths shorter than λ_{max} and 75 percent is emitted at wavelengths exceeding λ_{max}. Solar radiation received at earth is most intense at 0.48 μm, in the blue-green band of the color spectrum. By inserting 0.48 μm into the displacement equation and rearranging gives the apparent temperature of the sun:

$$\overset{\circ}{T} = \frac{5216}{0.48 \, \mu m} = 10,800°R = 6000°K$$

Photographers will recognize this value, as most daylight-type color films are balanced for color temperature sources of 5500-6000°K.

Most radiation emitted by natural sources on earth falls in a range centered around 60F, or 520°R. Substituting this into Wien's Law yields an average wavelength for terrestial radiation of 10 μm. Environmental thermal radiation occurs in two distinct wavelength ranges which are denoted as *shortwave* (0.3 - 4.0 μm) for solar radiation, and *longwave* (4-80 μm) for sources of terrestial origin, for example, heated building interior materials.

The solar spectrum ranges from within the ultraviolet (about 5% of total solar energy received on earth) at a wavelength of approximately 0.3 μm, and terminates, for practical purposes at about 1.4 μm. Roughly half (47%) of the sun's radiation is visible, occupying the range 0.4 - 0.7 μm. The other half (48%), is invisible, but is felt as warming *infrared* (IR) radiation. The infrared spectrum extends from 0.7 - 25 μm for the "near" infrared, and from 25 μm to 1000 μm for the "far" infrared. Because longwave radiation falls entirely within the infrared spectrum, "longwave" and "infrared" are often used interchangeably to designate invisible thermal radiation.

Note that an ordinary 100-watt tungsten filament lamp operates at a color temperature of about 2900°K (5220°R). Use of Wien's Law indicates that the greatest emission of the lamp occurs at 1.0 μm —well outside the visible spectrum. From this, we deduce the widely known fact that while an incandescent lamp is an inefficient light source, it is an effective radiant heater. The actual distribution of the lamp's radiant emission over the visible and infrared spectrum can be predicted by Planck's Law, discussed in most physics textbooks.

Emissive Power.

The rate of emission, or *emissive power,* of a body radiating to the hemisphere it "sees" is related to its absolute temperature by the *Stephan-Boltzmann Law,*

$$E = \varepsilon\sigma\mathring{T}^4$$

where E = hemispherical emissive power (Btuh/ft^2)
     ε = hemispherical emittance, dimensionless ratio
    $\mathring{T}$ = absolute temperature of radiator (°R)
    σ = Stephan-Boltzmann constant, 0.1714 x 10^{-8} Btu/ft^2(hr)°R^4

For an ideal ("black body") material and surface, the emittance ε has a value of unity. No such bodies exist on earth, so ε is always less than one, but greater than zero. The sun, however, can be assumed to have ε =1, so its radiant power can be estimated by assuming it has an absolute temperature of 10,800°R:

$$E_{sun} = (1)(0.1714 \times 10^{-8})(10,800)^4 = 23.3 \text{ M Btuh/ft}^2$$

All bodies radiate to one another, and in building heat transfer problems, heat is lost by radiators through convection and conduction, as well as by radiation. For cooling buildings by radiation, of primary interest is the *net* radiant exchange between devices on earth and the sky.

Radiation to Clear Sky.

Although deep space can be considered an infinite sink for radiant emissions from earth, atmospheric carbon dioxide and water vapor intervene by absorbing much of the earth's longwave radiation. As a result, the atmosphere takes on a fictitious, or effective, temperature which the earth "sees" as very close to its own. The sky then radiates longwave heat energy back to the earth at a rate similar in magnitude, but slightly less, than that of earth. The earth's overall heat balance is maintained through this continual heat loss by daily charging of heat by solar radiation. The role played by atmospheric carbon dioxide and water vapor is often (somewhat erroneously) likened to the glazing in a hothouse; the phenomenon is, therefore, commonly described as the "global greenhouse effect."

Because the longwave absorptance of the atmosphere is related to the water vapor content of the air (the CO_2 content remains constant, for our purposes), the effective temperature of the sky, or simply, *sky temperature* $\mathring{T}_{sky}$ can be expressed as a function of absolute humidity (dew point temperature) and the dry-bulb temperature of the air near the ground. The following correlation between sky temperature and dew point and dry-bulb air temperatures has been offered by Clark [1981], for use with data obtained from the U.S. Weather Service (the error in $\mathring{T}_{sky}$ may be as great as ±7F):

$$\mathring{T}_{sky} = [0.742 + 0.0015T_d]^{1/4} (T_o + 459.7)$$

where $\mathring{T}_{sky}$ = effective absolute temperature of sky (°R)
    T_d = dew point air temperature near the ground, coincident with T_o (F)
    T_o = outdoor dry-bulb air temperature near the ground, coincident with T_d (F)

Given an assumed operating temperature T_r of a horizontal radiating panel facing the hemispherical sky vault, such as a building roof surface, and estimating T_{sky} from the relation given above, expressions can be written for the emissive power of both the radiator and the sky through use of the Stephan-Boltzmann Law (because of the way in which T_s has been derived, we need not include an expression for sky emittance in the equation for E_{sky}).

$$E_r = \varepsilon_r\sigma(\mathring{T}_r)^4 \quad \text{where } E_r = \text{emissive power of radiator}$$

$$E_{sky} = \sigma(\mathring{T}_{sky})^4 \quad \text{where } E_{sky} = \text{emissive power of sky}$$

For most naturally-occurring materials, thermal absorptance equals thermal emittance. The amount of sky radiation absorbed by the rooftop radiator, therefore, can be assumed to equal $\varepsilon_r E_{sky}$. The net radiative heat exchange ΔIR at the radiator is simply the difference between its emitted and absorbed heat flux (in Btuh/ft^2):

$$\Delta IR = (E_r - \varepsilon_r E_{sky})$$
$$\Delta IR = \varepsilon_r\sigma(\mathring{T}_r)^4 - \varepsilon_r\sigma(\mathring{T}_{sky})^4$$

Collecting terms and substituting in the value of the Stephan-Boltzmann constant, one can write

$$\Delta IR = 0.1714 \, \varepsilon_r \left[\left(\frac{\mathring{T}_r}{100} \right)^4 - \left(\frac{\mathring{T}_{sky}}{100} \right)^4 \right]$$

Radiation to Clear Sky: Rules of Thumb.

Although the cooling-rate equation deals with fourth power differences in temperature, the radiant temperature differences that one deals with in building problems (typically 20-30°R) are very small compared to the absolute temperatures themselves (typically 500-540°R). Consequently, the relationship between cooling rate ΔIR and temperature difference ($T_r - T_{sky}$) is practically linear. Noting this, Clark [1981] has stated that *the cooling rate of a radiator in Btuh/ft² is equivalent to the Fahrenheit (or Rankine) temperature difference between radiator temperature* T_r *and the virtual sky temperature* T_{sky}:

$$\Delta IR = (\mathring{T}_r - \mathring{T}_{sky}) \quad Btuh/ft^2$$

The major difficulty in estimating cooling rate is in making suitable approximations of sky temperature. Analysis of weather data reveals that sky temperature roughly parallels the daily cycle of dry-bulb air temperature (because dew point is comparatively stable throughout the day). A second general rule of thumb can be stated: *the sky temperature on an average summer night ranges 15 to 20F below the dry-bulb air temperature measured near the ground.* The magnitude of difference between sky and dry-bulb temperature varies from one climate region to another in relation to the moisture content of the air. The larger (20F) differences can be found in arid regions, while the smaller differences are characteristic of humid climates. Atwater and Ball [1978] have mapped estimated sky temperatures for the four seasons of the year, and Clark [1981] has mapped net nocturnal cooling rates for horizontal radiating surfaces of different temperatures for the month of July.

Radiation to Cloudy Sky.

The radiant cooling power equations given thus far are maximum or potential values which can be achieved under clear skies. Clouds increase the thermal absorptance of the atmosphere, and so its temperature increases proportionately. Cooling rates of radiators on earth must be adjusted downward to compensate for this. This may be done by applying a minimum *cloud correction factor* C_n to the clear sky cooling rate ΔIR [Clark 1981]:

adjusted cooling rate $= C_n \Delta IR$

where C_n = (1-0.059n) for low level clouds
(1-0.039n) for mid-level clouds
(1-0.029n) for high level clouds
(1-0.056n) as best estimate when cloud level is unknown
n = opaque cloud cover value in tenths, as reported by U.S. Weather Service.

Convection Effects.

The overall heat loss or gain of a radiator depends upon its interaction with the surrounding air. Total system heat flux due to both longwave exchange and convective exchange is given (Btuh/ft²):

$$\frac{q}{A} = \Delta IR - h_c(T_o - T_r)$$

where h_c = coefficient of convective heat transfer (Btu/ft² (hr)F).

When the radiator comes into equilibrium with its surroundings so that its losses to the sky and gains or losses from the air are balanced, its net heat flux q/A will come to rest at zero. This condition is referred to as *stagnation*. At stagnation, the temperature difference ($T_o - T_r$) becomes constant and this difference is termed the *stagnation temperature depression* $(T_o - T_r)_{stag}$:

$$(T_o - T_r)_{stag} = \frac{\Delta IR}{h_c}$$

The radiator's temperature at equilibrium is the lowest temperature it can attain under the ambient conditions. This temperature is known as the radiator's *stagnation temperature*, $T_{r(stag)}$:

$$T_{r(stag)} = T_o - \frac{\Delta IR}{h_c}$$

Note that the stagnation temperature is similar in both form and concept to the sol-air temperature (see "Minimize solar gain").

Unfortunately, determination of suitable values for the coefficient of convective heat transfer h_c is not so simple as the formulations in which it is necessary. A long-used approximation originally offered by Hottel and Woertz states, $h_c = (1 + 0.3V)$, where h_c is given in Btu/ft^2(hr)F and V is wind velocity in miles per hour. The constant 1 accounts for "free" convective transfer in the absence of wind. More recent work suggests this correlation may overestimate h_c [Clark, et al, 1981].

Radiant Cooling Strategy

The sky is the most problematic heat sink to exploit in building design. The difficulties may be described as environmental on one hand, and architectural on the other. Environmental constraints are described below.

1) Even at maximum clear sky cooling potential, the sky is not a powerful sink; except under very arid conditions, its temperature rarely drops more than 20F below ambient air temperature.

2) The difference between sky and ambient air temperature is related to the moisture content of the atmosphere, and is smaller in humid regions than in arid zones, under clear skies. Cloud cover severely reduces the sky cooling rate, and a heavy overcast can effectively shut off the thermal "view" of the sky altogether.

3) Heat gains from convection can erase much of the radiator's temperature depression. According to the relation, $h_c = (1 + 0.3V)$, a 3-mph breeze is sufficient to halve the maximum clear sky flux loss, and a 7.5 mph wind will reduce clear sky loss by a factor of 8.5.

With these considerations in mind, it can be said that radiant cooling generally is not well suited to warm, humid climates in which nighttime breezes prevail. The greatest radiant cooling potential is found in dry, northern regions where summer skies are clear and night sky temperatures are low. Radiant cooling is also an important strategy in the U.S. Southwest, where very large stagnation temperature depressions are possible; the stagnation temperatures themselves will not be as low, however, as in equally dry, more northerly regions.

Architectural constraints revolve around the difficulty of coupling the interior to the night sky without subjecting it to the solar load of the day. Two design approaches are used in practice; these can be described in the same terms used to classify passive solar heating systems. In *isolated loss* systems, a low mass radiator is placed on top of an insulated roof, and is coupled to the interior by means of a heat transfer fluid, this usually being either water or air. Heat is dissipated from the fluid as it passes over or behind the radiator. Because the fluid becomes denser as it is chilled, it can drain by gravity directly into the space beneath or to an underfloor storage plenum. *Indirect loss* systems couple the interior to the sky through the roof, which is left uninsulated, and which serves as the radiator. There are two extremes to this approach: 1) the roof has very high conductance and very little mass; 2) the roof has a great deal of mass and as much conductance as allowable. *Direct loss* systems consist of conditions where the space has a direct, uninterrupted view of the sky. Outdoor courtyards and open-air rooftop sleeping decks are examples.

High conductance, low mass construction enables the roof to respond almost instantaneously to ambient sky and outdoor air conditions; it cools the interior by absorbing heat from the indoor air and by absorbing radiation emitted by interior occupants and furnishings. Such a roof offers the greatest radiant cooling potential at night, but is most likely to be unacceptable because of poor daytime performance. In short, low mass radiators are most capable of approaching sky temperatures, and are most adversely affected by convective heat gain.

High mass roofs have typified the vernacular approach to radiant cooling in arid regions of the world. One effect of the mass is to store heat absorbed during the daytime in the roof, which results in a greater $(T_r - T_{sky})$ temperature difference and, therefore, high nighttime cooling rates. The roof's underside surface temperature will remain higher than for the low mass roof, however. In very massive roofs, the radiating surface temperature is likely to remain above the nighttime air temperature, so nocturnal breezes will enhance the overall cooling effect. In contrast to low mass roofs, high mass construction sacrifices lower nighttime temperatures for more comfortable daytime conditions. It is most suitable in predominantly overheated arid regions where there is no significant winter heating load, or where the heating load is easily met by solar design.

The most thermally successful indirect loss radiant cooling systems make use of movable insulation so that the interior can be coupled through the roof with the night sky, but isolated from the solar load of the day. The best known of selectively-insulated roof cooling concepts is the "Skytherm System," developed by Harold Hay [Watson, 1977]. It is designed to exploit the additional cooling effect of evaporation.

10. Promote Evaporative Cooling: Roof Systems

Vaporization of water describes the process by which liquid water is transformed into water vapor. Water vaporizes through *boiling*, in which vapor evolves within the volume of the mass and passes to the surface in bubbles, and through *evaporation*, in which the transformation occurs only at the air-water surface. Whereas water has to be heated to 212F (at normal atmospheric pressure) in order to boil, evaporation can occur at any temperature, including below freezing (when it is called *sublimation*).

A large amount of heat is absorbed by water in its change of phase, and the specific quantity required to effect the change from liquid at a given temperature to vapor at the same temperature is called the *latent heat of vaporization, $\mathscr{L}$*. It is referred to as "latent heat" because the heat absorbed in the transformation is held in the vapor, and is released only when the vapor reverses the phase change through condensation back into liquid. The heat of vaporization is related almost linearly to temperature. At the boiling point, 970 Btu are absorbed by one pound of water in its change of phase; this value increases to 1075 Btu/lb at 32F. Intermediate values are given in *Table 10a*.

The cooling effect of evaporation is easily calculated from the latent heat of vaporization if the quantity of water evaporated and its temperature are known. The density of water is about 62.4 lb/ft^3, so one pound of water will cover an area of 1 ft^2 at a depth of 0.1923 in (a little more than 3/16 in). This represents a cooling effect of 1050 Btu/ft^2 for water at 76F. We usually assume that this heat is absorbed from within the water body and from the substrate it is carried upon, although some of the heat effecting the change of phase may be drawn from the air. Absorption of heat from the air is known as *advection*, and the evidence of its activity is a depression in the dry-bulb temperature of air over the evaporating surface. Advection is a meteorological term, and both it and its effects are normally disregarded (whether rightly or wrongly) in building surface evaporative cooling calculations. The air cooling process which occurs in advection is the basis for evaporative space conditioning systems, although advection is not used to describe the process.

Evaporative Surface (Roof) Cooling.

While the exact physics of evaporation is complicated, the process can be described by semi-empirical relationships which can be expressed in a form similar to those used in heat conduction calculations. The rate of evaporation depends primarily upon the difference in *vapor pressure* between the ambient outdoor air and the very thin film of vapor-saturated air that resides at the surface of the water. These pressures can be expressed in numerous ways, including the expressions *humidity ratios, absolute humidities, dew point temperatures,* or pressures in dimensions of mm Hg (mercury), in Hg, in H$_2$O, and psi, among others. Different researchers have chosen to work with different expressions, so a variety of formulas for estimating evaporative cooling effect have been devised.

The most straightforward formula has been attributed to Willis Carrier [Yellott 1981a], who developed it from empirical studies of cooling ponds:

$$\frac{q_{evap}}{A} = \mathscr{L}(0.093)h_c(P_s - P_a)$$

where q_{evap}/A = evaporative heat flux loss (Btuh/ft^2)
$\mathscr{L}$ = latent heat of water at surface temperature (Btuh/lb)
h_c = coefficient of convective heat transfer, approximated $h_c = (1 + 0.38V)$
V = wind velocity (mph)
P_s = vapor pressure at saturation, at temperature of surface (in Hg)
P_a = atmospheric vapor pressure (in Hg)

The saturation vapor pressure P_s is solely a function of temperature at atmospheric pressure; values for it are given in *Table 10b*. P_a must be found from observed weather data; it is the product of P_s and the relative humidity.

Sodha, Khatry, and Malik [1978] offer the following approach, adapted from R. V. Dunkle:

$$\frac{q_{evap}}{A} = \mathscr{L}(0.2)h_c[P_{s(s)} - rP_{s(o)}]$$

where h_c = coefficient of convective heat transfer (Btu/ft²(hr)F)
$P_{s(s)}$ = vapor pressure of saturation at temperature of evaporating surface (psi)
$P_{s(o)}$ = vapor pressure of saturation at temperature of outdoor air (psi)
r = relative humidity (dimensionless)

Over the temperature range 85F < T < 120F, P_s can be adequately represented by the relation,

$$P_s = (0.0309T - 2.11)$$

The average daily evaporative heat loss rate, therefore, can be written,

$$\frac{q_{evap}}{A} = \mathscr{L}(0.2)h_c[(0.0309\bar{T}_s) - (0.0309\bar{T}_a)r - 2.11(1 - r)]$$

where $\bar{T}_s$ = daily average surface temperature (F)
$\bar{T}_a$ = daily average outdoor air temperature (F)

Another formulation, described by Loxsom and Kelly [1980], begins with a conventional textbook statement of the problem using humidity ratios:

$$\frac{q_{evap}}{A} = \frac{\mathscr{L}}{(Le)\rho c} h_c(W_s - W_o)$$

where W_s = saturation humidity ratio at the temperature of the surface T_s (dimensionless)
W_o = humidity ratio of ambient air (dimensionless)
Le = Lewis number (dimensionless)
ρc = heat capacity of moist air (Btu/lb)
h_c = coefficient of convective heat transfer (Btu/ft²(hr)F)

This equation is combined with the expression for convective heat gain or loss,

$$\frac{q_{conv}}{A} = h_c(T_s - T_o)$$

to yield a composite expression of evaporative cooling and convective gain or loss,

$$\frac{q_{evap + conv}}{A} = 3h_c\left[\frac{T_s + T_w}{62.4} - 1\right](T_s - T_w)$$

where T_s = temperature of evaporating surface (F)
T_w = ambient wet-bulb temperature (F)
h_c = (1 + 0.3V) (Btuh/ft²)
V = wind velocity (mph)

Surface evaporation process and calculations are discussed in detail by M. M. Shah [1981].

Evaporative Roof Cooling Potential.

While the rate of evaporation is governed by the difference in vapor pressure between saturated air at roof surface temperature and ambient outdoor air, the lowest limit to which the roof can theoretically be cooled by evaporation alone is given by the ambient *wet-bulb temperature* T_w, the "temperature of evaporation." Wet-bulb temperature is defined as the temperature at which water, by evaporating into air, can bring the air to saturation "adiabatically" at the same temperature. ("Adiabatically" means that no heat is gained or lost from the water-air system.) It is the lowest temperature that a surface film of water can assume when it evaporates freely. This process is mimicked by a wet-bulb thermometer which, for practical purposes, gives a sufficiently accurate reading of the true wet-bulb temperature.

How closely roof surface temperature approaches the wet-bulb temperature depends on what other sources and sinks the roof is coupled to. Increasing air speed always increases the rate of evaporation by whisking away the highly humidified air that clings to the evaporating surface, but the air itself may either supply heat to or extract heat from the surface, depending on whether T_o is greater or less than $\bar{T}_s$. The net heat flow due to combined convective and evaporative effects, therefore, depends both on the temperature and humidity of the media involved in the energy transaction. This balance normally becomes critical only during nocturnal cooling applications, since roof surface temperature can always be assumed to exceed air temperature during daylight hours.

An index of evaporative cooling potential is the *wet-bulb temperature depression*, which is defined as the difference between ambient air dry-bulb and wet-bulb temperatures $(T_o - T_w)$. July daytime wet-bulb temperature depressions fall in the range of 30-40F in the arid Southwest, and 10 − 15F along the Gulf Coast (*Table 10c*). A more meaningful index for daytime roof wetting treatments is the (*sol-air*) − (*wet-bulb*) temperature depression,

$$(SAT - T_w) = \left[\frac{I\alpha}{h_c} + \frac{\Delta IR}{h_c} + T_o \right] - T_w$$

Some computations of $(SAT-T_w)$ are presented in *Table 10c*. The very large depressions resulting from calculating solar effects reveals enormous potential for evaporative roof treatments in both arid and humid regions. Some of the early experimentation with roof spray was, in fact, carried out in Florida, where substantial reductions in interior cooling loads were measured [Sutton 1950].

It is important to note that the wet-bulb temperature is only an index of the evaporative cooling resource, and that the $(SAT-T_w)$ depression is merely an index of potential. The performance of evaporative roof treatments depends on the design of the structure to which they are applied.

Evaporative Cooling Strategy.

There are two very different approaches to evaporative roof cooling. On one hand, roof sprays and ponding can be considered as a means of dissipating solar heat absorbed at the surface; that is, they provide a method of reducing the amount of heat conducted through the roof into the interior. In this sense, sprays and ponding are compatible with the conduction control strategy. On the other hand, evaporative treatments can be used to extract heat from the interior. In this sense, outdoor air provides a heat sink which evaporative treatments serve to exploit. In order to achieve this, the interior must be coupled with the evaporating surface through a highly conductive roof structure.

Heat Dissipation.

The heat dissipation value of evaporative cooling depends on what other mechanisms of resisting heat gain are provided in the design. Among these are maximizing the solar reflectivity of the roof surface, ventilating the underside of the roof, and resisting the conductive intrusion of heat through insulation (provision of mass in the roof will reduce the maximum intensity of heat gain, but does little to reduce the accumulated daily heat load). It can be noted that whereas the cooling resource for evaporative treatment is indicated by the wet-bulb temperature, the comparable potential of a ventilated roof is indicated by the dry-bulb temperature. The dry-bulb temperature is also the index of the potential for perfectly reflective and shaded roofs. Wetted roofs, therefore, are always potentially cooler than dry ones. In opposition, it may be said that a well-insulated, well-ventilated roof receives little additional benefit from evaporative cooling; conversely, the greatest benefit of evaporative treatments will be realized with poorly-insulated roofs. This supports the popularity of installing roof spray systems as a remedial retrofit. Roof sprays are well suited in general to humid southern regions where small winter heating loads require little in the way of insulation, but where solar loads are always high.

Little measured data exists which can be used to assess the practical value of evaporative treatments, especially on a regional basis. The old ASHRAE *Guides* used to include a table, reproduced here in part as *Table 10d*, which indicates "Equivalent Temperature Differentials" for calculating heat gains through different roof systems. The Equivalent Temperature Differential (ETD) is an expression of the difference between sol-air temperature and a design room air temperature, here selected by ASHRAE as 80F. It can be seen that, during maximum insolation, the wetted roof temperature differentials are at most about one-third of those for light-weight and medium construction roofs. In experimental work in Phoenix, Yellott [1966] demonstrated that use of about 0.3 lb. of spray water per hour per square foot of roof reduced the roof temperature to a point where the day-long average difference between the roof and ambient (dry-bulb) air temperature was close to zero. Under the same conditions, the day-long average difference between roof and air temperatures ranged between 30 to 40F.

An interesting additional benefit of roof spray systems can be realized on greenhouse exteriors. Research with experimental greenhouses [Hess 1950] has shown that a film of water 1/16 inch thick will absorb 19 percent of incident solar energy—most of which falls in the spectrum of the invisible infrared. This shortwave radiant heat is intercepted and rejected before it reaches the glass. The combined additional reflection and interception and evaporative dissipation effects of a water film applied to greenhouse glazing has been found to remove 44 percent of incident radiation without significant losses in daylight transmission.

Heat Extraction.

The problem inherent in using the roof as an evaporative cooling surface to extract heat from the interior is identical to that of coupling the interior to the roof as a radiator. There are, however, additional alternative solutions, since it is neither necessary nor desirable for the wetted roof—or subroof—to "see" the sky. The evaporating surface can be shaded, and the devices utilized for shading could serve also to direct the air-stream of natural breezes over the wetted roof. Because of their similarities, heat extraction by evaporative cooling and by radiative loss to the sky are often combined in systems employing movable insulation or thermosiphoning circuits linking the roof with the interior (see discussion of radiative cooling).

Table 10a   Latent Heat of Vaporization $\mathscr{L}$ of Water at Atmospheric Pressure

Temperature (F)	Btu/lb	Btu/gal[a]	Btu/in (ft²)[b]
32	1075.1	8961.0	5586.1
40	1070.6	8924.5	5563.3
50	1065.0	8875.5	5532.8
60	1059.3	8822.1	5499.5
70	1053.7	8766.7	5465.0
80	1048.0	8706.5	5427.4
90	1042.2	8646.7	5390.2
100	1036.7	8575.6	5345.8
110	1030.9	8510.1	5305.0
120	1025.2	8444.6	5264.2
130	1019.4	8376.4	5221.7
140	1013.6	8308.1	5179.1
150	1007.7	8237.4	5135.0
160	1001.8	8166.1	5090.6
170	995.8	8092.9	5044.9
180	989.8	8018.9	4998.8
190	983.7	7916.2	4934.8
200	977.5	7838.5	4886.4
212	970.0	7749.7	4831.0

a   One gallon of water weighs 8.336 lb at 40F, decreasing to 7.989 lb at 212F
b   One gallon = 1.604 in (ft²)

Table 10b   **Saturation Vapor Pressure P_s of Water at Atmospheric Pressure**

Temperature (F)	in Hg	psi
32	0.18050	0.08865
40	0.24784	0.12173
50	0.36264	0.17811
60	0.52193	0.25635
70	0.73966	0.36329
80	1.03302	0.50737
90	1.42298	0.69890
100	1.93492	0.95034
110	2.59891	1.2764
120	3.45052	1.6947
130	4.53148	2.2256
140	5.88945	2.8926
150	7.57977	3.7228
160	9.6648	4.7469
170	12.2149	5.9994
180	15.3097	7.5194
190	19.0358	9.3495
200	23.4906	11.5375
212	29.9493	14.7097

Source: ASHRAE *Fundamentals* [1981] pp. 6.13-6.16
Note: P_s in kilopascals can be computed from the following relation [Campbell 1977] when T is in °K:

$$P_s = \exp\left(52.57633 - \frac{6790.4985}{T} - 5.02808 \ln T\right)$$

Table 10c   **Wet-Bulb Temperature Depression Indices, July 21**

Location	Max T_o[a]	Max T_w[a]	SAT[b]	$(T_o - T_w)$	$(SAT - T_w)$
Albuquerque	91.4	55.4	169	36	114
Phoenix	102.2	65.6	176	36.6	110
Oklahoma City	92.3	70.0	159	22.3	89
San Antonio	94.9	72.6	158	22.3	85
Atlanta	87.8	72.1	153	15.7	81
Indianapolis	84.4	70.5	150	13.9	79
Miami	87.9	75.5	151	12.4	75
Houston	88.6	77.3	151	11.3	74

a Daily maximum dry-bulb and wet-bulb temperatures
b SAT computed using the constants h_c = 3; $\Delta IR/h_c$ = 7; α = 0.85

Total Equivalent Temperature Differentials for Calculating Heat Gain Through Sunlit and Shaded Roofs

Description of Roof Construction[a]	Sun Time								
	A.M.			P.M.					
	8	10	12	2	4	6	8	10	12
Light Construction Roofs — Exposed to Sun									
1″ Wood[b] or 1″ Wood[b] + 1″ or 2″ insulation	12	38	54	62	50	26	10	4	0
Medium Construction Roofs — Exposed to Sun									
2″ Concrete or 2″ Concrete + 1″ or 2″ insulation or 2″ Wood[b]	6	30	48	58	50	32	14	6	2
2″ Gypsum or 2″ Gypsum + 1″ insulation 1″ Wood[b] or 2″ Wood[b] or + 4″ rock wool 2″ Concrete or in furred ceiling 2″ Gypsum	0	20	40	52	54	42	20	10	6
4″ Concrete or 4″ Concrete with 2″ insulation	0	20	38	50	52	40	22	12	6
Heavy Construction Roofs — Exposed to Sun									
6″ Concrete	4	6	24	38	46	44	32	18	12
6″ Concrete + 2″ insulation	6	6	20	34	42	44	34	20	14
Roofs Covered with Water — Exposed to Sun									
Light construction roof with 1″ water	0	4	16	22	18	14	10	2	0
Heavy construction roof with 1″ water	−2	−2	−4	10	14	16	14	10	6
Any roof with 6″ water	−2	0	0	6	10	10	8	4	0
Roofs with Roof Sprays — Exposed to Sun									
Light construction	0	4	12	18	16	14	10	2	0
Heavy construction	−2	−2	2	8	12	14	12	10	6
Roofs in Shade									
Light construction	−4	0	6	12	14	12	8	2	0
Medium construction	−4	−2	2	8	12	12	10	6	2
Heavy construction	−2	−2	0	4	8	10	10	8	4

a Includes 3/8 in. felt roofing with or without slag. May also be used for shingle roof.
b Nominal thickness of the wood.

Notes: 1. Estimated for about August 1 in 40 deg north latitude. For typical design day where the maximum outdoor temperature is 95 F and minimum temperature at night is approximately 75 F (daily range of temperature, 20 F) mean 24 hr temperature 84 F for a room temperature of 80 F. All roofs have been assumed a dark color which absorbs 90 percent of solar radiation, and reflects only 10 percent.

2. Application. These values may be used for all normal air conditioning estimates; usually without correction, in latitutde 0 deg to 50 deg north or south when the load is calculated for the hottest weather. Note 3 explains how to adjust the temperature differential for other room and outdoor temperatures.

3. Corrections. For temperature difference when outdoor maximum design temperature minus room is different from 15 deg. If the outdoor design temperature minus room temperature is different from the base of 15 deg, correct as follows: When the difference is greater (or less) than 15 deg. add the excess to (or subtract the deficiency from) the above differentials.

For outdoor daily range of temperature other than 20 deg. If the daily range of temperature is less than 20 deg., add 1 deg. for every 2 deg. lower daily range; if the daily range is greater than 20 deg., subtract 1 deg. for every 2 deg. higher daily range. For example, the daily range in Miami, Florida is 12 deg. or 8 deg. less than 20 deg., therefore, the correction is +4 deg. at all hours of the day.

Source: ASHRAE 1960 Guide, p. 190; also see Houghten, Olson and Gutberlet [1940].

11. Promote Conductive Cooling

The only sink to which a building can continuously lose heat by means of conduction during overheated seasons is the ground. Problems of heat exchange between the ground and buildings buried in it are much more complicated than those of heat exchange between a building interior and the out of doors. The temperatures of the sun, sky, and air are unaffected by their heat exchange with any individual building, whereas the temperature of the ground—and, consequently, its potential as a heat sink—is significantly altered by the thermal presence of a building. As yet, there exist no simple analytical techniques for predicting the cooling potential of the ground adjoining a building. It is, on the other hand, a relatively simple matter to estimate the natural, or undisturbed, temperature of the ground, from which it is possible to determine the potential, although unrealizable, value of the ground as a cooling sink.

Ground Surface Temperature.

Ground temperature analysis has been a subject of investigation within many disciplines for over a century. Lord Kelvin (Sir W. Thomson) published a sophisticated work on "The Reduction of Observations on Underground Temperature" as early as 1861. For most purposes, it is sufficient to describe variations in ground temperature as a cosine function. This is possible because the ground is heated by the sun, and the intensity of solar radiation as it falls on a horizontal surface varies sinusoidally over the course of a year. If one neglects seasonal differences in cloud cover and other atmospheric effects, solar intensity I_h received on the horizontal ground plane can be described by the simple relation,

$$I_h = I_{mh} - A_h \cos \left[\frac{360}{365} (t - t_o) \right]$$

where I_h = average daily insolation on a horizontal surface on day t of the calendar year (Btu/(ft^2)day)
I_{mh} = mean annual insolation on a horizontal surface (Btu/(ft^2)day)
A_h = amplitude of annual insolation on a horizontal surface (Btu/(ft^2)day)
t = time of year, beginning January 0 (days)
t_o = a phase constant (days)
360/365 = conversion of 365 calendar days into 360°

The solar amplitude A_h represents the difference between the maximum or minimum radiation intensity and the annual mean intensity; it is equal to 0.5 [(I_h on June 21) − (I_h on December 21)].

Because the cosine function begins with its lowest value on $t = 0$, and because the least intensity of radiation occurs on December 21 (day 355 of the calendar year), the phase correction term t_o is necessary in order to reconcile the simplified mathematical model with fact. Setting $355 = t_o$ achieves the desired reconciliation by shifting the entire curve to the right until the minimum falls on December 21. The same end result is accomplished by shifting the curve backward in time by 11 days, by setting $t_o = -11$.

Ground surface temperature follows the sinusoidal pattern of solar heat flux. Annual variation in ground surface temperature can therefore be represented by a similar, idealized formula,

$$T_s = T_m - A_s \cos \left[\frac{360}{365} (t - t_o) \right]$$

where T_s = temperature of the ground surface on day t (F)
T_m = mean annual ground surface temperature (F)
A_s = amplitude of annual ground surface temperature (F)

The term T_s represents the long-term average temperature of the ground surface on day t, and is not necessarily that to be expected on any given day of the year. The *mean annual ground temperature* T_m is the long-term annual average temperature of the ground, which is essentially constant from the surface through the first few hundred feet of the earth. Its value can be approximated very closely by taking the temperature of local well water; a less exact approximation is made by adding, typically, 2 or 3F to the average annual air temperature. T_m ranges from almost 80F in the most southern regions of the U.S. to about 45F in the northern tier of states; it is not everywhere 55F, as is often mistakenly supposed.

The *surface ground temperature amplitude* A_s represents the maximum departure of temperature from the annual mean. It would, like the amplitude of solar radiation on a horizontal surface, increase with latitude, all other things being equal. "All other things," however, include the influences of oceans, mountains, and atmospheric clarity, among others, so A_s has only an underlying relationship to latitude. It ranges between 15 and 30F for the U.S., with the lowest values appearing throughout Florida and the Gulf Coast region, and along the Pacific Coast. The greatest values of A_s are found in the northern Midwest. Approximations of A_s can be made by taking one-half the difference between the July and January monthly average air temperatures, and adding about 2F.

Measured ground temperatures reveal that the cyclical wave of surface temperature follows, on the average, about 1/8 cycle, or 46 days, after the phase of the solar radiation wave. The value of t_o for the T_s equation, therefore, can be represented by 35 days (46 + [− 11]), plus or minus a week or so. A time-lag of 1/8 cycle is predicted by periodic heat conduction theory [Labs 1981a], so that good agreement with observations lends validity to the mathematical formula. This time lag is reflected in the annual cycle of average daily air temperature. Because the ground heats the air, air temperature acts as a mirror of ground temperature.

Subsurface Temperatures.

The annual wave of temperatures occurring at the ground surface is propagated into the soil, creating a sinusoidal pattern of temperature variation at every depth at which a fluctuation occurs. A model of temperature variation at any depth x can be constructed using the ground-surface temperature variation equations as a basis:

$$T_x = T_m - A_s e^{-xr}\cos\left[\frac{360}{365}(t - t_o - xL)\right]$$

where T_x = temperature at depth x at time t (F)
  x = depth (ft)
  e = Euler's number, 2.71828 (exponential function)
  r = logrithmic decrement (ft^{-1})
  L = lag time (days/ft)
  α = thermal diffusivity of soil (ft^2/day)

The *logrithmic decrement* r represents the quantity,

$$r = \sqrt{\frac{\pi}{365\,\alpha}}$$

and the expression for *lag time* L is given by the relation,

$$L = 0.5\sqrt{\frac{365}{\pi\alpha}}$$

The only variable introduced in the equation for T_x is *thermal diffusivity* α which is defined as the thermal conductivity k divided by the heat capacity ρc . Thermal diffusivity is very difficult to predict with accuracy in soils, because both k and ρc vary with moisture content. Typical annual average values of α can be given for soils characterized as dry, mean, and wet; these, along with related values of x and L are given in *Table 11a*.

The term (e^{-xr}) is called the *decrement factor*, as it indicates the fraction by which the amplitude is reduced at depth x. The amplitude at depth x, denoted A_x, is simply the product of the surface amplitude and the decrement factor,

$$A_x = A_s e^{-xr}$$

At the surface, x = 0, so the value of (e^{-xr}) is one, leaving $A_o = A_s$. As x approaches infinity, (e^{-xr}) goes to zero, indicating that temperature fluctuation has ceased, and $T_x = T_m$.

If one wishes to find the depth at which the subsurface temperature fluctuation is reduced to 10 percent of that of the surface, then set (e^{-xr}) = 0.10, and solve for (−xr):

$$e^{-xr} = 0.10, \text{ therefore } -xr = \ln 0.10 = 2.3; \quad -x = 2.3/r$$

The depth x is dependent on the value of r and, since r is defined in terms of thermal diffusivity α, the range of temperature fluctuation in the soil is a function of its thermal diffusivity. Solutions for x for the three representative soils are given in *Table 11a*.

The *lag time* term L represents the fact that the temperature wave at any depth in the soil lags behind the waves at all depths above it. Lag time is a linear function of depth in homogeneous soils and is also related to thermal diffusivity.

The ground temperature model described here assumes that the soil is homogeneous with depth, that conduction is the only heat transfer process acting in the soil, and that the temperature wave at the surface is exactly sinusoidal over the course of the year. None of these assumptions is strictly true in soils. Nevertheless, experimental work confirms the general validity of the model, which can be expected to yield estimates within 2 or 3F of measured temperatures if accurate values for T_m, A_s, and α can be chosen. The model will not apply well to regions characterized by significant seasonal snow cover or large seasonal fluctuations in groundwater level near the surface, for which more sophisticated calculation methods must be used.

A simplified form of the equation for T_x can be written for the moment in mid-summer when air temperature and ground surface temperature are both at maximum, designated as T_x':

$$T_x' = T_m + A_s(e^{-xr})\cos(xL)$$

where T_x' = ground temperature at depth x at the time when $T_{x=o} = (T_m + A_s)$
xL = lag time, expressed in degrees (the conversion factor (360/365) is technically necessary, but can be disregarded)

Summer Cooling Potential.

Any device which extracts coolth from the ground in turn raises the temperature of surrounding soil above the natural, undisturbed values predicted by the equation for T_x. Although methods for describing this disturbance and the loss of cooling potential that it produces are only now being researched, an approach to relating cooling rate to undisturbed ground temperature is easily described in principle. In lieu of considering the temperature T_x at a single depth x, integrate the equation for T_x with respect to depth over the range x = a to x = b, and divide by the difference (b–a). This yields an average temperature of the soil profile, designated $\bar{T}_{(a-b)}$:

$$\bar{T}_{(a-b)} = T_m + \frac{A_s}{(b-a)r\sqrt{2}}(e^{-xr})\cos\left[\frac{360}{365}(t-t_o-xL-45.6)\right] \Big|_a^b$$

where $\bar{T}_{(a-b)}$ = average temperature of profile extending from depth a to b (F)
a = upper bounds of profile (ft)
b = lower bounds of profile (ft)

The heat loss from an underground wall extending from depth a to b can then be represented by the expression [Akridge, 1981],

$$\frac{q_{(a-b)}}{A} = [T_i - \bar{T}_{(a-b)}]\frac{\mathcal{F}}{R_w}$$

where T_i = indoor air temperature (F)
$\bar{T}_{(a-b)}$ = undisturbed average temperature of soil profile corresponding to wall depth range (F)
R_w = overall thermal resistance of wall in contact with the soil (ft^2 (hr) F/Btu)
$\mathcal{F}$ = adjustment factor (dimensionless)

The reconciliation factor $\mathcal{F}$ (termed the "decrement factor" f_d by Akridge, not to be confused with the expression e^{-xr}) is a function of soil and wall properties. It has an initial, theoretical value of 1.0, at which it would remain if the wall were perfectly insulated. In reality, $\mathcal{F}$ diminishes with time, approaching some constant value greater than zero as the building comes into thermal equilibrium with its surroundings. The influence of varying soil and wall thermal properties on $\mathcal{F}$, as well as a theoretical framework for its determination, have been discussed by Akridge. Poulos [1982] has subsequently calculated values of $\mathcal{F}$ by relating undisturbed ground temperature predictions to computer-simulated performance of subgrade walls. His $\mathcal{F}$ factors apply to fully subgrade walls extending in depth from 2 to 10 feet (Table 11b).

The value of the quantity $(T_i - \bar{T}_{(a-b)})$ is of interest as a driving potential regardless of the value of $\mathcal{F}$. Of even more general interest as a climatic index is the difference between daily average air temperature T_o and the soil profile temperature $\bar{T}_{(a-b)}$. This difference $(\bar{T}_{(a-b)} - T_o)$ can be termed the *ground temperature depression*, in nomenclature consistent with that used to describe the potential of other heat sinks. Ground temperature depressions have been computed [Labs 1981a] for a depth range of 2'-12' on July 21 for a number of locations in the U.S. Some of these are presented in *Table 11c*. It should be noted that when air and ground surface temperature are at maximum, the deeper layers have not yet reached their peaks; no single value of ground temperature depression, therefore, represents the continually variable conditions in the ground.

Earth Cooling Strategy.

In most regions of the United States, the ground acts as a heat sink throughout the entire year, so that ground cooling becomes, first, a problem of how to dissipate heat to a low grade sink in the summer without incurring the penalty of large winter heating loads. Two secondary problems may also arise: 1) how to maximize the effectiveness of ground coupling without risking condensation occurrence; 2) how to situate and design the structure for useful ground coupling without sacrificing the effective use of natural ventilation.

Because the soil acts as an extension of the building envelope, the indoor space of an underground building can be thought of as *directly coupled* to the ground. Some designers have attempted to exploit the cooling potential of the ground without actually building underground by *indirectly coupling* the interior to the ground by means of earth-air heat exchangers or "earth pipes" [see Abrams *et al.* 1980; Scott *et al.* 1965; Ingersoll *et al.* 1954]. *Isolated coupling* with the ground has also been achieved through heat pump mechanical systems, although these work more effectively with ground water.

The problem of balancing winter and summer heat losses so far has been solved only by compromise in the thickness and configuration of insulation placement. In northern regions, selective coupling with uninsulated structural walls could be achieved by furring out an insulated interior panel, and venting the air space thus formed during the summer. In more southerly regions, carrying roof insulation out over the backfill seems to offer adequate protection against heat exchange with the surface, while maximizing coupling with the subsoil [see Labs 1982].

The thermal value of earth cover is closely related to the importance of roof radiant and evaporative cooling, both of which must be viewed in regional context. The likelihood of condensation occurrence also varies from region to region. It can be prevented by uncoupling the wall from the soil with insulation or by dehumidifying the indoor air. If most of the moisture responsible for condensation is generated indoors, the problem can be alleviated by ventilation. If outdoor air dew point temperatures are below indoor wall temperature, however, ventilation will only exacerbate the problem. Tradeoffs between optimum earth coupling and effective ventilation rest with the designer but, like the foregoing, this issue is also of a regional nature.

In very broad terms, the earth cooling potential is greatest in northern regions; arid conditions favor earth coupling, whereas earth coupled cooling design for humid zones will always be accompanied by compromise between benefits and liabilities [Labs and Watson 1981]. The value of ground-coupled cooling can always be enhanced by ground temperature modification techniques [Givoni 1979b; Labs 1981b].

Table 11a   Generalized Soil Properties*

Property	Dry	Mean	Wet
α (ft^2 / day)	0.33	0.52	0.75
r (ft^{-1})	0.16150	0.12865	0.10713
L (days / ft)	9.38	7.47	6.22
$A_x / A_s = 0.1$; x =	14.2 ft	17.9 ft	21.5 ft

*based on a cluster analysis performed by Labs and Harrington [1982] of thermal diffusivity values reported by Kusuda [1968].

Table 11b $\mathscr{F}$ **Factors as a Function of Soil Thermal Conductivity**
and Wall Thermal Resistance

Wall R	Soil Conductivity (Btu/ft (hr) F)					
	.25	.50	.75	1.0	1.25	1.5
Uninsulated*	.11	.20	.28	.34	.39	.44
R7	.31	.50	.60	.67	.72	.77
R12	.43	.61	.70	.78	.83	.86
R17	.52	.69	.77	.83	.87	.90
R22	.59	.75	.82	.87	.90	.93

*Has an R value of 2.0
Source: Poulos [1982]

Table 11c   **Ground Temperature Depression Indices, July 21**

Location	$\overline{T}_{(2\text{-}12)}$	T_a	$[\overline{T}_{(2\text{-}12)} - T_a]$
Phoenix	72	91	−19
Boston	56	73	−17
Albuquerque	62	78	−16
Oklahoma City	66	82	−16
Indianapolis	60	75	−15
Atlanta	68	78	−10
San Antonio	76	84	−8
Houston	79	84	−5
Miami	78	82	−4

Source: Labs [1981a]

12. Appendices

Temperature is defined as the thermal state of matter with reference to its tendency to communicate heat to matter in contact with it. Temperature is an index of the thermal energy content of materials, disregarding energies stored in chemical bonds and in the atomic structure of matter.

Fahrenheit temperature (F) refers to temperatures measured on a scale devised by G. D. Fahrenheit, the inventor of the alcohol and mercury thermometers, in the early 18th century. On the Fahrenheit scale, the freezing point of water is 32F and its boiling point is 212F at normal atmospheric pressure. It is said that Fahrenheit chose the gradations he used because it divides into 100 units the range of temperatures most commonly found in nature; the Fahrenheit scale, therefore, has a more humanistic basis than other temperature scales.

Rankine temperature ($^\circ$R) refers to temperatures measured on the Rankine scale, which uses the same gradations as the Fahrenheit scale but which takes its origin at absolute zero (-459.69F); it is, therefore, an absolute scale of temperature.

Celsius temperature ($^\circ$C) refers to temperatures measured on a scale devised in 1742 by Anders Celsius, a Swedish astronomer. The Celsius scale is graduated into 100 units between the freezing temperature of water (0°C) and its boiling point at normal atmospheric pressure (100°), and is, consequently, commonly referred to as the *Centigrade scale*.

Kelvin temperature (K) refers to temperatures measured on the Kelvin scale, which uses the same graduations as the Celsius scale but which takes its origin at absolute zero (-273.16°C); it is, like the Rankine, an absolute scale of temperature.

F	$^\circ$R	$^\circ$C	K
F	= ($^\circ$R $-$ 459.69)	= (9/5) $^\circ$C + 32	= (9/5) K $-$ 459.69
(F + 459.69)	$^\circ$R	(9/5) $^\circ$C + 491.69	(9/5) K
(F $-$ 32) (5/9)	$^\circ$R (5/9) $-$ 273.16	$^\circ$C	(K $-$ 273.16)
F (5/9) + 255.38	$^\circ$R (5/9)	$^\circ$C + 273.16	K

Dry-bulb temperature (T_d, DBT) is an indicator of sensible heat, or the heat content of perfectly dry air. It is the temperature measured by an ordinary (dry-bulb) thermometer, and is independent of the moisture content and insensitive to the latent heat of the air.

Wet-bulb temperature (T_w, WBT) is an indicator of the total heat content (or enthalpy) of the air, that is, of its combined sensible and latent heats. It is the temperature measured by a wet-bulb thermometer, a thermometer having a wetted sleeve over the bulb from which water is able to evaporate freely. A wet bulb thermometer is easily made by slipping a sleeve cut from a light cotton shoestring over the bulb of an ordinary outdoor or photo thermometer.

Dew point temperature (T_d, DPT) is an indicator of the moisture content of the air, with specific reference to the temperature of a surface upon which moisture contained in the air will condense. Stated differently, it is the temperature at which a given quantity of air will become saturated (reach 100% relative humidity) if chilled at constant pressure. Dew point temperature is not easily measured directly; it is conveniently found on a psychrometric chart if dry-bulb and wet-bulb temperatures are known.

Humidity refers to the water vapor contained in the air. Like the word "temperature," however, "humidity" must be qualified as to its type for it to have quantitative meaning.

Absolute humidity is defined as the weight of water vapor contained in a unit volume of air; typical units are pounds or grains of water per cubic foot. Absolute humidity is also known as the water vapor density (d_v).

Relative humidity (RH or r) is defined as the (dimensionless) ratio of the amount of moisture contained in the air under specified conditions to the amount of moisture contained in the air at saturation at the same (dry bulb) temperature. Relative humidity can be computed as the ratio of existing vapor pressure to vapor pressure at saturation, or the ratio of absolute humidity to absolute humidity at saturation existing at the same temperature and barometric pressure.

Specific humidity is the weight of water vapor contained in a unit weight of air. It may be expressed as a dimensionless ratio of pounds of water per pounds of (moist) air, or in terms of grains of moisture per pound of (moist) air.

Humidity ratio (W), or *moisture content* or *mixing ratio,* is defined as the (dimensionless) ratio of the weight of the water vapor to the weight of the dry air contained in a given volume of (moist) air.

Water vapor pressure (P_v) is that part of the atmospheric pressure ("partial pressure") which is exerted due to the amount of water vapor present in the air. It is expressed in terms of absolute pressure as inches of mercury (in. Hg) or pounds per square inch (psi). The vapor pressure of ambient air is conventionally associated with the dry bulb temperature at which the air would become saturated if it were chilled to; it is then termed the *water vapor saturation pressure* (analogous to dew point temperature).

Note: One pound contains 7000 grains

Appendix 4 — Relation Between Pressure Units

psi	in. Hg[a]	atm	mm Hg[b]	Pa[c]	in. H$_2$O[d]
1	2.0360	0.06805	51.715	6894.8	27.678
0.4911	1	.03342	25.400	3386.4	13.595
14.696	29.921	1	760.0	101,325	406.78
0.01934	0.03937	0.001316	1	133.32	0.5353
1.4504×10^{-4}	2.953×10^{-4}	9.869×10^{-5}	0.00750	1	0.004015
0.03613	0.07355	0.002458	1.8681	249.08	1

a at 32F

b at 32F; also known as torr

c Pascal = Newton/meter^2 (SI)

d at 39.2F (4°C)

Note: 1 bar = 10^5 Newton/meter^2, or approximately 1 atm (.9869 atm)

Appendix 5 — Relation Between Wind Speed Units

knot	m/sec	mph	km/hr	ft/sec	ft/min
1	= .515	=1.152	=1.853	=1.689	=101.3
1.943	1	2.237	3.600	3.281	196.9
0.868	0.447	1	1.609	1.467	88.0
0.540	0.278	0.621	1	0.911	54.67
0.592	0.305	0.682	1.097	1	60.0

SITE
PLANNING

WIND BREAKS

PLANTS & WATER

BUILDING
ENVELOPE

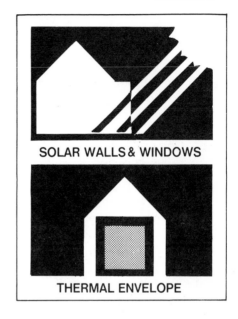

SOLAR WALLS & WINDOWS

THERMAL ENVELOPE

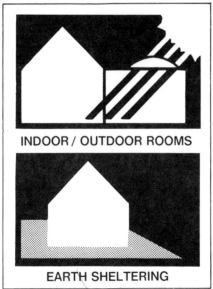

INDOOR / OUTDOOR ROOMS

EARTH SHELTERING

BUILDING
MASSING / PLAN

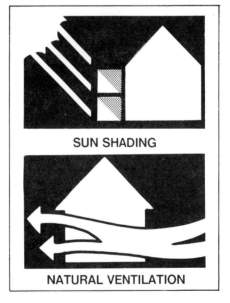

SUN SHADING

NATURAL VENTILATION

BUILDING
OPENINGS

Part II: Practices

79

Concept 1: **WIND BREAKS** (winter)

Two climatic design techniques serve the function of minimizing winter wind exposure:

2   Use neighboring land forms, structures, or vegetation for winter wind protection.

10   Shape and orient the building shell to minimize winter wind turbulence.

Concept 2: **PLANTS AND WATER** (summer)

Several techniques provide cooling by the use of plants and water near building surfaces for shading and evaporative cooling.

6   Use ground cover and planting for site cooling.

7   Maximize on-site evaporative cooling.

36   Use planting next to building skin.

38   Use roof spray or roof ponds for evaporative cooling.

Concept 3: **INDOOR / OUTDOOR ROOMS** (winter and summer)

Courtyards, covered patios, seasonal screened and glassed-in porches, greenhouses, atriums and sunrooms can be located in the building plan for summer cooling and winter heating benefits, as in these three techniques:

14   Provide outdoor semi-protected areas for year-round climate moderation.

20   Provide solar-oriented interior zone for maximum solar heat gain.

21   Plan specific rooms or functions to coincide with solar orientation.

Concept 4: **EARTH-SHELTERING** (winter and summer)

Techniques such as using earth against the walls of a building or on the roof, or building a concrete floor on the ground, have a number of climatic advantages for winter insulation and wind protection and for summer cooling. These techniques are often referred to as earth-contact or earth-sheltering design:

11   Recess structure below grade or raise existing grade for earth-sheltering effect.

24   Use slab-on-grade construction for ground temperature heat exchange.

33   Use sod roofs.

Concept 5: **SOLAR WALLS AND WINDOWS** (winter)

Using the winter sun for heating a building through solar-oriented windows and walls is covered by a number of techniques:

1   Maximize reflectivity of ground and building surfaces outside windows facing the winter sun.

9   Shape and orient the building shell to maximize exposure to winter sun.

19   Use high-capacitance materials to store solar heat gain.

29   Use solar wall and roof collectors on south-oriented surfaces.

41   Maximize south-facing glazing.

42   Provide reflective panels outside of glazing to increase winter irradiation.

43   Use skylights for winter solar gain and natural illumination.

Concept 6: **THERMAL ENVELOPE** (winter)

Many climatic design techniques to save heating energy are based upon isolating the interior space from the cold winter climate.

8   Minimize the outside wall and roof areas (ratio of exterior surface to enclosed volume).

15   Use attic space as buffer zone between interior and outside climate.

16   Use basement or crawl space as buffer zone between interior and grounds.

17   Provide air shafts for natural or mechanically assisted house-heat recovery.

18   Centralize heat sources within building interior.

22   Use vestibule or exterior "wind-shield" at entryways.

23   Locate low-use spaces, storage, utility and garage areas to provide climatic buffers.

25   Subdivide interior to create separate heating and cooling zones.

28   Select insulating materials for resistance to heat flow through building envelope.

30   Apply vapor barriers to control moisture migration.

31   Develop construction details to minimize air infiltration and exfiltration.

32   Select high-capacitance materials for controlled heat flow through the building envelope.

39   Provide insulating controls at glazing.

40   Minimize window and door openings on north, east, and/or west walls.

44   Detail window and door construction to prevent undesired air infiltration.

45 Provide ventilation openings for air flow to and from specific spaces and appliances.

Concept 7: **SUN SHADING** *(summer)*

Because the sun angles are different in summer than in winter, it is possible to shade a building from the sun during the overheated summer period while allowing it to reach the building surfaces and spaces in winter. Thus the concept to provide sun shading does not need to conflict with winter solar design concepts, as shown in the following techniques:

3 Minimize reflectivity of ground and building surfaces outside windows facing the summer sun.

4 Use neighboring land forms, structures, or vegetation for summer shading.

12 Shape and orient the building shell to minimize exposure to summer sun.

34 Provide shading for walls exposed to summer sun.

35 Use heat reflective materials on surfaces oriented to summer sun.

46 Provide shading for glazing exposed to summer sun.

Concept 8: **NATURAL VENTILATION** *(summer)*

Natural ventilation is a simple concept by which to cool a house using the following techniques:

5 Use neighboring land forms, structures, or vegetation to increase exposure to summer breezes.

13 Shape and orient the building shell to maximize exposure to summer breezes.

26 Use "open plan" interior to promote air flow.

27 Provide vertical air shafts to promote interior air flow.

37 Use double roof and wall construction for ventilation within the building shell.

47 Orient door and window openings to facilitate natural ventilation from prevailing summer breezes.

48 Use wingwalls, overhangs, and louvers to direct summer wind flow into interior.

49 Use louvered wall for maximum ventilation control.

50 Use roof monitors for "stack effect" ventilation.

Minimize Conductive Heat Flow

Promote Solar Gain

Minimize External Air Flow

Minimize Infiltration

Promote Earth Cooling

Minimize Solar Gain

Promote Ventilation

Promote Evaporative Cooling

Promote Radiant Cooling

Recommended Design Concepts

WIND BREAKS

PLANTS & WATER

INDOOR / OUTDOOR ROOMS

EARTH SHELTERING

SOLAR WALLS & WINDOWS

THERMAL ENVELOPE

SUN SHADING

NATURAL VENTILATION

81

Maximize reflectivity of ground and building surfaces outside windows facing the winter sun.

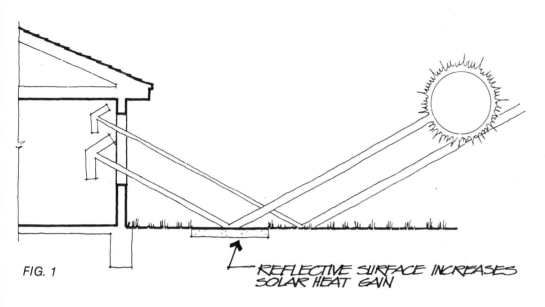

FIG. 1

REFLECTIVE SURFACE INCREASES SOLAR HEAT GAIN

Solar radiation transmitted through windows may be either *direct, diffuse,* or *reflected* from exterior surfaces. The most significant of these surfaces are the ground plane and materials immediately surrounding the window opening.

Increasing the reflectivity of surface materials outside of the south-facing windows will slightly increase the amount of total radiation entering the interior (*FIG. 1).* Snow increases ground reflectivity at a time when this is needed, but in most U.S. climates snow cover is transient, and could be blown away from the immediate vicinity (if the wind is from the north, the south side of the building would be in the wind eddy).

All materials except mirrors and highly polished metals are characterized by *diffuse* reflectivity, that is, light striking the surface is bounced off in all directions. Because of this, no precise size or extent of the reflective plane from the building can be specified. For the same reason, highly reflective exterior surfaces that promote increased transmission through glazing will likely lead to summertime overheating. In most U.S. locations, it is therefore most desirable to utilize temporary or seasonal techniques for increasing outside reflectivity as detailed in Technique 42 rather than

as the permanent building or landscape features. However, if the ground and building surfaces to the south are shaded in summer, it may be possible to select materials for those surfaces that will increase reflected solar irradiation (See *Table 1* for reflectance values of various surface materials).

Design Concept

Table 1. Reflectance Values of Various Ground Covers*	
Material	Reflectance in %
Snow cover, fresh	75-95
Snow cover, old	40-70
Light sand dunes	30-60
Concrete surfaces	30-50
Snow, dirty firm	20-50
Grass	20-30
Brick, various colors	23-48
Soil, sandy	15-40
Meadow	12-30
Woods	5-20
Soil, dark cultivated	7-10
Blacktop	<10-15
Slate, dark clay	7+
Bluestone (sandstone)	18
Dry grass	32
Bark	23-48
Green fields	3-15
Green leaves	25-32
Water surfaces, sea	3-10

*Reflectance values must be considered generalized estimates. They will vary with conditions such as moisture content and angle of incidence.

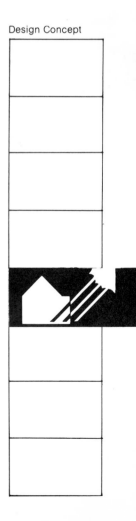

2

Use neighboring land forms, structures, or vegetation for winter wind protection.

FIG. 2c
General Rules of Windbreak Design

1. The range of protected area downwind is proportional to the height of the windbreak—the higher the barrier, the longer the "wind shadow". Angle of the windward edge is also important—the more vertical, the greater the effect.

2. The maximum length of wind shadow is developed only when the width of the windbreak is at least 11-12 times its height.

Design Concept

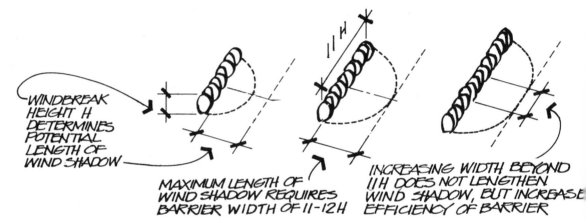

WINDBREAK HEIGHT H DETERMINES POTENTIAL LENGTH OF WIND SHADOW

MAXIMUM LENGTH OF WIND SHADOW REQUIRES BARRIER WIDTH OF 11-12H

INCREASING WIDTH BEYOND 11H DOES NOT LENGTHEN WIND SHADOW, BUT INCREASE EFFICIENCY OF BARRIER

3. The permeability or density of the barrier affects the length of the downwind protected zone—dense and solid barriers offer greatest reduction in wind speed, but only for a short distance immediately behind the barrier. Further downwind turbulence quickly restores wind speed. Wind permeable barriers pass some wind at reduced speed, creating an "air cushion" that reduces turbulence and extends the length of the sheltered zone downwind.

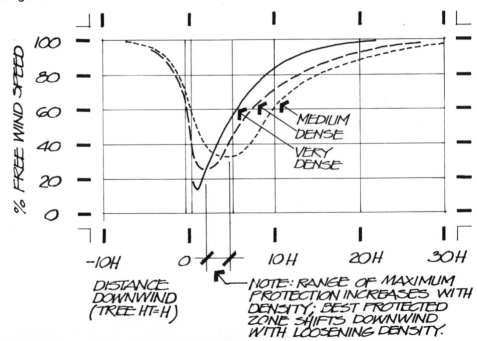

% FREE WIND SPEED

100 80 60 40 20 0

MEDIUM DENSE

VERY DENSE

-10H   0   10H   20H   30H

DISTANCE DOWNWIND (TREE HT=H)

NOTE: RANGE OF MAXIMUM PROTECTION INCREASES WITH DENSITY; BEST PROTECTED ZONE SHIFTS DOWNWIND WITH LOOSENING DENSITY.

Use neighboring land forms, structures, or vegetation for winter wind protection.

An analysis of the building site should be made to determine if there are existing wind protected areas. In siting a house, the builder should avoid open areas, hilltops and valley floors that are directly exposed to prevailing winter winds. Existing hedgerows, hillocks and tree stands may be available as wind-shields (FIG. 2a). The arrangement of housing units in cluster plans and planned unit developments can be used to take advantage of these.

Cold air flows downhill just like water, so that although a valley may seem to be shielded from cross winds, it may actually be in the midst of a steady downstream nighttime cold air current. The best protected sites, occur on leeward slopes (FIG. 2b).

Windbreaks

In the absence of, or to augment, existing land forms and tree masses, barriers can be created for wind control. Trees and shrubs are the most common of these, but berms, fences and walls can also provide benefit. Usually, the higher the barrier, the larger the protective "wind shadow" (FIG. 2d). Solid or wind-impenetrable barriers are undesirable, as they create turbulent areas on the leeward side (FIG. 2c). The energy effectiveness of wind protection follows directly from the amount of windspeed reduction. The economic benefit of landscaping has been studied with the aid of computer simulation at the University of Wisconsin. Researchers have concluded: [Grist 1977]

> The most desirable windbreaks, from a heat loss point of view, are those that range in porosity from 25% to 60%. High porosity (about 50%) usually means more protection 5H to 20H downwind where velocity is reduced to about 30% of free stream velocity. Lower porosity (about 25%) usually indicates more protection up to 4H range, where velocity may be reduced to 10%, but less protection 4H to 20H downwind where velocity may be 60% of the free stream velocity.

Climatically planned landscaping can have significant energy benefits for homeowners, in addition to its aesthetic and property value. As a cautionary consideration, however, it should be noted that windbreaks can possibly impede summer ventilating breezes. This should be taken into account when planting plan is prepared.

Minimize Conductive
Heat Flow

Minimize External
Air Flow

Minimize Infiltration

Design Concept

FIG. 2a
Increasing the width of shelterbelt does not increase the length of its wind shadow. Contrarily, a wide belt (such as woods) consumes much of the area of its own protection.

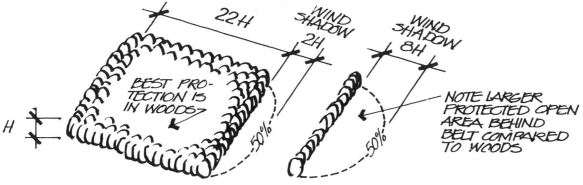

22H WIND SHADOW 2H WIND SHADOW 8H

BEST PROTECTION IS IN WOODS?

H

50%

50%

NOTE LARGER PROTECTED OPEN AREA BEHIND BELT COMPARED TO WOODS

FIG. 2b
Cross ridge winds skip over leeward slope, leaving protected region desirable for building since winter winds often come from the north and northwest. Slope in the "shadow" of the wind will often enjoy a sunny southern exposure. Look for them during site selection and lotting.

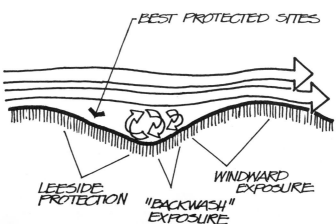

BEST PROTECTED SITES

LEESIDE PROTECTION

"BACKWASH" EXPOSURE

WINDWARD EXPOSURE

Use neighboring land forms, structures, or vegetation for winter wind protection.

FIG. 2d.
Man-made Features: Fences & Walls

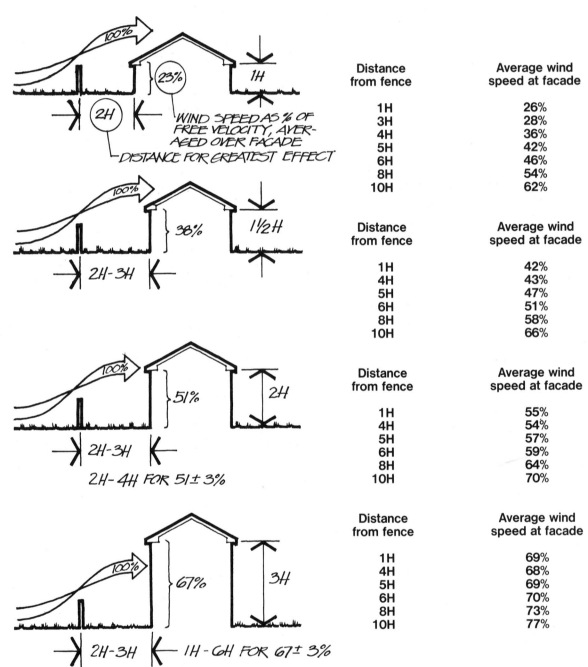

Distance from fence	Average wind speed at facade
1H	26%
3H	28%
4H	36%
5H	42%
6H	46%
8H	54%
10H	62%

Distance from fence	Average wind speed at facade
1H	42%
4H	43%
5H	47%
6H	51%
8H	58%
10H	66%

Distance from fence	Average wind speed at facade
1H	55%
4H	54%
5H	57%
6H	59%
8H	64%
10H	70%

Distance from fence	Average wind speed at facade
1H	69%
4H	68%
5H	69%
6H	70%
8H	73%
10H	77%

Design Concept

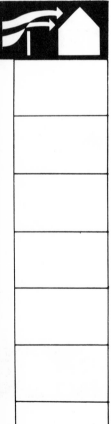

Fence-facade relationships: wind impact on a structure is related to height of facade and distance from windbreak. Recommendations given here are derived from observations in the open (not at a building surface) behind a windbreak of 15-25% permeability. Note that with decrease in relative fence height, the horizontal range of effectiveness (±3%) expands: although low fences provide less benefit, their location is less critical.

Minimize reflectivity of ground and building surfaces outside windows facing the summer sun.

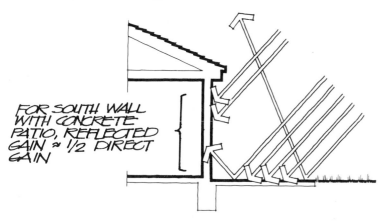

FOR SOUTH WALL WITH CONCRETE PATIO, REFLECTED GAIN ≈ 1/2 DIRECT GAIN

FIG. 3a. For a perfectly difusing horizontal surface the amount of radiation reflected onto a vertical wall will be one-half of the radiation reflected from the horizontal surface.

Considering that on a summer day in the middle U.S. latitudes about 2½ times as much radiation falls on the ground as on a south wall, the reflected heat gain at the wall will be 1½r x 2½r=1¼r times the direct gain. Values of r typically range from 10-50% (see *Table 1*).

Since the amount of solar irradiation received by the ground surface during summer months is about twice that received by either east or west walls, the radiation reflected from the ground into windows and external walls can add significantly to the cooling requirements of the structure. Selecting exterior surface materials of low reflectivity, therefore, is a means of minimizing solar heat load.

Some reflectivity values of typical exterior ground surfaces are presented in *Table 1*. Because reflection from outdoor surfaces is diffuse—that is, nondirectional—it cannot be controlled as easily as direct radiation through louvers, sunscreens and other geometric shading devices. Nonetheless, the heating effect of diffuse reflected radiation

can generally be assumed to follow the rule that "angle of reflectance equals angle of incidence." The greatest concern, therefore, will be the area immediately outside the building, determined as if the surface were a mirror (*FIG. 3a*).

One cautionary note: although locating an asphalt driveway outside a wall will keep reflected radiation at a minimum, the heat absorbed by the paving will make air temperatures outside the wall much higher than over lawn and most other surfaces. Shrubs and ground cover make effective sun absorbers, and also provide an evaporative air-cooling effect. *FIG. 3b* depicts how broadleaf planting and geometric manipulation of the reflecting surface can be used to minimize heat loading on house walls.

Design Concept

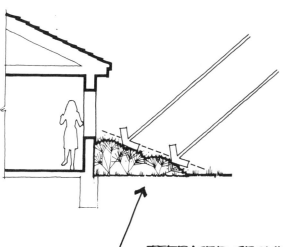

FIG. 3b. The irregular or "rough" surface of shrubbery helps intercept sun giving a lower reflectivity value than even a planted planar surface such as grass.

TERRACED GEOMETRY OF SHRUBS OR BERM MINIMIZES REFLECTION AT LOW SUN ANGLES — WEST EXPOSURES, FOR EXAMPLE

Use neighboring land forms, structures, or vegetation for summer shading.

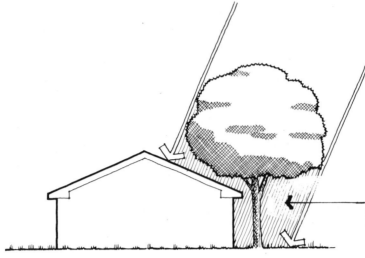

KEEP UNDERSTOREY CLEAR SO AS NOT TO DISRUPT AIRFLOW FOR VENTILATION

FIG. 4b. **Plant tall canopy trees on south side of house to shade roof and walls.**

SHADE PLANTING ON WEST AND NORTHWEST SIDES OFTEN CAN DOUBLE AS WINTER WIND-BREAK. CONSIDER EVERGREENS, FENCES, AND WALLS.

Design Concept

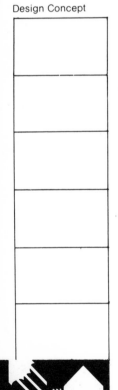

FIG. 4c. **Plant dense trees, shrubs, hedges on west side of house to intercept afternoon sun.**

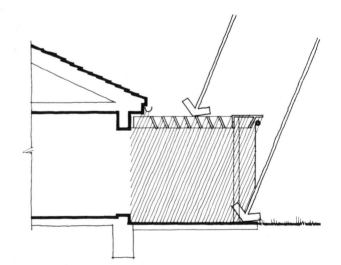

FIG. 4d. **Attached overhead shading structures can provide multiple benefits. Not only does this patio cover shade the wall, it also reduces reflected gain from loading on the wall.**

Use neighboring land forms, structures, or vegetation for summer shading.

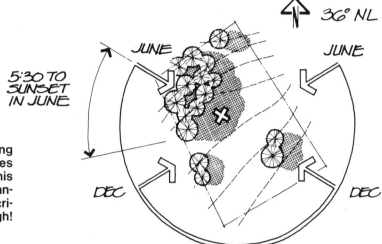

FIG. 4a "X" marks the spot for existing on-site sun protection. Look for sites shrouded by trees on west side. In this example, trees are on upslope, enhancing their shading ability. Don't sacrifice winter southern exposure, though!

Opportunities for sun control and shading exist in almost all aspects of site planning and development, from the initial selection of a building site or the plotting of the lots in a subdivision to the final selection of plant materials and details in landscaping.

Siting

For a custom house on property that is large enough to offer several building sites, the choices should be analyzed for natural shading from existing trees and land masses. These can be utilized to reduce solar gain from low afternoon summer sun by siting the building to the east of such features (*FIG. 4a*).

In plotting subdivisions, streets should be laid out to create lots having the best possible solar orientation. Generally, this means streets should run east-west; with this orientation, tightly clustered units will shade each other from afternoon summer sun, without obstructing desirable southern exposures.

On-lot development

Inasmuch as air temperatures usually peak during the end of July or in early August in most regions of the U.S., and since solar impact on *walls* is greater at this time than at the date of maximum insolation (June 21), the afternoons of this period can be considered as the selected design condition for determining a shade planting plan. When the site of the house has been determined, planning of other exterior shading devices can begin.

The best known, of course, are shade trees (*FIG. 4b*), which protect the house in summer and shed their leaves in winter to allow the house to receive solar gain. On the south side of the house, tall growing species should be selected that will shade the roof as well as the wall.

West elevations will benefit most from lower, more compact trees, and tall, dense shrubs that provide screening from the low afternoon summer sun (*FIG. 4c*). Because the winter sun does not reach around as far as it does in summer, shading vegetation for western walls may be evergreen. If winter winds are westerly or northwesterly, the planting may double as a windbreak.

Other shading devices for south and west walls include overhead trellises and shade walls and fences. Trellises can be built with louvers or slats to block the sun (*FIG. 4d*), or may consist of little more than a light framework to support climbing ivy.

The garage can be located on the western side of the house and may be sited to create a breezeway or shaded patio area between it and the house. There are many approaches to providing outdoor shading structures. The most suitable will depend on the interior organization of the plan, the intended areas for outdoor living space (patios, porches, etc.), the site of the lot and its relationship to other dwellings.

Minimize Solar Gain

Promote
Radiant Cooling

Design Concept

Use neighboring land forms, structures, or vegetation to increase exposure to summer breezes.

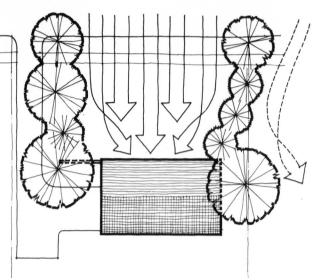

FIG. 5b. **Wind Funnels**

Tree planting can be used to guide wind into unit. Here tree funnel lines are "disguised" as driveway and property line planting to better blend with siting.

FIG. 5c. **Wind Dams**

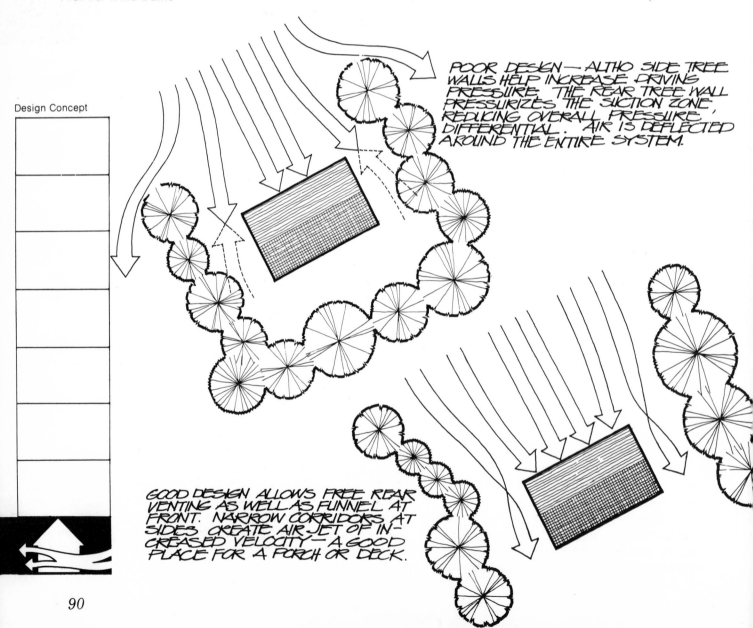

Design Concept

POOR DESIGN — ALTHO SIDE TREE WALLS HELP INCREASE DRIVING PRESSURE, THE REAR TREE WALL PRESSURIZES THE SUCTION ZONE, REDUCING OVERALL PRESSURE DIFFERENTIAL. AIR IS DEFLECTED AROUND THE ENTIRE SYSTEM.

GOOD DESIGN ALLOWS FREE REAR VENTING AS WELL AS FUNNEL AT FRONT. NARROW CORRIDORS AT SIDES CREATE AIR JET OF INCREASED VELOCITY — A GOOD PLACE FOR A PORCH OR DECK.

Use neighboring land forms, structures, or vegetation to increase exposure to summer breezes.

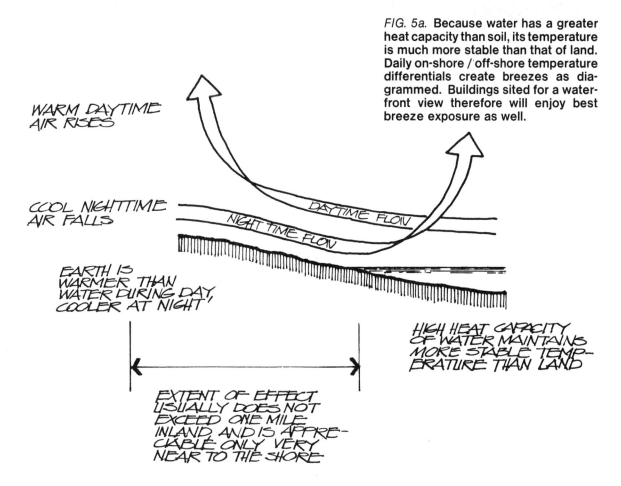

WARM DAYTIME
AIR RISES

COOL NIGHTTIME
AIR FALLS

DAYTIME FLOW

NIGHT TIME FLOW

EARTH IS
WARMER THAN
WATER DURING DAY,
COOLER AT NIGHT

HIGH HEAT CAPACITY
OF WATER MAINTAINS
MORE STABLE TEMP-
ERATURE THAN LAND

EXTENT OF EFFECT
USUALLY DOES NOT
EXCEED ONE MILE
INLAND AND IS APPRE-
CIABLE ONLY VERY
NEAR TO THE SHORE

FIG. 5a. Because water has a greater heat capacity than soil, its temperature is much more stable than that of land. Daily on-shore / off-shore temperature differentials create breezes as diagrammed. Buildings sited for a water-front view therefore will enjoy best breeze exposure as well.

Promote Ventilation

Promote Evaporative Cooling

Design Concept

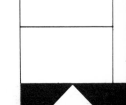

The direction and velocity of flow of summer breezes are influenced considerably by local land forms, tree masses, and existing structures. The resulting air flow patterns may be quite different from that of prevailing breeze directions given in published (airport) climatic data.

Although building on the crest of a hill will maximize exposure to prevailing breezes, this practice also maximizes the structure's vulnerability to winter wind chill. Unless winters are mild and the ventilating season is long (as in the Gulf Coast States), it will be more advantageous to locate site development along hillsides, rather than at the ridgeline.

On slopes and in valleys, cool air flows downhill, washing along the slope and settling in depressions or following the valley downstream near large water bodies. Daily on-shore and off-shore breezes are created which are largely independent of regional air patterns (FIG. 5a). A topographic analysis of the area is thus necessary to determine probable on-site wind flow patterns and the most desirable building locations. (FIG. 5e).

Trees and shrubs can be used to channel air flow toward the structure, and may even be used to increase air velocity through the building by "funnelling" air into openings (FIG. 5b). Fences, walls, and adjacent structures can create air dams that increase the inflow pressures (FIG. 5c). Caution should be exercised in landscaping, however, since undesirable air dams can also be created that will cause air to pool rather than flow. Also, the location of planting outside of the house can either aid in deflecting air into the structure or hurt by deflecting it away from open windows (FIG. 5d).

Use neighboring land forms, structures, or vegetation to increase exposure to summer breezes.

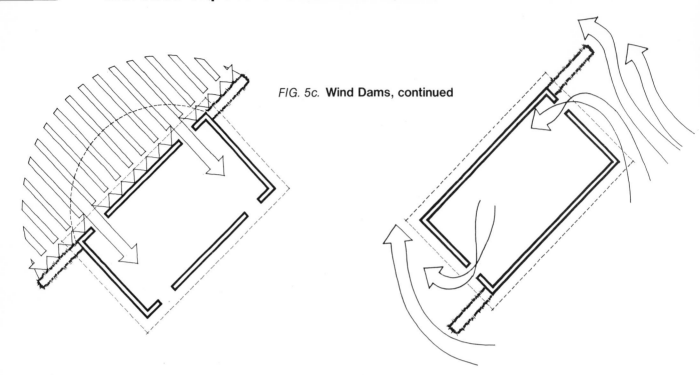

FIG. 5c. **Wind Dams, continued**

Design Concept

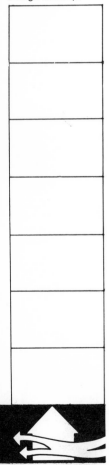

FIG. 5d. **Wind Deflectors**

Hedge and shrub planting outside window relieves unwanted pressure component, fosters downward deflections of air stream. Effect will be produced for distances D up to 15 to 20 ft.

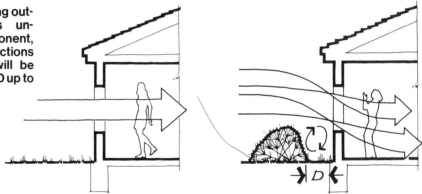

Influence of tree canopy outside the window is to "lift" or warp the airstream upward by relieving downward pressure (opposite of shade-effect). If tree is immediately outside window it will produce a ceiling wash flow. At a distance from the house, canopy may warp the airstream sufficiently to miss the house altogether.

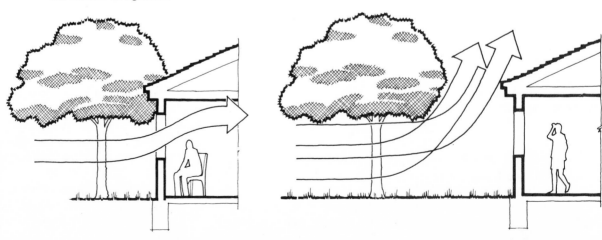

Use neighboring land forms, structures, or vegetation to increase exposure to summer breezes.

FIG. 5e.

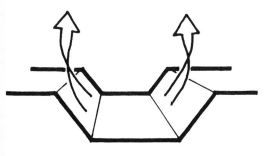

MORNING—BREEZE RISES UP FACE OF SLOPE DUE TO SURFACE WARMING

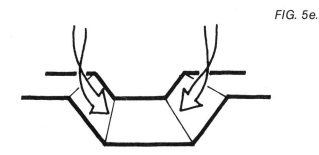

LATE EVENING—SURFACES COOL, START DOWNWARD FLOW OF COOL AIR

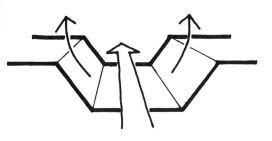

NOON—AIR POOL IN VALLEY HAS BEEN WARMED, CREATES UP-VALLEY BREEZE

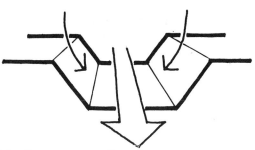

NIGHT—COOL, HEAVY AIR COLLECTED IN VALLEY BEGINS TO FLOW DOWNSTREAM

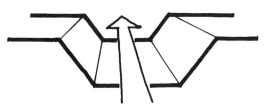

AFTERNOON—UP-VALLEY BREEZE PREDOMINATES AS SURFACES REACH & PASS MAXIMUM TEMPERATURE

DAWN—DOWNSTREAM FLOW PRE-DOMINATES AS COOL AIR WASHES DOWN SLOPES

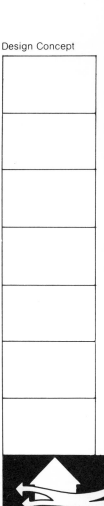

Design Concept

6   Use ground cover and planting for site cooling.

FIG. 6b Neighborhood air temperatures can be kept low by minimizing the expanse of paving, and by shading paved areas. Many of the guidelines above save dollars or boost sales appeal.

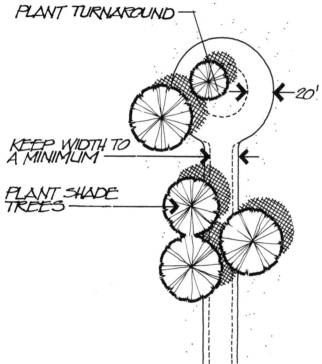

PLANT TURNAROUND

20'

KEEP WIDTH TO
A MINIMUM

PLANT SHADE
TREES

Site Planning Suggestions:

Keep paved area to a minimum—an 8ft. dia. turnaround with a 20 ft. ring road is recommended.

If spillover parking areas are required use a porous paving block instead of asphalt.

Plant shade trees to shade paving.

Use 18-20 ft. street width for large lot (34 acre or more) developments.

Use 26 ft. street width for 14 acre lots on cul-de-sacs and short loops.

Avoid 34-36 ft. street widths—these are never warranted in well planned new developments.

Design Concept

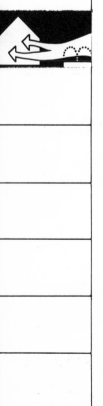

FIG. 6c. Porous concrete paving can be precast or cast-in-place with forms made for this purpose ("grasstone," shown below, and "grasscrete" are examples) use it for stabilizing shoulders and for spillover parking spaces both on and off lot.

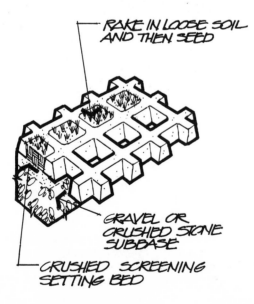

RAKE IN LOOSE SOIL
AND THEN SEED

GRAVEL OR
CRUSHED STONE
SUBBASE

CRUSHED SCREENING
SETTING BED

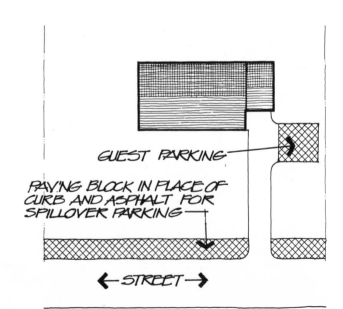

GUEST PARKING

PAVING BLOCK IN PLACE OF
CURB AND ASPHALT FOR
SPILLOVER PARKING

← STREET →

Use ground cover and planting for site cooling.

The climatic benefit of landscaped ground cover is seldom considered. On a sunny summer day, an acre of turf may evaporate about 2400 gallons of water. At this rate, the rear yard of a typical 1/4 acre lot will have the cooling effect of 2 million Btu per day. This has a significant influence on air temperature. In similar terms, the daily evaporation from a mature beech tree is said to provide an air cooling effect of one million Btu—the equivalent of 10 room-sized air conditioners operating 20 hours a day.

The difference in surface temperature between grass and asphalt can easily exceed 25°F (*FIG. 6a*). The air temperature in the "microclimate zone" (one to four feet) above these surfaces also differs appreciably, registering on the order of 10°F or more. The relationship of lawn and other living ground cover surfaces to non-evaporating surfaces (driveways, streets, roofs) will in part determine neighborhood air temperatures. These, in turn, will influence the cooling load on houses in the area, as well as the suitability of natural ventilation as a cooling strategy. Additionally, plants help create fresh air.

Stated a different way, vegetation should be maximized, and where possible, man-made surfaces such as streets and roofs should be shaded by trees (*FIG. 6b, 6c*).

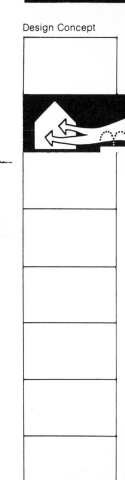

Promote Ventilation

Promote
Evaporative Cooling

Promote
Radiant Cooling

Design Concept

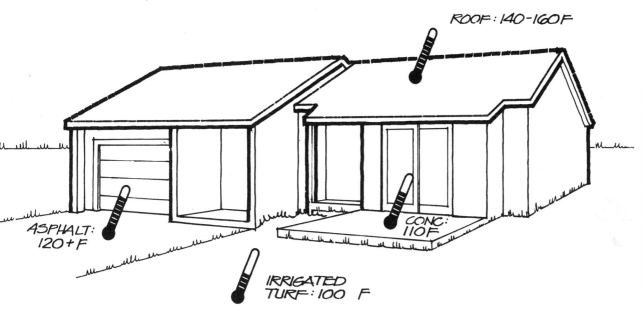

ROOF: 140-160 F

ASPHALT: 120+ F

CONC: 110 F

IRRIGATED TURF: 100 F

FIG. 6a. **Non-living surfaces are much hotter than grass (which would be cooler yet, if well irrigated) since they don't dissipate heat through evaporation. A black roof is hotter than an asphalt driveway, because the ground underneath the paving stores heat. The hottest roof will be one with insulation right under the roofing—having negligible mass. The coolest roofs will be sprayed, ponded, or covered with irrigated sod.**

Maximize on-site evaporative cooling.

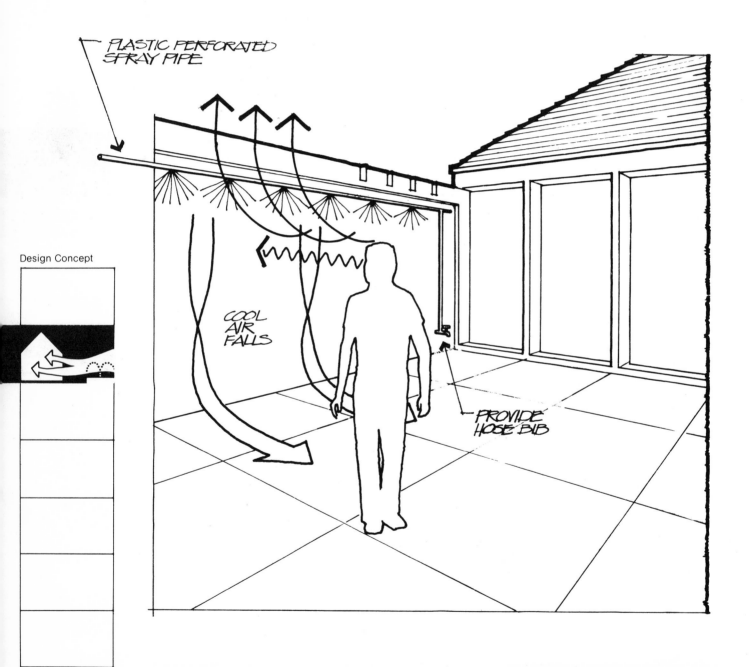

PLASTIC PERFORATED
SPRAY PIPE

COOL
AIR
FALLS

PROVIDE
HOSE BIB

Design Concept

FIG. 7b. **Spray-pipe on atrium walls is an excellent and inexpensive way of providing evaporative cooling for courtyard. Spray cools air as well as wall surface, thereby reducing both ambient and mean radiant temperatures. More dramatic (and expensive) treatment is a "wall wash" of water flow from a top-mounted trough.**

Maximize on-site evaporative cooling.

Outdoor evaporative cooling mechanisms can help to provide outdoor comfort as well as to lower indoor cooling costs by lowering air temperature surrounding the building. This reduces the cooling load transmitted through the building shell and makes natural ventilation more desirable and more effective.

Since cool air is denser than warm air, it will tend to drain away, flowing downhill. The traditional response to retaining spray-treated air is to enclose the outdoor space with a wall or fence, in effect creating an open-topped "tank" of air. In some parts of the world, atrium or courtyard house design is standard practice, often with fountains or spray jets (*FIG. 7a*). In some cases, dual courtyard design has been employed, wherein a shaded, spray-cooled courtyard provides a cool ventilation air supply, while the heat trapping effect of a sunny courtyard on the other side of the unit propels an upward flow of warmed air, drawing the cool air through the house.

The same dual courtyard principle will work in many areas of the United States, but even simpler designs can achieve useful benefits. Small-lot patio houses can have cool yard areas with the aid of little more than a lawn sprinkler–provided that the fence or wall isn't too "leaky". Other low cost devices include running a perforated garden hose or pipe around the fencetop, creating a "wet wall". Spray-mist type area "foggers" can cool a large air mass instantly — and benefit the plants as well. The water that falls out of the air will help keep ground and patio surfaces cool by evaporation as well, so little water need be wasted (*FIG. 7b*).

Promote Ventilation

Promote
Evaporative Cooling

Design Concept

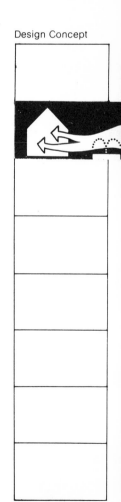

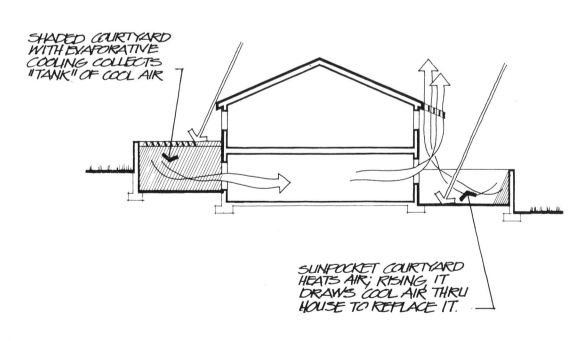

SHADED COURTYARD
WITH EVAPORATIVE
COOLING COLLECTS
"TANK" OF COOL AIR

SUNPOCKET COURTYARD
HEATS AIR; RISING IT
DRAWS COOL AIR THRU
HOUSE TO REPLACE IT.

FIG. 7a. **Dual courtyard design can be used to drive flow-through ventilation, when one is used to cool air, and the other to heat it.**

Minimize the outside wall and roof areas (ratio of exterior surface to enclosed volume).

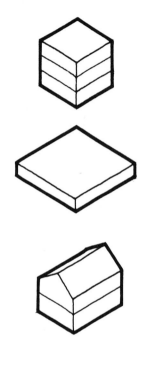

TOTAL VOLUME= 32 000
SURFACE AREA= 5 039   SVR = 0.157

3rd FLOOR   = 1 007
2ND FLOOR   = 1 007
BASE AREA   = 1 007   SFAR = 5.0
Combined Area = 3 021   SFAR = 1.67

TOTAL VOLUME= 32 000   SVR = 0.17
SURFACE AREA= 5 429

FLOOR AREA   = 3 200   SFAR = 1.7

TOTAL VOLUME= 32 000
SURFACE AREA= 4 754   SVR = 0.148

UPPER FLOOR = 1 350
BASE AREA   = 1 350
Combined Area = 2 700   SFAR = 1.76
(includes attic)

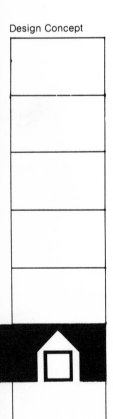

Design Concept

TOTAL VOLUME= 32 000
SURFACE AREA= 3 864   SVR = 0.12

UPPER FLOOR = 1 618
BASE AREA   = 1 932   SFAR = 2.0
Combined Area = 3 550   SFAR = 1.09

TOTAL VOLUME= 32 000
SURFACE AREA= 4 435   SVR = 0.138

UPPER FLOOR = 1 600
BASE AREA   = 1 600   SFAR = 2.77
Combined Area = 3 200   SFAR = 1.38

TOTAL VOLUME= 32 000
SURFACE AREA= 4 800   SVAR = 0.15

UPPER FLOOR = 1 600
BASE AREA   = 1 600   SFAR = 3.0
Combined Area = 3 200   SFAR = 1.5

FIG. 8

Minimize the outside wall and roof areas (ratio of exterior surface to enclosed volume).

Surface-to-volume ratio (SVR) is one way to express the relationship between outside building surface area and the enclosed space that can be used in order to compare differently shaped buildings containing equal volumes of space. For any given building volume, the more compact the shape, the less wasteful it is in losing heat. Only the above-ground surface area of the structure is significant in determining the SVR, since the most severe climatic stresses occur by exposure to fluctuating air temperatures and to winter winds.

The surface-to-volume ratio indicates a measure of the efficiency of enclosing space, but does not express the efficiency of the *use* of the enclosed space for living. Consider, for example, a cubical structure: if it contains three floors, it will have three times as much usable space than if it had a single floor, yet both will contain the same volume and will consume roughly the same amount of heating energy. The *surface-to-floor area ratio* (SFAR), therefore gives a more critical evaluation of different building plans. As in the case of SVR, the lower the SFAR value for a given floor area, the better is the performance of the house.

The relative benefit of different building configurations can be generalized by comparing the SVR of alternatives that contain the same interior volume of space. The hemisphere, for example, encloses the most space with least surface area, hence can be considered the most efficient. The minimum SVR for conventional building construction is obtained with a square floor plan with a flat roof of a height of one-half the length of its sides. It can easily be recognized that this is the approximate shape and dimension of a two story house. If the attic floor is well insulated, the *effective* shape of the structure will be that of having a flat roof, regardless of the roof configuration. If the house has a cathedral ceiling, the volume under the roof is an important part of the SVR calculation.

Surface-to-volume ratios can be minimized by making use of compact geometric forms. Minimum surface-to-floor area ratios are achieved by partitioning these as efficiently as possible by avoiding excessive interior ceiling heights. (See *FIG. 8*).

Design Concept

Shape and orient the building shell to maximize exposure to winter sun.

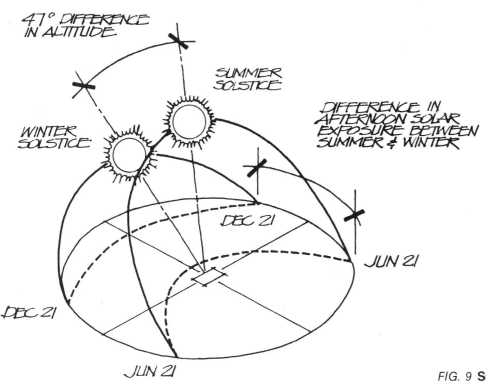

47° DIFFERENCE IN ALTITUDE

SUMMER SOLSTICE

WINTER SOLSTICE

DIFFERENCE IN AFTERNOON SOLAR EXPOSURE BETWEEN SUMMER & WINTER

DEC 21

JUN 21

DEC 21

JUN 21

FIG. 9 **Sunpath diagram illustrates that the only meaningful orientation for winter sun exposure is south. In summer, east and west exposures are the worst villains for overheating.**

Design Concept

Absorption of solar energy by an opaque building surface will not be a significant winter benefit to heating in cold weather climates, since the building should be well insulated against conductive heat flow. However, the shape and orientation of the building does determine the area for favorable window and solar collector placement. Building shape and orientation is, therefore, a significant design consideration.

The winter sun path is much shorter and lower than the summer sun path (*FIG. 9*). Because the winter sun rises and sets south of east and west respectively, the east and west facades do not receive any significant amount of irradiation in winter throughout much of the U.S. East and west facades are major recipients of unwanted summer heat, however. The appropriate strategy for winter solar heating, therefore, is to face the major wall and window areas of the house to the south.

An orientation slightly east of south is often favored (typically 15° east of south), in that this exposes the unit to more morning than afternoon sun and enables the house to begin to heat earlier in the day. Orienting the unit west of south means the house will retain its morning chill later into the day but will carry over afternoon heat into the evening. These choices are discussed with reference to specific solar techniques which follow.

10 Shape and orient the building shell to minimize winter wind turbulence.

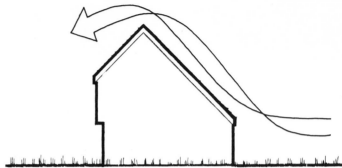

FIG. 10b. Traditional New England "saltbox" turned its back to the wind— long, windowless roof plan deflects air flow over house in streamline fashion, avoiding air dam effect that its height would otherwise have. Note also "paper facade"—flat planar surfaces devoid of protuberances.

Design Concept

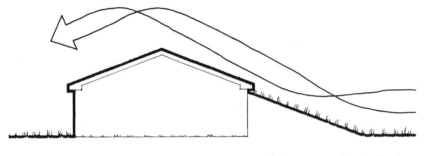

FIG. 10c. Berming or earth sheltering on the windward side of the house eliminates infiltration at the band joist, and reduces infiltration at the top plate as well. The streamlined roof shape also reduces conduction-convection losses through the roof.

FIG. 10d. A compact form—in plan as well as section—is the first rule in minimizing wind exposure. Orientation is equally important: plan B has the same configuration and area as plan A, yet orientation increases its apparent width to the same as C when rotated 45°.

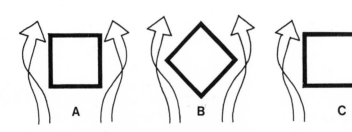

Shape and orient the building shell to minimize winter wind turbulence.

BUILDING
MASSING

10

Minimize Conductive
Heat Flow

Minimize External
Air Flow

Minimize Infiltration

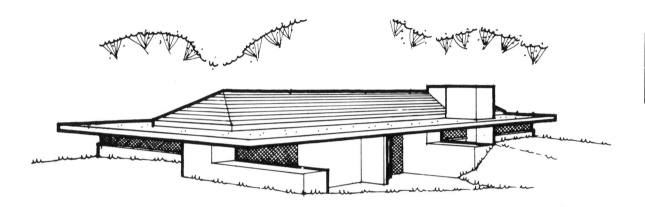

FIG. 10a. This berm type house was designed by architect Frank Lloyd Wright for a Minnesota climate. Note low profile all around—including flatness of roof. This "well-pocketed" plan would minimize the projection of the house into the wind stream; numerous overhangs, fins, and baffles, however, are wind catchers (see Technique 49) and work against shedding of wind flow. A smooth facade would be better suited to windy site.

Winter windflow against a structure affects the rate of heat loss by increasing infiltration rates and by increasing conductive heat flow through the shell by the rate at which heat is convected away from the exterior shell surface. In a well-insulated building, infiltration will be the greater of these two, so the more effective strategy generally should be to minimize the pressure differential between opposing sides, rather than merely attempting to minimize the velocity of air movement at the shell surface.

This technique is the opposite of that recommended to capture the flow of summer breezes: the facade of smallest area should face into the direction of prevailing winter winds, and windows and doors (openings vulnerable to infiltration) should be located in zones of minimum pressure.

Obviously, it is necessary to know the direction of prevailing winter winds before designing or siting a house to minimize its wind exposure. In areas where winter wind direction changes frequently, the best approach is to keep the building and roof pitches low. This reduces the "air dam" effect of the building and permits relatively smooth air flow (*FIGS. 10b, 10c*).

Where winter winds do come from a predictable direction, the shape of the building shell can be streamlined with respect to orientation. An idealized geometric form for shedding the wind would probably be a teardrop or low-rise dome. In conventional construction, wind exposure can be reduced by manipulating roof pitches and orientation, or by integrating the shell with the ground surface to minimize the exposed area of the building. The effectiveness of shaping the shell can be enhanced a great deal by planting and windbreaks, some possible approaches for which are illustrated in *FIGS. 10c* and *10d.*

Rounded corners reduce pressure build-up by "leading" the air stream around the structure to promote laminar flow of wind. Similarly, smooth wall surfaces offer least resistance to the flow, thereby keeping pressures at minimum. Angling a building with the idea that the "knife edge" of a corner will help the wind slip by can be counterproductive, despite the initial sense this seems to make: the resulting increase in width of the structure seen by the oncoming wind creates a larger wind shadow, so that the resulting suction and driving force of indoor air movement can be expected to be larger.

Design Concept

11 Recess structure below grade or raise existing grade for earth-sheltering effect.

Buildings constructed on the surface are exposed to the full brunt of climatic stresses—a wide range of temperature fluctuations, constantly changing humidity levels, rapid expansion and contraction due to direct exposure to the sun, and of course, driving rain, sleet, snow, high winds and whatever other abuses the weather may deliver. Most of these stresses are appreciably damped out or stabilized below grade, and this benefit can be exploited through subsurface construction.

Building below grade may take on a variety of forms, most common of which is the basement. A number of euphemistic terms describing undersurface building are used, including "earth-integrated," "earth-contact" and "earth-tempered" design, among others. A categorization of basic types is illustrated in FIG. 11a.

Subsurface construction fulfills requirements of many climatic design strategies, including thermal control, wind protection, stability of moisture effects on materials, and protection from other extraordinary stresses such as brush fire, tornadoes, and excessive noise (from airports and freeways, for example).

Underground thermal environment
The thermal mass of the earth damps out and delays the fluctuations of the annual temperature cycle. Below the depth of 1½ to 2 feet, daily fluctuations are virtually eliminated.

FIG. 11b illustrates the annual fluctuation of earth temperatures related to depth for a typical soil and surface temperature range. From this graph it can be seen that earth temperature is not "constant" at a depth of 6 feet, as is commonly supposed. Instead, the temperature at this depth will usually swing about 10 or 11°F above and below the annual average ground temperature.

Another common misconception is that ground temperature is 55°F everywhere in the United States. Regional ground temperature averages can be estimated for general purposes from NWWA's well water map, FIG. 11c. Local ground temperature averages can be determined readily by taking the temperature of well water with an ordinary thermometer.

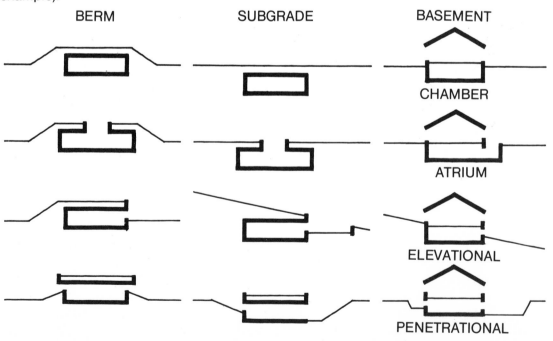

BERM   SUBGRADE   BASEMENT

CHAMBER

ATRIUM

ELEVATIONAL

PENETRATIONAL

FIG. 11a. **Earth tempering benefits may be captured through a variety of design approaches, ranging from conventional basements to turf-topped houses with sunken courtyards.**

Recess structure below grade or raise existing grade for earth-sheltering effect.

Heat loss and gain

Soil is not a very good insulating material. Four feet of an average soil are required to attain the equivalent of one inch of extruded polystyrene foam. The *effective thickness* of a soil as insulation, however, quickly adds up, since heat flow from subgrade walls in the winter follows a near-radial path to the surface. The method of calculating maximum winter heat loss from a subgrade wall is illustrated in *FIG. 11d*. Except in extreme northern climates, where the ground itself is at a low temperature, it is usually considered necessary to insulate only the first three to six feet of wall below grade. Beneath this depth, the additive resistance of the soil is sufficient to take over the insulating job by itself.

Since average ground temperatures for most of the U.S. are below the comfort zone, most underground portions of buildings constantly lose some heat to the subsoil through the lower parts of the wall and floor slab. In winter, these losses are usually considered negligible (due to the small temperature differential) compared to the losses near the surface, and not worthy of insulation. In summer these interior surfaces provide a desirable source of cooling, familiar to basement owners. Only when summer humidity levels are high and condensation becomes a potential problem does insulation become desirable.

Solar orientation

Undersurface construction is often thought to be somehow incompatible with passive solar heating. Like above ground houses, the solar heating potential of underground construction is a matter of orientation and design; it is not an intrinsic property of being either above- or below-grade. Nonetheless, the foundation and floor slab and the earth in contact with it provides an inexpensive form of thermal storage for solar-oriented house designs.

Costs of underground building shells are normally higher than that of frame construction due to additional structure required to resist soil loads and hydrostatic pressure. This may be partly off-set by elimination of exterior surface finishes. Underground houses have greater resistance to damage from windstorms than conventional construction and are therefore of great interest throughout tornado zones in the midwest and south.

Minimize Conductive Heat Flow

Minimize External Air Flow

Minimize Infiltration

Promote Earth Cooling

Minimize Solar Gain

Design Concept

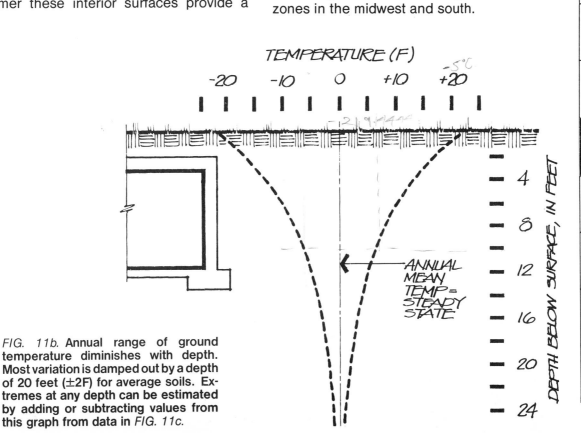

FIG. 11b. Annual range of ground temperature diminishes with depth. Most variation is damped out by a depth of 20 feet (±2F) for average soils. Extremes at any depth can be estimated by adding or subtracting values from this graph from data in *FIG. 11c*.

Recess structure below grade or raise existing grade for earth-sheltering effect.

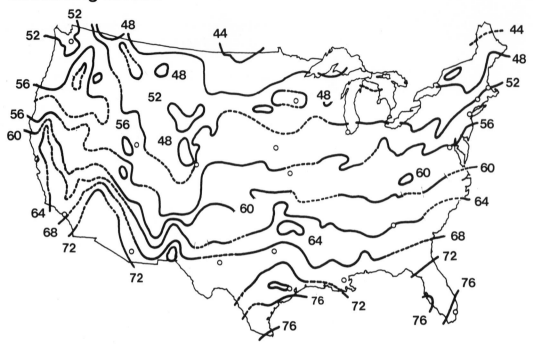

FIG. 11c. Approximate steady state of earth temperatures can be estimated from this map of shallow nonthermal well water temperature distribution (redrawn from a map prepared by the National Water Well Association).

FIG. 11d. For near-surface underground heat transfer calculations, heat can be assumed to follow circular flow paths centered on the interior wall face at outside grade level.

Design Concept

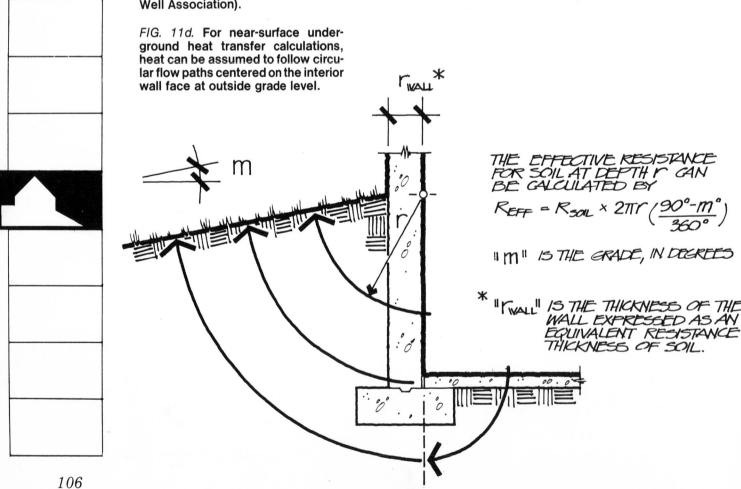

THE EFFECTIVE RESISTANCE FOR SOIL AT DEPTH r CAN BE CALCULATED BY

$$R_{EFF} = R_{SOIL} \times 2\pi r \left(\frac{90° - m°}{360°}\right)$$

"m" IS THE GRADE, IN DEGREES

* "r_{WALL}" IS THE THICKNESS OF THE WALL EXPRESSED AS AN EQUIVALENT RESISTANCE THICKNESS OF SOIL.

Shape and orient the building shell to minimize exposure to summer sun.

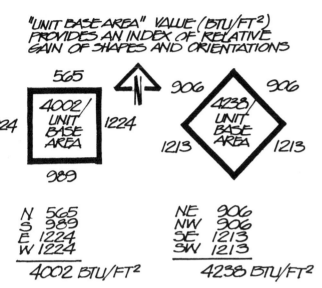

"UNIT BASE AREA" VALUE (BTU/FT²) PROVIDES AN INDEX OF RELATIVE GAIN OF SHAPES AND ORIENTATIONS

N	565		NE	906
S	989		NW	906
E	1224		SE	1213
W	1224		SW	1213

4002 BTU/FT²   4238 BTU/FT²

FIG. 12a. **Solar irradiation in BTU/sq/ft. of vertical wall area is illustrated here for Dodge City, Kansas area (about 38° N. latitude). Daily values are June-July averages.**

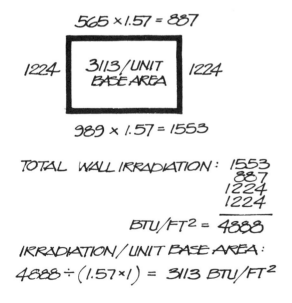

$565 \times 1.57 = 887$

$989 \times 1.57 = 1553$

TOTAL WALL IRRADIATION: 1553
887
1224
1224
BTU/FT² = 4888

IRRADIATION / UNIT BASE AREA:
$4888 \div (1.57 \times 1) = 3113$ BTU/FT²

FIG. 12b. **Plan proportions which yield least irradiation per unit base area at Dodge City is 1:1.57, or 30.3′ x 47.5′ for a 1440 square foot house.**

The amount of solar heat received by the surfaces of a house can be minimized for any period of the year through manipulation of the (1) shape and orientation of the building plan with respect to the sun; (2) height of the building exposed to the sun; (3) shape and pitch of the roof. With continuous shading controls at the facades, the plan shape, orientation, and building height will be rendered immaterial, inasmuch as solar gain will be blocked prior to receipt. This section, therefore, applies to buildings without overhangs and other wall shading devices, and assumes, for purposes of discussion, either equal or no fenestration.

Plan shape and orientation

For all regions of the United States, summer solar gain can be minimized by orienting the facades to face due N, S, E, & W. The proportions of the plan should be selected to equalize the sum of gains at north and south elevations with those received at the east and west facades at the most overheated time of the year. Although incoming solar radiation is greatest on June 21, air temperatures usually do not peak until late July or August. The design period should be selected in consideration of the two effects. For this discussion, radiation values averaged over the months of July and August will be utilized.

FIG. 12a. represents daily solar heat received per square foot of vertical wall at Dodge City, KS, during the maximum overheated period. The length-to-width ratio that produces the least solar receipt is given by L/W=(E+W)÷(N+S). Following this formula, the plan yielding the least solar gain for the selected period in Dodge City is given by 2440:1554=1.57:1. Applied to a rectangular, 1440 square foot floor area, this yields dimensions of 48′ x 30′ with the long axis running east-west. (*FIG. 12b*).

The optimum ratio will vary with latitude, with longer plans being favored in the south (1.64:1 in Miami, for example), and squarer plans in the north (1.30:1 for Glasgow, MT). Optimum length-to-width ratios can be determined by the builder or designer for east-west orientation simply by applying this formula to solar radiation data contained in

Minimize Solar Gain

Promote Radiant Cooling

Design Concept

12 Shape and orient the building shell to minimize exposure to summer sun.

Kusuda [1977]. The method presented here is an approximating method tailored to use the data in this reference.

Application of the same method to diagonally oriented plans reveals that the most favorable relationship will be a square, regardless of latitude; this follows from SE and SW exposures, and NW and NE facades, receiving identical solar loads. It can be seen from totaling the unit solar gains for *FIGS. 12a & 12b* that the diagonal plan receives greater heat gain.

Building height

Building height has considerable influence on solar gain by the dwelling, since increasing height increases wall area, hence exposure to the sun. Of primary interest is the relationship between enclosing a given floor area on one level versus two or more levels. Since summer irradiation is always most intense on a horizontal surface, it will be advantageous to decrease roof area and increase building height; walls, too, are more easily shaded than roofs. As a general rule of solar control, a building in the south should be taller than a building in the north that encloses the same interior volume.

A more rigorous analysis of solar loading has been performed by D.H.K. Lee [1970] for five different building shapes and three different orientations at latitudes 0° and 30°N (Table 12a). Lee's conclusion is that in extremely southern latitudes (of the northern hemisphere), "the least proportional solar load is experienced in hot season by the tall, slab-like building, with the long axis running east-west", and moreover, "this type of building has the further advantage of acquiring a high radiation load in the cool season of mid-latitudes". Lee also points out that "the single story building, whether square or rectangular, has the highest solar load under hot conditions when it is undesirable, but fails to retain this under cold conditions when it would be welcome."

If, however, roof gains are largely restricted or dissipated through attic ventilation, roof spray, or considerable insulation in the attic floor, the two story unit may have the greater transmitted solar load. This is possible because the wall solar gain of the two-story dwelling exceeds that of the single story unit.

The importance of such wall-area to roof-area relationship applies mostly in southern regions where cooling is the principal energy requirement.

Roof shape, roof pitch

The effect of roof shape and roof pitch on solar gain depends on the sun angle. When the sun is high, all roofs of the same horizontal area intercept the same amount of sunlight, as a simple shadow projection reveals (*FIG. 12c*). At lower sun angles, high pitched roofs increase exposure area, and therefore increase solar gain; this is especially true for east and west elevations.

Far more important than roof shape itself is what is under the roof. Flat roofs are difficult to ventilate and concentrate noonday radiation over a small surface area. A higher pitched roof, on the other hand, can be ventilated by both cross ventilation and stack effect. Roof shape becomes less important still if the exterior surface is a reflective color, if it is evaporatively cooled, or when the roof or ceiling is well insulated.

Design Concept

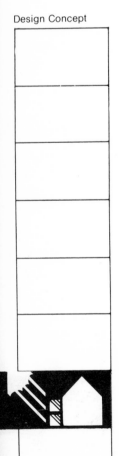

Shape and orient the building shell to minimize exposure to summer sun.

TABLE 12a
Incidence of solar radiation on buildings per sq. ft. of floor space for 30° north latitude (BTU/hr). [Lee 1970] .

Type	[1]Orientation	Summer			Winter		
		12.00 Hours	15.00 Hours	Average	12.00 Hours	15.00 Hours	Average
Cube 2 story	N	193	272	233	205	172	189
	NW	199	305	252	259	139	199
	W	193	262	228	221	174	198
Square 1 story	N	365	388	377	254	179	217
	NW	365	405	385	311	159	235
	W	365	359	362	281	179	230
Rectangle 1 story	N	339	339	339	292	167	230
	NW	339	398	[2]369	302	132	[2]265
	W	339	382	361	240	172	206
Square 4 story	N	113	252	183	211	201	206
	NW	113	305	209	278	152	215
	W	113	239	176	216	203	210

1. Buildings have simple gable roofs, with ridge at right angles to orientation.

2. Averages for types with rectangular base at intermediate orientation include projections for 9.00 hours (although not given in table).

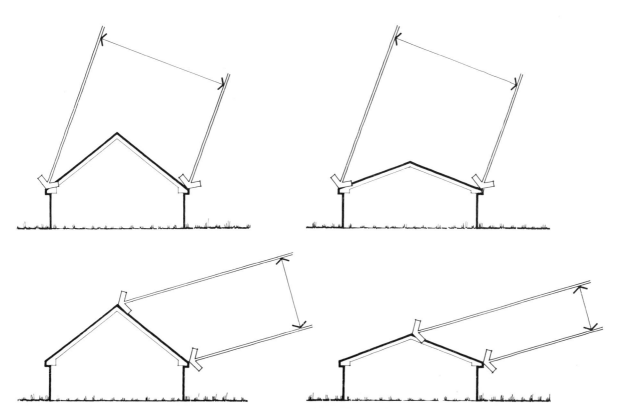

FIG. 12c. Roof shape has little effect on mid-day gain when sun is high.

Design Concept

13 Shape and orient the building shell to maximize exposure to summer breezes.

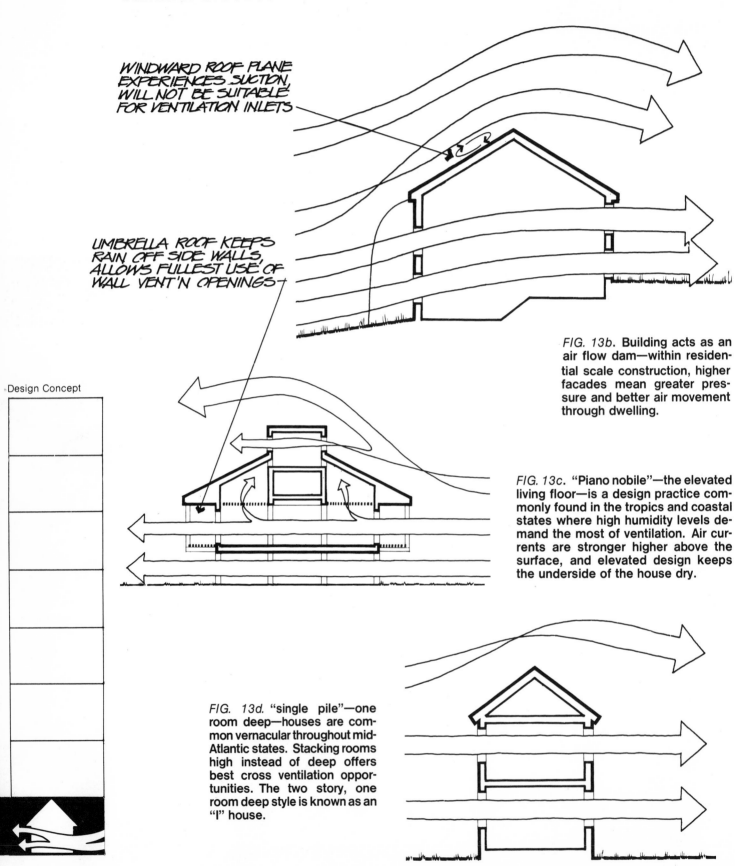

WINDWARD ROOF PLANE EXPERIENCES SUCTION, WILL NOT BE SUITABLE FOR VENTILATION INLETS

UMBRELLA ROOF KEEPS RAIN OFF SIDE WALLS, ALLOWS FULLEST USE OF WALL VENT'N OPENINGS

Design Concept

FIG. 13b. Building acts as an air flow dam—within residential scale construction, higher facades mean greater pressure and better air movement through dwelling.

FIG. 13c. "Piano nobile"—the elevated living floor—is a design practice commonly found in the tropics and coastal states where high humidity levels demand the most of ventilation. Air currents are stronger higher above the surface, and elevated design keeps the underside of the house dry.

FIG. 13d. "single pile"—one room deep—houses are common vernacular throughout mid-Atlantic states. Stacking rooms high instead of deep offers best cross ventilation opportunities. The two story, one room deep style is known as an "I" house.

Shape and orient the building shell to maximize exposure to summer breezes.

For simple rectangular shapes, typical of housing, summer wind currents are best exploited for natural ventilation by orienting the building shell so that its longer facade is approximately perpendicular to the flow of prevailing breezes (*FIG. 13a*). Although good ventilation can be obtained with the windward side facing somewhat askew from perpendicular, the fact that the breeze may deviate from its prevailing direction justifies the general rule of facing squarely into the breeze. When a building shape is complex, high and low pressures can build up along a wind-exposed facade which could be used to increase air flow into window openings.

If no predominant breeze direction can be identified—that is, currents come from nearly all directions when ventilation is needed—the building must be designed to ventilate efficiently along both its axes. In this case, a square floor plan would be suitable, with windows placed so that openings exist on windward and leeward sides.

Where a prevailing breeze direction does exist, the building can be shaped, and oriented to capture or funnel the flow of wind through its interior. In plan, for example, an L-shaped configuration will act as an effective air dam. In section, a high facade on the windward side of the building also will increase surface air pressure, which is the driving force of ventilation. Stacking rooms vertically also facilitates good cross-ventilation design (*FIG. 13b*). A traditional response to maximizing ventilation is to elevate the building to be exposed to higher wind velocities (*FIG. 13c*).

Flat and low pitched roofs are subject to suction (or uplift), even if they face the direction of the wind. These roof surfaces, therefore, are good locations for exhaust vents, but rarely good locations for air inlets (*FIG. 13d*). Steep roofs receive some pressure on the windward plane, but this will never be as great nor as effective for ventilation openings as in the windward wall itself. Roof-mounted "wind catchers" are ventilating devices commonly used in the Middle East and are often the primary means of admitting air flow. These are effective, however, only if the inlets are located high enough above the roof surface to project into the streamline flow zone.

Providing roof overhangs and porches on the windward side of the building is a means of utilizing roof shape to enhance ventilation inflow. It does this by damming the airstream in a pocket at the wall, thereby increasing the pressure outside the windows in the wall.

Finally, traditional techniques for capturing wind currents are worth consideration. In the southern "dog-trot" house, for example, the dwelling is organized around a horizontal air corridor to increase the exposure of at least two sides of each room to passing wind flow. Its modern counterpart is the ranch house with a breezeway. Central hallway plans utilize this same flow-through principle to ventilate flanking rooms, with the hallway acting as an interior breezeway.

Promote Ventilation

Promote Evaporative Cooling

Design Concept

FIG. 13a. **As a general rule long facade should be centered to face into direction of prevailing breezes. Plan may be rotated off axis up to 20° or 30° without seriously impairing ventilation performance.**

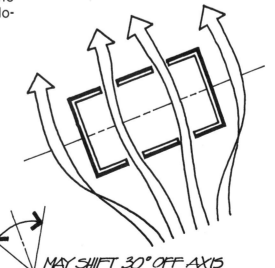

MAY SHIFT 30° OFF AXIS

Provide outdoor semi-protected areas for year-round climate moderation.

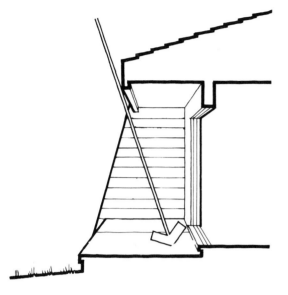

SIZE OVERHANG FOR SUMMER SUN PROTECTION

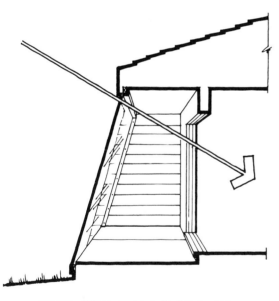

IF PROPERLY PLANNED IN ADVANCE, A SEASONAL GREENHOUSE CAN BE ADDED BY "STITCHING" IN A PLASTIC FILM TO SOFFIT, CURB, AND WINGWALL NAILING STRIPS.

FIG. 14 **Outdoor spaces can be designed for both summer sun protection and winter sun collection with demountable glazing panels or films.**

Patios, porches, courtyards and other protected outdoor living areas can contribute to the comfort of the indoors as well as providing pleasant, private living space in themselves.

In summer, porches and patio covers, for example, shade the house wall and openings and surrounding ground and outdoor floor surfaces. This helps to keep the temperature of the outdoor air low, making both natural ventilation more suitable and minimizing conductive heat gain through walls. A variety of shading treatments is possible; options should be selected with consideration for openness to ventilating breezes.

When patios or courtyards are well planted and watered, or when they contain fountains or other misting devices, they provide a positive source of cooling through the "clothesline effect" of evaporation. This also will depress the temperature of the outdoor space. In fact, by air-drying laundry in such a space, the surrounding air is cooled appreciably – 8,100 Btu of cooling for every gallon of water absorbed by the air. When held within a courtyard, this can enhance the cooling ability of natural ventilation and can reduce the house's cooling load. Well designed outdoor spaces, therefore, create an amenity in themselves, as well as preconditioning the exterior climate to make indoor comfort control more easily achieved.

The same outdoor spaces can provide a winter benefit: the house itself can be shaped to form sun "pockets" that make use of winter sun to warm areas immediately outside the dwelling. Not only will this expand the length of usefulness of outdoor living areas, but the heating effect will continue to provide benefit by creating a warmer climate outside part of the house, thereby reducing winter heat loss. Planting, overhangs, wing walls, and even pseudo-greenhouse effects like that illustrated in *FIG. 14* can help make warmer surroundings for the house.

Promote Solar Gain

Minimize Infiltration

Promote Evaporative Cooling

Promote Radiant Cooling

Design Concept

Use attic space as buffer zone between interior and outside climate.

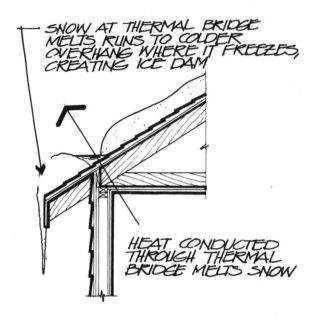

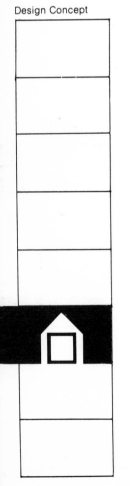

Design Concept

FIG. 15b. Ice dams are a result of an improperly insulated attic and insufficient attic ventilation. The floor of the attic should be insulated to keep heat from escaping into the attic and warming the roof, causing snow melt. Attic temperature can also rise as a result of south slope solar heat absorption, causing snow melt and ice damming on the north side. To prevent this, and condensation in the attic, the space should be vented.

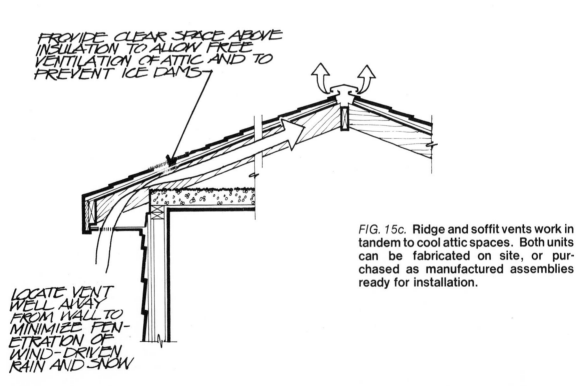

FIG. 15c. Ridge and soffit vents work in tandem to cool attic spaces. Both units can be fabricated on site, or purchased as manufactured assemblies ready for installation.

Use attic space as buffer zone between interior and outside climate.

Attics serve as intermediate spaces between the temperature-controlled portion of the house and the part of the building shell (the roof) that takes the brunt of both summer and winter stresses. With little added construction cost, the attic shelters the spaces below from direct exposure to summer sun and winter cold. By following two simple strategies – attic ventilation and ceiling insulation (e.g. in the floor of the attic) – the buffering ability of the attic space can be much enhanced.

Attic insulation

Heat that is lost through the roof in winter first enters the attic space through the attic floor from the ceiling beneath. Either the roof or the attic floor may be insulated to restrict this flow, but insulation in the attic floor is the more advantageous location. This has the effect of keeping the surface-to-volume ratio small, whereas insulating the roof makes the attic space itself become part of the heated structure beneath. In short, insulating the attic floor avoids heating an uninhabited space (*FIG. 15a*). Proper insulation also prevents problems of ice damming, which are common in older houses in northern zones (*FIG. 15b*).

Attic ventilation

There are three basic techniques of attic ventilation: cross ventilation, stack effect, and mechanical (fan) power. All serve the purpose of flushing out air that is heated by contact with the underside of a hot roof.

Cross ventilation is normally provided for by use of gable end louvers (*FIG. 15a*). With adequate outdoor breeze flow, cross ventilation technique, effecting the greatest number of air changes per hour. Unfortunately, cross ventilation is not a reliable mechanism, being dependent on wind direction and speed.

Since attic ventilation is frequently most needed when there is no wind, chimney effect ventilating devices often are installed to draw off hot air under still conditions. Stack effect ventilation requires hot air exhausts at high points in the attic and intake vents located as low as possible, in order to maximize the distance and resulting temperature differential between them. The use of ridge- and soffit-vents is one familiar technique (*FIG. 15c*), although a variety of other intake/outlet combinations are possible, such as roof vents, gable and rake vents and soffit vents in projecting gable eaves.

The thermal driving potential of the stack effect is increased with the distance between intakes and outlets. Raising the pitch of the roof, therefore, will generally enhance the ventilating ability of the attic space. The same effect may be achieved by projecting the outlet above the roofline. Cupolas and "dummy flues" may be used to this end.

Fan power can be coupled to most of the aforementioned ventilating devices to augment their natural processes, but is not cost effective except in very warm climates with inadequate ceiling insulation. Even so, it should be noted that in many poorly insulated attics, the primary mechanism of heat transfer from roof to attic floor is by thermal radiation. Ventilation cannot affect this transfer, although it can aid in cooling the floor once the heat has been received.

Minimize Conductive Heat Flow

Minimize Infiltration

Minimize Solar Gain

Promote Ventilation

Design Concept

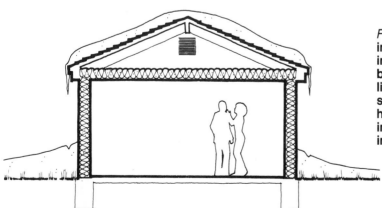

FIG. 15a. In winter, the attic serves as insulation space between the heated interior and the outside. To maximize benefit, heat should be confined to living areas by insulating attic floor. In summer, ventilate the attic to exhaust heat gained through the roof. Floor insulation keeps heat from transferring into living space beneath the attic.

Use basement or crawl space as buffer zone between interior and ground.

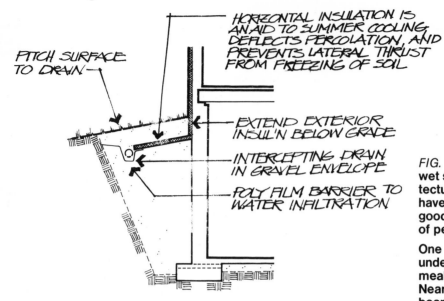

HORIZONTAL INSULATION IS AN AID TO SUMMER COOLING, DEFLECTS PERCOLATION, AND PREVENTS LATERAL THRUST FROM FREEZING OF SOIL

PITCH SURFACE TO DRAIN

EXTEND EXTERIOR INSUL'N BELOW GRADE

INTERCEPTING DRAIN IN GRAVEL ENVELOPE

POLY FILM BARRIER TO WATER INFILTRATION

FIG. 16c. The old farmer's adage "A wet soil is a cold soil" applies to architecture as well as agriculture. Soils have higher R-value dry than wet, so a good strategy is to keep the water out of perimeter soils.

One approach is to make surface & undersurface soils relatively impermeable to precipitation infiltration. Near-horizontal impervious insulation board can serve as a water deflector, and keeps backfill soil mass warm.

Design Concept

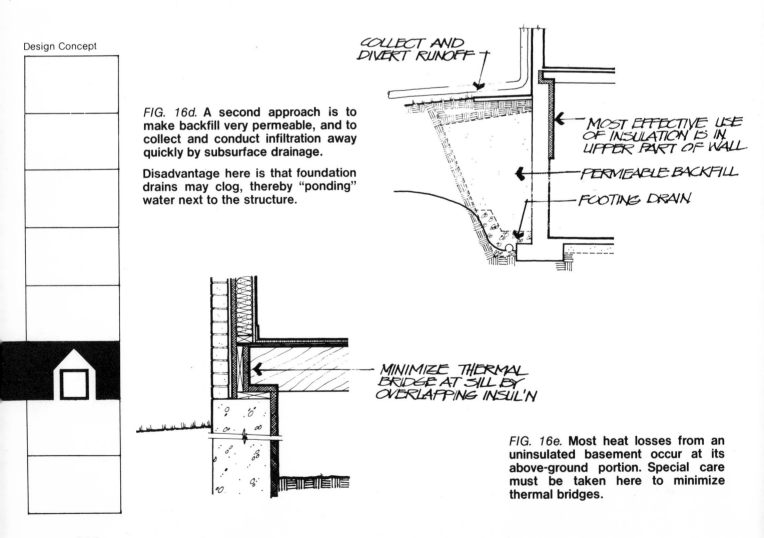

FIG. 16d. A second approach is to make backfill very permeable, and to collect and conduct infiltration away quickly by subsurface drainage.

Disadvantage here is that foundation drains may clog, thereby "ponding" water next to the structure.

COLLECT AND DIVERT RUNOFF

MOST EFFECTIVE USE OF INSULATION IS IN UPPER PART OF WALL

PERMEABLE BACKFILL

FOOTING DRAIN

MINIMIZE THERMAL BRIDGE AT SILL BY OVERLAPPING INSUL'N

FIG. 16e. Most heat losses from an uninsulated basement occur at its above-ground portion. Special care must be taken here to minimize thermal bridges.

Use basement or crawl space as buffer zone between interior and ground.

One way to reduce conductive losses from the floor to the ground is to raise the body of the house off the surface by constructing it over a basement or crawl space. This area then serves as a "dead air" insulation zone.

Basements and crawl spaces should be insulated regardless of whether or not they are heated. Most heat loss occurs through the perimeter walls of the upper few feet of the space and in particular through the above-grade portion of the wall. The most economical placement of insulation within the basement or crawl space is around the perimeter. This will only be effective, however, if the space is tightly sealed against infiltration (*FIG. 16a & 16e*).

If the space beneath the structure cannot be made airtight, or if it is necessary to ventilate the space to prevent excessive moisture levels, insulation should be placed in the crawl space ceiling, under the floor of the structure.

In closed crawl spaces and basements, the purpose of perimeter insulation is to isolate the lower walls and floor from the ground surface, which serves as the primary sink for subsurface heat. In northern regions where average ground temperatures are very low, it may be advisable to provide some form of floor insulation within the crawl space or basement.

Generally, a basement wall requires insulation around the upper portion, especially the band exposed to the exterior. Since perimeter heat losses are influenced by the conductivity of surrounding soils, drainage provisions can aid in reducing heat losses from basement or crawl space. As with slabs on grade, basement or crawl space insulation may be placed at either the interior or the exterior of the foundation wall. In the case of heated basements, exterior insulation is preferable so that the exposed interior wall mass will help to dampen internal temperature swings. (See *FIGS. 16c & d*).

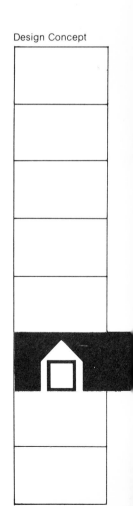

Minimize Conductive Heat Flow

Promote Earth Cooling

Design Concept

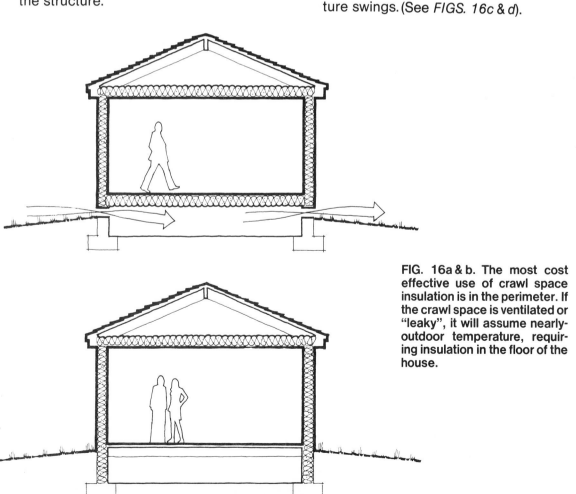

FIG. 16a & b. The most cost effective use of crawl space insulation is in the perimeter. If the crawl space is ventilated or "leaky", it will assume nearly-outdoor temperature, requiring insulation in the floor of the house.

Provide air shafts for natural or mechanically assisted house-heat recovery.

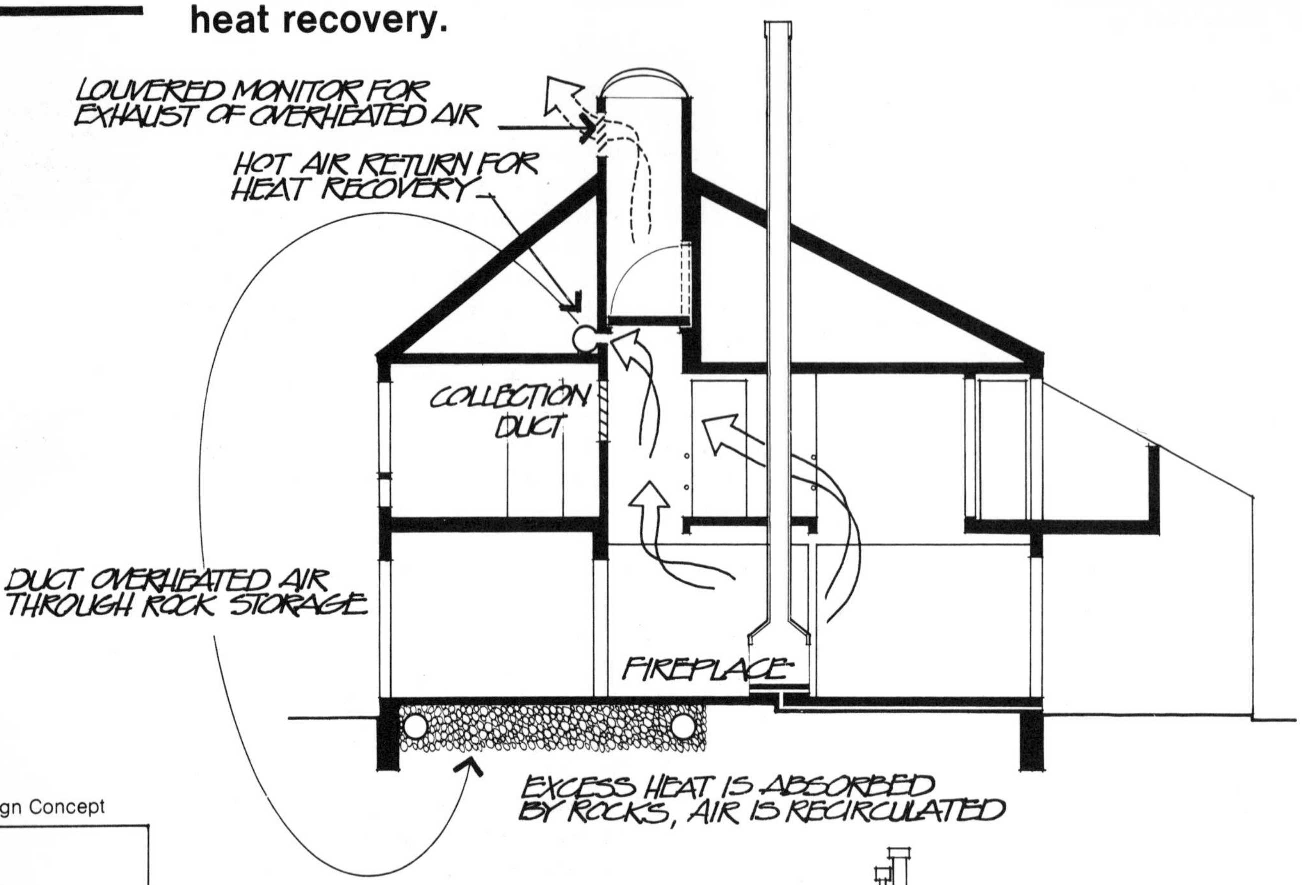

LOUVERED MONITOR FOR
EXHAUST OF OVERHEATED AIR

HOT AIR RETURN FOR
HEAT RECOVERY

COLLECTION
DUCT

DUCT OVERHEATED AIR
THROUGH ROCK STORAGE

FIREPLACE

EXCESS HEAT IS ABSORBED
BY ROCKS, AIR IS RECIRCULATED

Design Concept

FIG. 17a. **Overheated indoor air can be used to positive advantage in winter by collecting it and piping it through a rock bin where it is available "on call" for underheated periods.**

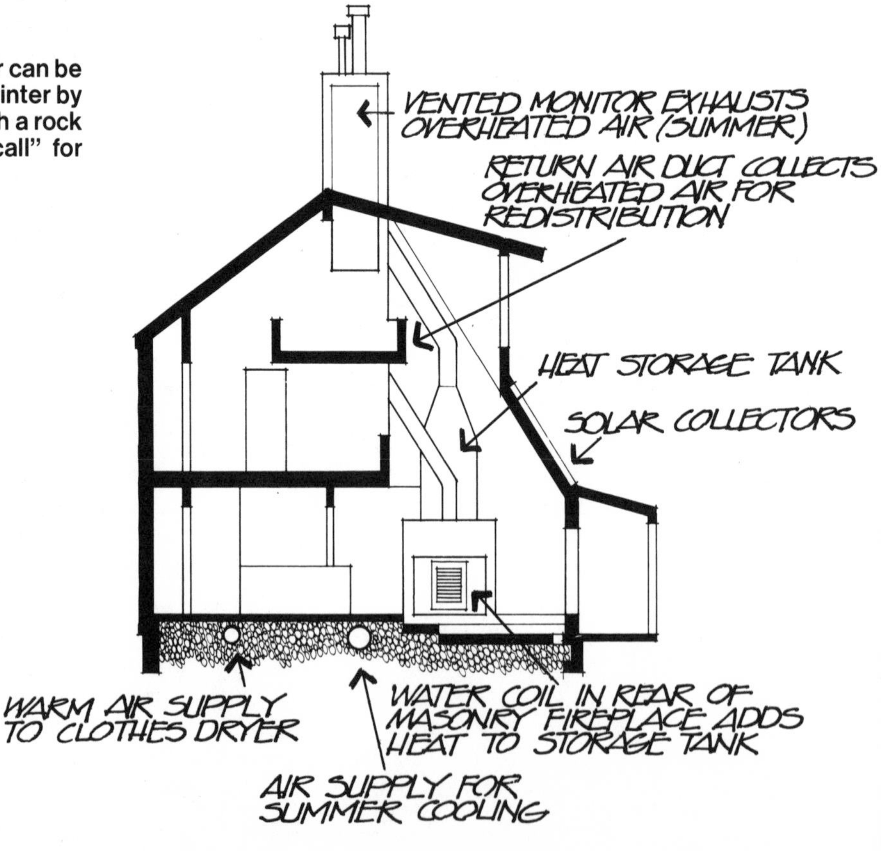

VENTED MONITOR EXHAUSTS
OVERHEATED AIR (SUMMER)

RETURN AIR DUCT COLLECTS
OVERHEATED AIR FOR
REDISTRIBUTION

HEAT STORAGE TANK

SOLAR COLLECTORS

WARM AIR SUPPLY
TO CLOTHES DRYER

WATER COIL IN REAR OF
MASONRY FIREPLACE ADDS
HEAT TO STORAGE TANK

AIR SUPPLY FOR
SUMMER COOLING

FIG. 17b. **House designed by solar engineer Everett Barber and architect Charles Moore makes use of multiple heat recovery and pre-tempering systems. Open plans allows central collection of overheated air, which is stored in both water tank and rock bed beneath radiating slab.**

Provide air shafts for natural or mechanically assisted house-heat recovery.

Minimize Conductive
Heat Flow

Promote Solar Gain

Solar energy utilization in buildings is distinguished from that of other energy sources by its low temperature and its irregular supply, so that it is desirable to store excess "low grade" solar and waste heat from internal sources. If the house is properly designed to store excess heat, "overheating" in a passive system can be made into a virtue, instead of a discomfort.

Heat recovery systems can be categorized as either space-heat recovery of overheated air or the "tapping off" of appliance-generated heat before being used to heat the space. A third type of heat recovery, in a seasonal storage sense, is the tempering of outdoor air by drawing it through deeply buried pipes since deep ground temperatures are warmer in winter than average outdoor air temperatures.

Space heat recovery

Recovery of heat from overheated interior air can be achieved through design practices similar to those recommended for exhausting hot air in summer. Instead of venting the air to the outside, however, a ducted system must be employed to route the air into a storage reserve, such as a rock bin. Providing a high internal space such as the stairway affords a central collection point that prevents overheating of any one room or level within the house. Since this lack of interior zoning can lead to low air temperatures at the first floor level, it may be advisable to use a storage system that also provides radiant heating effects (*FIG. 17a*).

Particular attention should be given to the detailing of dampers and insulating devices, so that hot air is not wasted through heat-leaky monitors, belvederes, or cathedral ceilings.

Appliance heat recovery

Fireplaces and furnaces are examples of devices that in themselves produce a great deal of heat that is directly wasted—up the chimney. Much of this heat can be captured by embedding ducts or pipes within or around the flue, so that heat can be transferred into storage. *FIG. 17b* illustrates a design in which water coils are placed in the rear of a masonry fireplace; heated water is conveyed from the fireplace to an insulated water tank, where it retains its heat long after the fireplace masonry itself has cooled. Like the design in *FIG. 17a*, this house features a high hot air return duct; reclaimed hot air is first streamed past the water storage tank, then routed into a rock storage-distribution plenum which also acts as a radiant floor, thereby creating a comfort effect even if heated only in the range of 80°F.

Some fireplaces are available with recirculating air systems that instantaneously recover and convect heat to several rooms. These prefabricated designs could be coupled to rock storage beds, or storage radiant-floor beds, so that excess fireplace heat could be more effectively spread over both time and area.

Combined systems

Heat recovery and storage systems make most economic sense when these are coupled to components that exist as part of other systems. This is exemplified in the house shown in *FIG. 17b* where the large water storage tank has been installed to store solar heat from roof-mounted liquid collectors.

Even if no form of storage is used, air shafts that redirect air from the top of the house back through the heat distribution system have the benefits of lowering temperatures against the upper floor ceiling thereby reducing house heat loss and of balancing the temperature throughout the house interior.

Design Concept

Centralize heat sources within building interior.

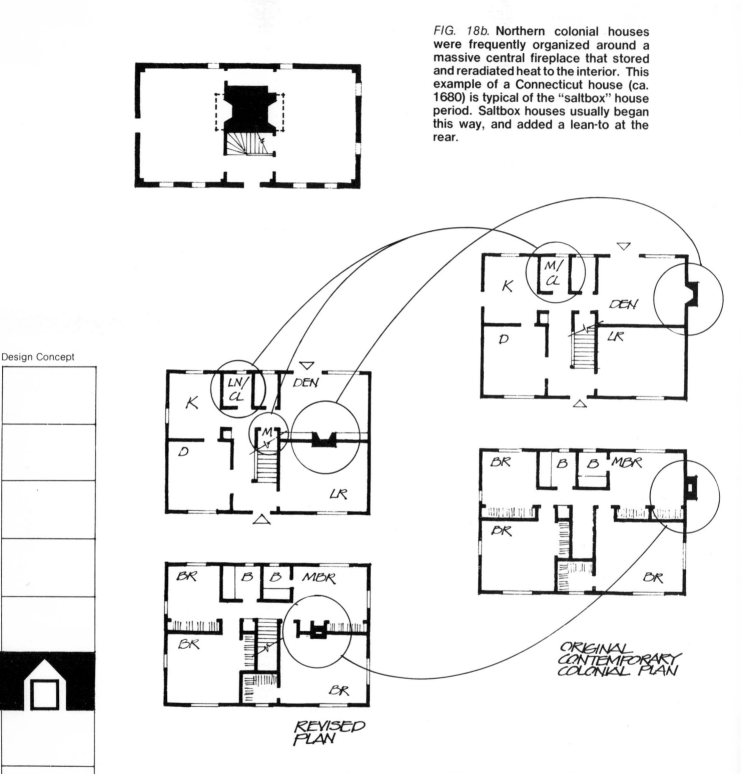

FIG. 18b. Northern colonial houses were frequently organized around a massive central fireplace that stored and reradiated heat to the interior. This example of a Connecticut house (ca. 1680) is typical of the "saltbox" house period. Saltbox houses usually began this way, and added a lean-to at the rear.

Design Concept

ORIGINAL
CONTEMPORARY
COLONIAL PLAN

REVISED
PLAN

FIG. 18c. Relocating exterior wall fireplace to the interior makes this popular colonial style plan better suited to northern regions. Switching furnace, hot water heater, laundry to interior areas also helps makes use of "waste heat" of these appliances.

Centralize heat sources within building interior.

From the time of the earliest colonial houses, a heat conservation principle has been to centralize heat-generating devices in the interior of the house. New England's "saltbox" houses exemplify this practice in the central location of fireplaces and cooking devices. Currently, wood-burning stoves are being re-introduced as major heating elements in northern regions of the United States. These can be sized to heat an entire house, in which case a central location on a lower floor is desirable.

Although contemporary houses don't rely on centrally located radiating fireplaces, the principles remain the same (*FIGS. 18a* and *18b*). Contrary to this, however, baseboard radiators and warm-air registers are located along external walls and hot air convectors are normally placed beneath windows. These locations are selected for heating sources for precise reasons: outside walls, windows, and corners are the coldest part of the house, thereby demanding additional heat to make them comfortable. They are the coldest areas because they lose heat faster than other parts of the house, and so they are also the least energy efficient location for

warming the house. If the house walls and windows are well insulated, the cold drafts usually associated with uninsulated walls and floors are not such a concern, so that other system layouts can be considered.

Locating air delivery registers and baseboards to more energy-efficient internal locations within rooms may result in less uniform temperature levels that residents are likely to find uncomfortable. However, in warm air systems, the returns can be located under windows, thereby eliminating cold down drafts across the floors. With other types of heating, the efficiency of heat delivery is improved by insulation near radiators, and use of heat reflective finishes (*FIG. 18c*).

Locating ovens and stoves and other heat sources such as mechanical rooms, against interior walls will also help contain heat within the interior instead of speeding conductive losses through outside walls. Similarly, rooms can be planned so that critical comfort zones occur near interior, rather than outside walls—locating bedroom closets along outside walls, for example, frees up interior partitions for the beds.

Design Concept

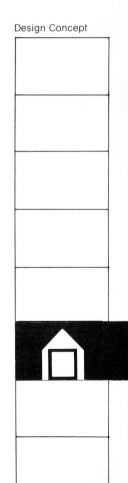

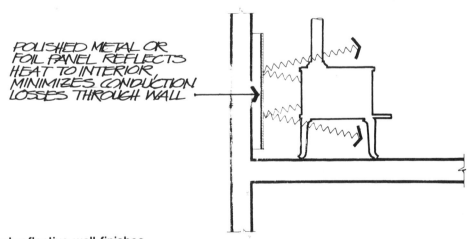

POLISHED METAL OR FOIL PANEL REFLECTS HEAT TO INTERIOR, MINIMIZES CONDUCTION LOSSES THROUGH WALL

FIG. 18a. **Heat-reflective wall finishes "bounce" heat back into room space, rather than allowing absorption and conduction at and out through the wall. This strategy is particularly appropriate in this "new era" of radiators—of wood burning stoves and free standing fireplaces.**

121

Use high-capacitance materials to store solar heat gain.

TABLE 19a

Thermal admittance expresses the "acceptability" a material has towards heat absorption and storage. Materials of high admittance quickly store and release heat (metals, for example) while materials of low admittance are relatively "indifferent" to heat presence—they respond slowly and hold little heat.

THERMAL ADMITTANCE = $\sqrt{\text{CONDUCTIVITY} \times \text{HEAT CAPACITY}}$

Design Concept

Material Description	Heat Capacity BTU/cu.ft ($^\circ$F)	Conductivity BTU /hr (ft)$^\circ$F	T. Admittance BTU/ft^2($^\circ$F) $\sqrt{\text{hr}}$
Acoustic tile	5.8	.033	.44
Adobe	19.6	.37	2.7
Aluminum	35.9	128.	67.8
Brick, common (120 pcf)	24.	.42	3.2
Brick, face (130 pcf)	26.	.75	4.4
Concrete	29.4	1.0	5.4
Copper	51.	227	108.
Corkboard	24.6	.023	.27
Glass (Pyrex)	26.8	.59	4.1
Gypsum	51.3	.25	2.2
Iron, cast	54.	27.6	38.6
Limestone	34.7	.54	3.5
Marble	18.	1.5	7.1
Paraffin	18.6	.14	2.3
Particleboard (160 pcf)	27.7	.1	1.36
Plasterboard	22.4	.43	3.1
Plywood	9.9	.067	.81
Polystyrene (Beadboard)	.3	.023	.083
Sand	18.	.19	1.85
Soil, light & dry (80 pcf)	18.	.2	1.9
Soil, average (damp, 131 pcf)	30.1	.75	4.75
Soil, wet (117 pcf)	35.1	1.4	7.0
Wood, hardwood	18.7	.09	1.3
Wood, white oak	26.8	.1	1.6
Wood, softwood	10.6	.067	.84
Wood, white pine	18.1	.063	1.07
Water, still*	62.4.	.35 *	4.67 *
Glass, cellular insulation	2.2	.033	.27
Lead	21.8	20.1	20.9
Ice	27.	1.35	6.04
Bakelite	20.4	9.7	16.6
Steel (mild)	58.7	26.2	39.2
Granite	31.7	1.40	6.6

*The "apparent" admittance of stirred water is much higher, due to increase in heat transfer (apparent conduction) by convection; it may range from 8 to 400 times greater than still water.

Use high-capacitance materials to store solar heat gain.

Interior spaces with large south-facing windows can overheat on a clear sunny day in winter. This effect, plus the fact that the sun shines for only one-third of the day during much of the heating season, makes it desirable to utilize interior materials to absorb solar heat as it arrives and to reradiate back to the interior after the sun has passed. These materials should be of high thermal capacitance and should be located wherever surfaces are exposed to direct solar radiation for extended periods of time.

Heat capacity alone, however, does not necessarily indicate a good heat storage medium. The time rate of acceptance and diffusion of heat throughout the material will determine its overall performance, and this is related to the material's conductivity as well as its heat capacity. An index of this property is "thermal admittance", and is equal to the square root of the heat capacity-conductivity product. Table 19a presents values of thermal admittance for a variety of materials.

Materials of low admittance are not effective for thermal storage. The heat of materials of low admittance absorbed at the surface will quickly be dissipated into the air, rather than in the material mass, producing large indoor temperature swings. Materials of high thermal admittance, in contrast, are quick to absorb heat to which they are exposed and are also quick to lose the heat they contain. Note that some materials listed in Table 19a have high heat capacity but also relatively low admittance values.

The construction systems most widely used for heat storage consist of concrete and masonry. Water, although less dense than brick and mortar, is superior as a storage medium because it can store more heat per unit of volume; it is frequently utilized in drums, containers, vertical columns, or even bottles placed in the sun path.

The location of interior storage materials has been used to classify some of the basic types of passive solar heating systems which are described as follows (FIGS. 19a & 19e).

Direct Gain

Direct gain systems make use of high mass materials that are exposed to the interior as ordinary room elements. Floors exposed to the sun should have tile, brick, or other heat absorbing material laid on a concrete or sand bed, with no carpeting or other insulative flooring surfaces applied. Similarly, masonry walls should be left with a brick, stone, concrete or tile facing when exposed to direct solar radiation.

For materials directly exposed to sun, a suggested rule of thumb for solar storage is that 30 Btu of storage mass be provided per square foot of solar-admitting glazed area. Thus, one hundred-twenty square feet of glazing would require 120 x 30 = 3600 Btu of storage. For an equivalent effect, four times as much storage capacity is required for storage materials *not* directly exposed to sun (materials exposed to the room interior, but not to the sun itself). Best results are obtained by distributing thermal mass locations throughout the interior, some exposed to direct sun, others to diffuse or reflected sunshine, in whatever combination of masonry floor, wall, fireplace, or ceiling materials that can be practically incorporated into the house construction.

One experimental solar structure built by the Massachusetts Institute of Technology utilizes a phase-change heat storage material sandwiched within a one-inch thick concrete tile. In a sense, the material is "preprogrammed" to evolve heat by its own internal chemical thermostat: the tile panels store excess day heat above 74°F and re-release it at night when room temperatures fall below 74°F. These lightweight storage cells can be used with conventional floor, wall, and ceiling construction, without the increased structural requirements for heavy masonry massing.

Indirect gain

Indirect gain systems admit solar radiation to a non-occupied collection space such as an attached greenhouse, skylit attic, or to

Design Concept

19 Use high-capacitance materials to store solar heat gain.

glazed storage mass. In each case, the surfaces exposed to the sun form a heat-storing and heat radiating partition with the interior living zone, usually a wall or ceiling. For glass-covered masonry walls in which the storage surface area approximately equals the aperture area, the storage mass requirement would be met by a concrete wall (heat capacity = 29.4 Btu/cubic foot) slightly greater than one foot in thickness. An adobe storage wall would necessarily be 50% thicker (since the heat capacity of adobe is about two-thirds that of concrete), while a brick storage wall (at 25 Btu/cubic foot) should be 20 percent thicker than concrete to provide the same storage. Only half the volume of water is necessary to store the same amount of heat as concrete.

Indirect gain, remote storage

The storage capacity of an indirect gain system can often be enhanced by ducting overheated air away from the sunspace itself into remote storage, such as a rock bin. When this is done, the storage capacity of the partition itself can be reduced and nearly eliminated. This is the basis of the "solar attic" concept in which case the attic is used as a solar collector.

Another variation is the attached greenhouse with a glazed wall between the greenhouse and the living space; if heat absorbing shades or blinds are placed on the outside of the interior wall, the system will function in direct gain, indirect gain, and/or remote storage modes of operation.

Design Concept

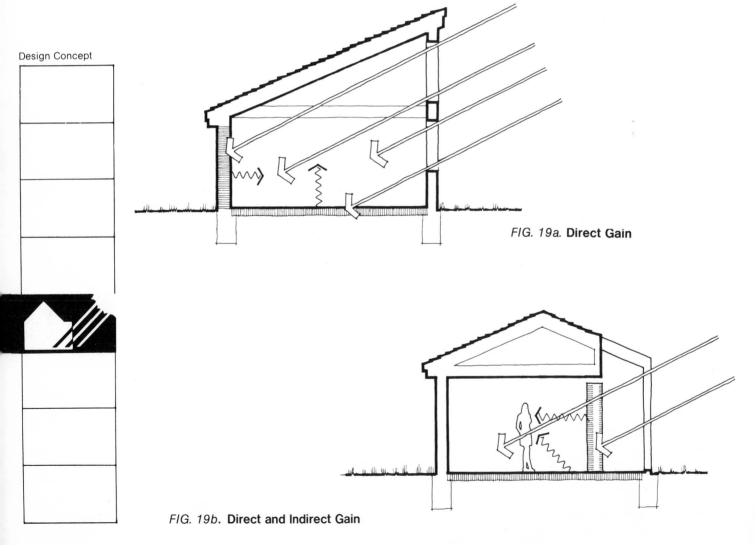

FIG. 19a. **Direct Gain**

FIG. 19b. **Direct and Indirect Gain**

Use high-capacitance materials to store solar heat gain.

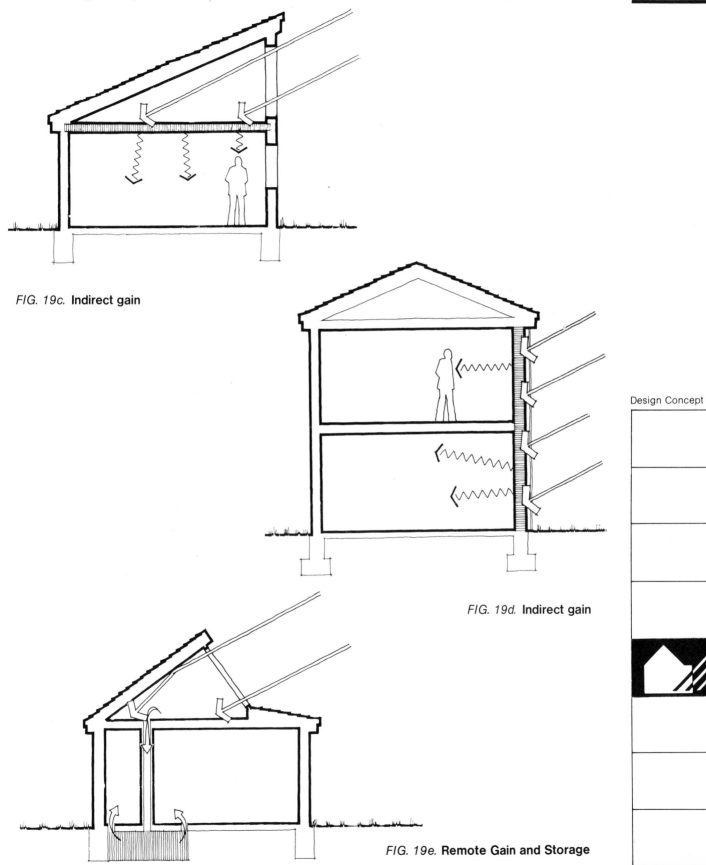

FIG. 19c. **Indirect gain**

FIG. 19d. **Indirect gain**

Design Concept

FIG. 19e. **Remote Gain and Storage**

Provide solar-oriented interior zone for maximum solar heat gain.

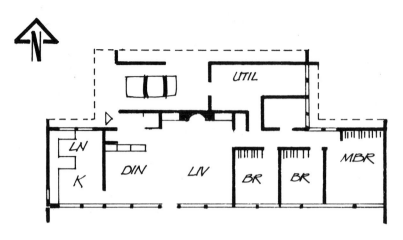

ONE STORY PLAN. Keck and Keck, Architects.

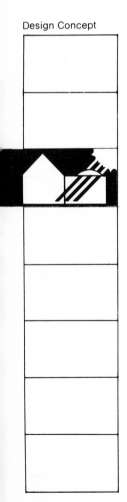

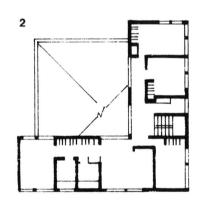

TWO STORY PLAN. Gelardin/Bruner/Cott, Architects.

Cross Section

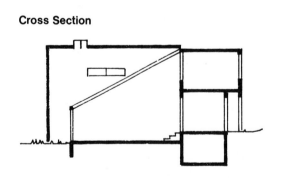

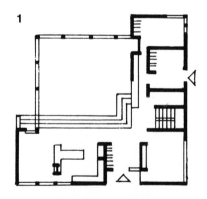

FIG. 20b. In lieu of or in addition to, orienting the majority of living areas for direct solar collection, a single "sunspace" can be used to fully maximize solar collection. Typical examples are attached greenhouses and internal glazed-roofed atria.

Provide solar-oriented interior zone for maximum solar heat gain.

The general concept of using south-facing room planning for passive heating can be carried one step further in design by creating an interior room such as a sunroom or atrium that maximizes its ability to function as a direct gain solar system (*FIG. 20a*). Such a room would not be well suited as a bedroom or kitchen, but may be ideal as a recreation area, sitting room, or working greenhouse. Heat gained from the sunroom can be removed from the room by natural convection or by a temperature-controlled fan for distribution throughout the rest of the house or to thermal storage.

The solarium and "Florida room" are familiar to most builders. Other "sunspaces" include skylit atriums, internal and attached greenhouses, and rooms with full wall glazing and/or large glazed roof areas (*FIG. 20b*). Since most glazing materials are losers of heat, some form of night insulation is essential in northern climates. In warmer climates the sunspace can be built with insulated glass, but otherwise left to cool at night. New glazing systems for sunspaces are currently under development, such as the "heat mirror". These materials are good solar "transmitters" but then "block" re-radiated heat from the interior.

A caution must be raised regarding the benefits of attached greenhouses: since most prefabricated, and even custom-built, attached greenhouse assemblies have little insulating value, their *net* solar gain is likely to be very small. The amount of heat contributed by the attached unit will be less than a window of size equivalent to the attachment area since it is in effect "shading" the actual wall between the greenhouse and house.

The advantage of the sunspace that is separated from the house is that on sunny days the temperature in the sunspace can be allowed to increase to above comfort limits so that it can be used more effectively for thermal storage, while at night the temperature can be allowed to drop below comfort limits and thus not require additional heating for a space that also adds to the living area of the house.

Minimize Conductive Heat Flow

Promote Solar Gain

Design Concept

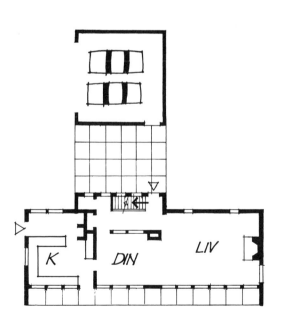

TWO STORY PLAN. Furno and Harrison, Architects.

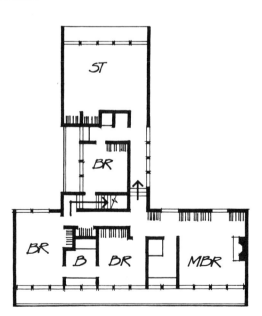

FIG. 20a. **A basic tenet of direct gain passive design is to orient major living areas to face south so that valuable radiation won't be wasted. Some "idea plans" are presented here.**

Plan specific rooms or functions to coincide with solar orientation.

TABLE 21a
Suggested room orientations

	N	NE	E	SE	S	SW	W	NW
Bedroom*	•	•	•	•	•			
Bath*	•	•	•	•	•	•	•	•
Kitchen			•	•	•			
Dining			•	•	•	•		
Living				•	•			
Family				•	•	•		
Utility / Laundry*	•	•						•
Workshop*	•	•						•
Storage*	•						•	•
Garage*	•					•	•	
Sun porch				•	•	•		
Outdoor space*				•	•	•		

*The most suitable location of those indicated will depend on local climate — whether largely too hot or too cold, direction of winter winds and summer breezes, etc.

A house can be made more energy efficient simply by designing the plan so that the order of rooms in which the normal daily sequence of activities occurs "follows" the path of the sun *(FIG. 21a)*. This strategy is complemented by, and most effective when combined with, partitioning the interior into separate heating and cooling zones, as recommended in subsequent techniques. By relating zones to sun movement, solar energy can be put to use when it is most available by direct orientation, and the call for mechanical heating or cooling can be minimized.

Examples of such favorable orientation include: an east- or southeast-facing window for bedrooms, kitchen and breakfast area to benefit from the earliest winter morning sunshine; a south orientation for daytime living areas; a southwest orientation for sunspaces or other indirect solar gain elements (such as the Trombe wall, *FIG. 19e*—the western sun would otherwise "overheat" a direct gain arrangement, particularly from spring through fall).

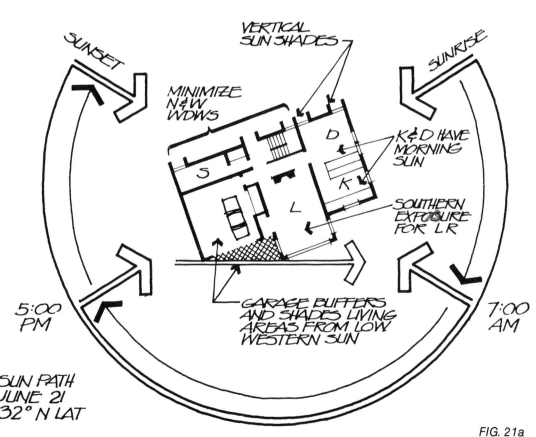

VERTICAL SUN SHADES

SUNSET

SUNRISE

MINIMIZE N & W WDWS

K & D HAVE MORNING SUN

SOUTHERN EXPOSURE FOR LR

5:00 PM

7:00 AM

GARAGE BUFFERS AND SHADES LIVING AREAS FROM LOW WESTERN SUN

SUN PATH JUNE 21 32° N LAT

FIG. 21a

Design Concept

Use vestibule or exterior "wind-shield" at entryways.

Design Concept

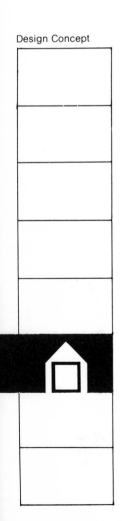

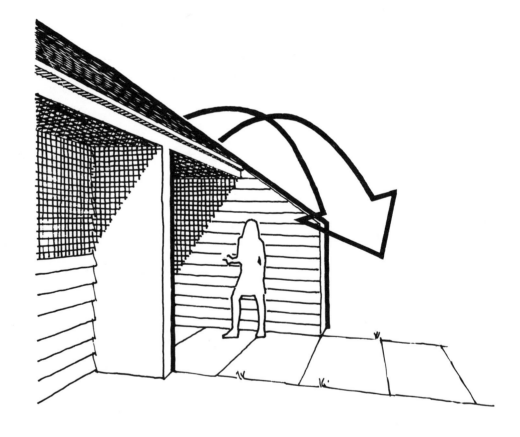

FIG. 22c. Wingwall, vertical fin panels can be incorporated into design or can be added to many conventional house style plans without visual discord. This is not a full substitute for vestibule, but can help deflect the wind.

Use vestibule or exterior "wind-shield" at entryways.

Vestibules control infiltration in two ways: 1) they function much better than a storm door in increasing the resistance to infiltration between door and jamb, 2) the sequence of closing one door before opening the second prevents direct air exchange between indoors and outdoors. ASHRAE *Fundamentals* [1981] indicates that the use of a vestibule can reduce infiltration amounts from 900 to 550 cubic feet per door opening when compared to a single outside swinging door.

Viewed in terms of overall air changes, infiltration tests conducted on houses near Princeton, NJ, indicate that under normal winter conditions with doors and windows closed, air infiltration rates contributed to approximately one-third of the total heating load. In situations where door openings and closings occurred for even short periods of time, however (one or two minutes each hourly interval), air infiltration heat losses were observed to increase to as much as 75 percent of the total heating load.

The effectiveness of vestibules in cutting energy costs will depend, of course, on how often the doorways are used, and whether one door is allowed to fully close before the second is opened (*FIG. 22a*). It should be apparent that vestibules will be of most value at entrances that are most frequently used.

Vestibules will be of greatest benefit at walls which are exposed to winter winds—although entrances at these locations should be avoided when possible (*FIG. 22b*). Exterior vestibules can often be added to existing house plans. An alternative to an enclosed vestibule is to provide some other form of exterior shelter or windshield at the entrance either as an extension of the building or in the form of plant materials (*FIG. 22c*). It follows that entrances should not be located at "wind funnel" areas that channel or otherwise increase wind flow past the entrance.

Minimize Infiltration

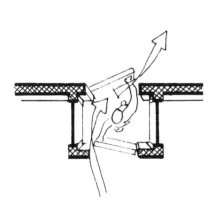

FIG. 22a. **Make sure dimensions are generous enough and door swings comfortable so that users allow one door to close before opening the second.**

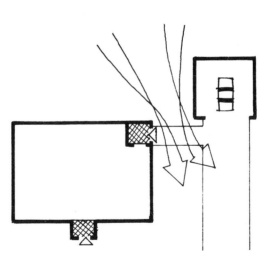

FIG. 22b. **Vestibules may be located within the plan or placed as a weather cover or closed porch outside of wall. Try to avoid locating doorways in wind channels, try to avoid creating wind funnels in site planning.**

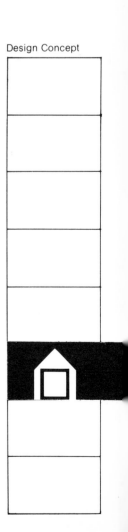
Design Concept

Locate low-use spaces, storage, utility and garage areas to provide climatic buffers.

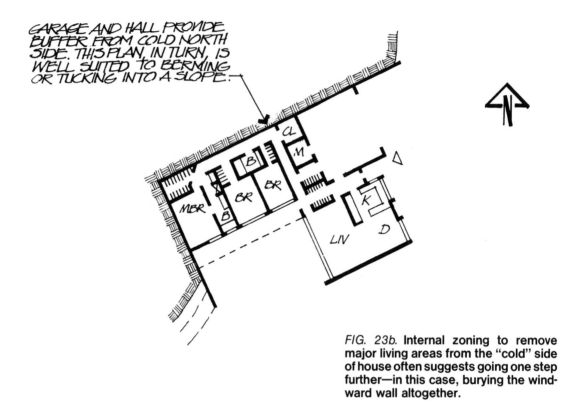

GARAGE AND HALL PROVIDE BUFFER FROM COLD NORTH SIDE. THIS PLAN, IN TURN, IS WELL SUITED TO BERMING OR TUCKING INTO A SLOPE.

FIG. 23b. Internal zoning to remove major living areas from the "cold" side of house often suggests going one step further—in this case, burying the windward wall altogether.

Design Concept

Locate low-use spaces, storage, utility and garage areas to provide climatic buffers.

Storage and other non-living areas of the house can be used as insulative dead-air zones of the house. These will be most effective when located on the windward side of the structure (*FIG. 23a*), especially if these areas are fully partitioned from the heated zones of the house. Under-utilized spaces can also be used to advantage on the west side of the house to absorb the full thermal impact of late afternoon sun in the summer.

Like basements and attics, heat lost from active portions of the house into unheated zones will increase the temperature of the latter, rather than being "washed away" by the wind (*FIG. 23b*). This increase in temperature decreases the temperature gradient across the partition between the heated and unheated spaces and thus retards the rate of heat loss. The rate of loss can be further slowed by making these partitions as insulative as practical. Another effect of this partitioning is to reduce the volume of the house that must be heated, which corresponds to the amount of energy necessary to maintain comfort levels.

Minimize Conductive Heat Flow

Minimize Infiltration

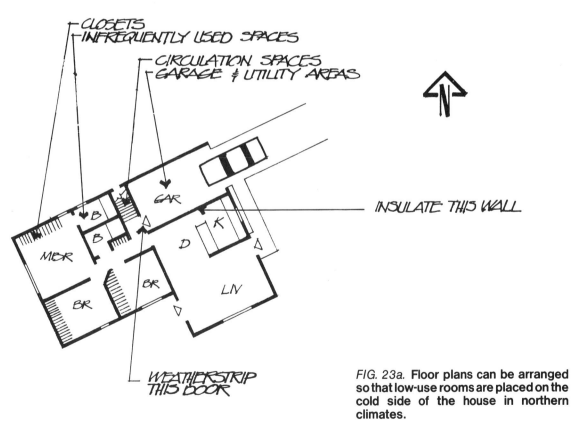

Design Concept

FIG. 23a. Floor plans can be arranged so that low-use rooms are placed on the cold side of the house in northern climates.

This same strategy can be applied to the hot side of the house in southern regions.

Use slab-on-grade construction for ground temperature heat exchange.

A slab-on-grade has two different zones of heat transfer: (1) direct losses and gains to the perimeter of the floor slab, and (2) fairly constant thermal exchange with the ground within the central area of the slab. The relative importance of these will vary with soil type, nearness of ground water level, building coverage area, and other factors. Each condition demands individual attention, as detailed below.

Perimeter heat exchange

Steady state (uni-directional) heat flow between a heated floor slab and a cold exterior ground surface near the perimeter will follow a series of nearly concentric radial paths (*FIG. 24a*).

Typical soils *in situ* have resistance values ranging between R1 and R2 per foot. Concrete falls in this same range, so grade beams and concrete foundations can be considered to behave much the same as soil.

Since the insulating effect (path length) of soil increases with depth, the use of perimeter insulation will exhibit diminishing returns as the insulating ability of the soil increases. Since the heat flow path is approximately circular, placing the insulation under the slab or at the outside face of the footing will have essentially the same effect in terms of thermal resistance. Vertical placement isolates the cold from penetrating through the foundation and therefore offers better protection from freeze-thaw stresses. Horizontal insulation, on the other hand, more effectively isolates the floor from all slab-ground interactions. Because of this, horizontal perimeter insulation is recommended when the ground water level is near the underside of the slab (less than 2 feet). When ground water is more than 4 feet below grade, the vertical placement is usually preferred.

Recommended R values for perimeter insulation (24″ extent) recommended by ASHRAE Standard 90-75 are given in *FIG. 24b*.

Center slab heat exchange

FIG. 24c illustrates theoretical paths of heat flow and lines of constant temperature (isotherms) occurring beneath a building under steady state conditions (one way flow, here shown as heat loss). The path lines, rate and even direction of heat flow will be affected by the seasonal changes in ground temperature, so the diagram represents only a simplified visualization of actual conditions.

Nevertheless, it does convey the fact that the center of the slab exchanges heat more directly with the deep subsurface rather than with the perimeter. This means that over the course of time a relatively stable temperature gradient will be established, and thermal interactions between ground and floor will be small. In many conventional frame houses, a slab-on-grade thus offers a heat storage mass that can moderate indoor temperature fluctuations.

Due to the immense thermal mass of the earth, the stability of this floor-ground temperature gradient can be an asset in maintaining indoor comfort by damping out internal swings due to both the "offs" and "ons" of the heating system. In northern climates, ground temperatures do impose some heating and cooling load on the house via an uninsulated slab, but this is usually regarded as negligible and not justifying subslab insulation. Two instances where underfloor insulation may be desirable are as follows:

1—In northern zones subject to high summer humidity levels, the surface temperature of the floor slab may fall below the dewpoint, causing condensation on the surface or within carpets and rugs. Continuous insulation

Design Concept

Use slab-on-grade construction for ground temperature heat exchange.

under the floor will correct this by "disconnecting" the slab from the thermal mass beneath, allowing it to be warmed by indoor temperatures. While this solves the condensation problem, it does so at the expense of a free cooling source. An alternative solution is to dehumidify the air and to enjoy the cool floor.

2 – Because ground water has enormous heat capacity, it is a very effective heat sink. In northern climates where ground water level is near the surface, this can result in a continuous washing away or bleeding off of heat from the underside of the slab (analogous to wind chill in the air). Even though high ground water levels are often seasonal and at a low in winter, the increase in soil moisture content due to capillary rise of water can increase the soil's conductivity. Continuous under-floor insulation may then be desirable. This situation demands special attention to provisions for underfloor drainage.

Solar Storage

The immense mass of earth beneath the uninsulated slab offers great potential for storage of solar energy via direct gain systems. In such applications, the foundation wall should be well insulated below the usual depth to keep stored heat from leaking out beneath the footing. A partial subsurface horizontal insulation layer may prove beneficial to trap heat under the floor (*FIG. 24d*).

Raising Frost Depth

Insulation placed horizontally at the exterior perimeter (see *FIG. 24e*) can be used to artificially "raise" the frost penetration depth, thereby allowing the use of shallower footings where permitted by local building codes. A standardized 2'6" deep foundation could be used in any region from 0 to 5000 degree days, with appropriate increases in insulation thickness and horizontal distance from the foundation.

Design Concept

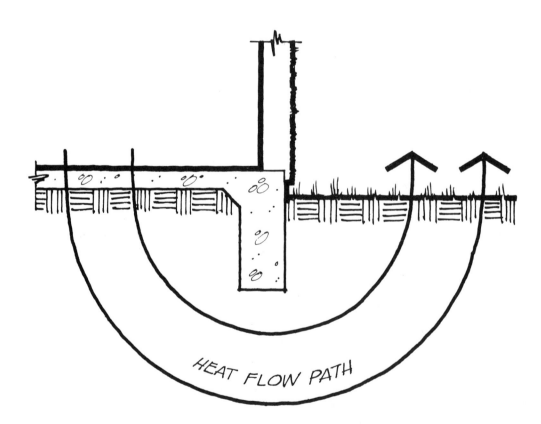

FIG. 24a. **The heat flow path length of 1/2 circle arc = π r for a soil of average conductivity r = 2' yields R9 resistance; Radius of 4' yields R18.**

Use slab-on-grade construction for ground temperature heat exchange.

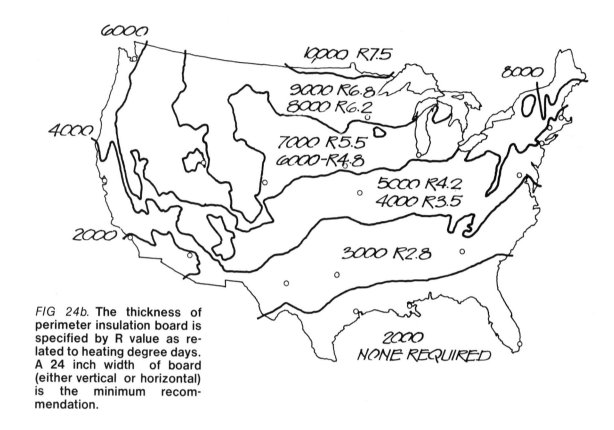

FIG 24b. **The thickness of perimeter insulation board is specified by R value as related to heating degree days. A 24 inch width of board (either vertical or horizontal) is the minimum recommendation.**

10,000 R7.5

9000 R6.8
8000 R6.2

7000 R5.5
6000-R4.8

5000 R4.2
4000 R3.5

3000 R2.8

2000
NONE REQUIRED

6000

10,000 R7.5

8000

4000

2000

Design Concept

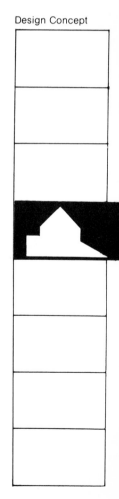

FIG. 24c. **The direction of heat flow from the central portion of slab is mostly into the subsoil, whereas at the perimeter, heat takes a path thru the soil to the surface.**

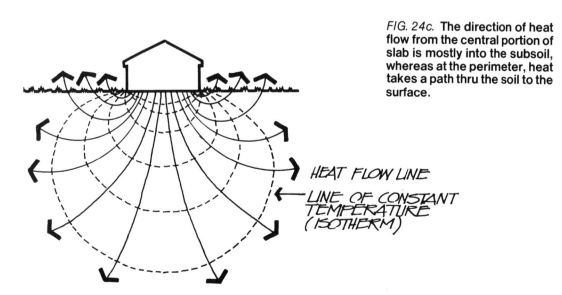

HEAT FLOW LINE

LINE OF CONSTANT TEMPERATURE (ISOTHERM)

Use slab-on-grade construction for ground temperature heat exchange.

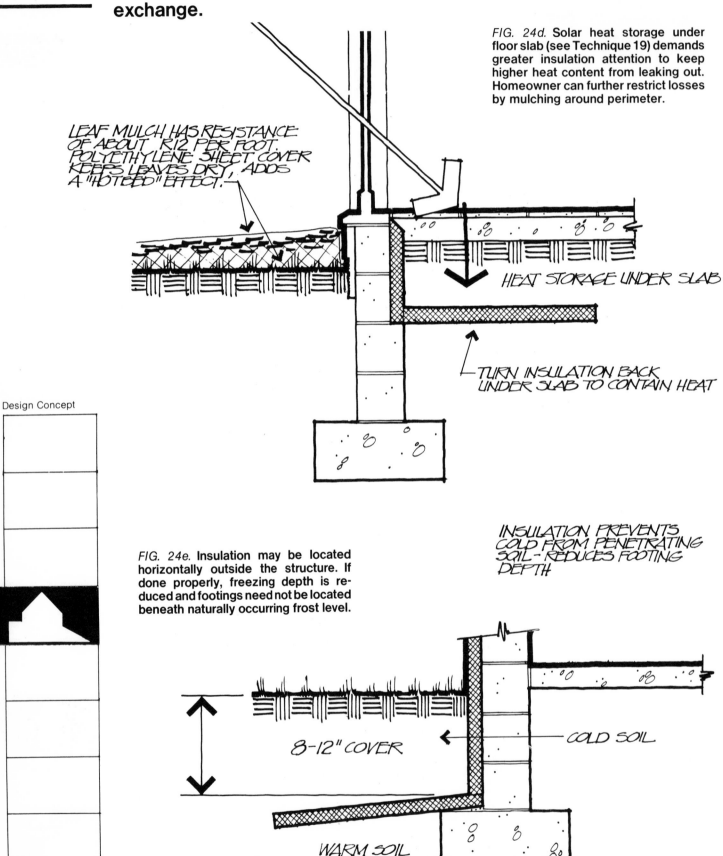

FIG. 24d. Solar heat storage under floor slab (see Technique 19) demands greater insulation attention to keep higher heat content from leaking out. Homeowner can further restrict losses by mulching around perimeter.

LEAF MULCH HAS RESISTANCE OF ABOUT R12 PER FOOT. POLYETHYLENE SHEET COVER KEEPS LEAVES DRY, ADDS A "HOTBED" EFFECT.

HEAT STORAGE UNDER SLAB

TURN INSULATION BACK UNDER SLAB TO CONTAIN HEAT

Design Concept

FIG. 24e. Insulation may be located horizontally outside the structure. If done properly, freezing depth is reduced and footings need not be located beneath naturally occurring frost level.

INSULATION PREVENTS COLD FROM PENETRATING SOIL - REDUCES FOOTING DEPTH

8-12" COVER

COLD SOIL

WARM SOIL

Subdivide interior to create separate heating and cooling zones.

Partitioning the interior of the house into different heating and cooling zones is one way of adjusting the interior volume of the house to meet daily living needs. If infrequently used and seasonal rooms can be closed off from the heated area of the house, less energy will be expended to maintain comfort in the remaining living area. The same principle applies to maintaining lesser used areas of the house at different temperatures than may be desirable for other zones, e.g., bedrooms can be kept cooler during winter days; also room air-conditioners can be used in bedroom areas only.

Interior zoning is achieved through use of standard partitioning and tightly sealed doors on those rooms that could be isolated from the heated areas. Weatherstripping is reasonable for such doorways. Rooms that are under-utilized should also be oriented to make best use of their buffering potential.

Horizontal partitioning is an important means of maintaining even temperature distribution in multi-storied dwellings. Hot air will rise through an open stairwell; when comfortable temperatures are maintained on the ground floor, upper floors may become overheated. This is wasteful during the daytime, when activities are often confined to the lower floors. Providing a door at the top or bottom of the stairway will help contain heat in the downstairs when closed and if left open—at nighttime, for example—will allow heat from downstairs where it is not needed to drift into bedrooms.

Two story spaces allow hot air to rise to the ceiling, leaving the living zone (up to 6 feet from the floor) relatively cold. Although the average temperature over the floor-to-ceiling height may fall within the comfort zone, upper levels may be overheated while lower levels may be uncomfortable. Some means of partitioning—whether by glazing, sliding panels, or other devices—can help equalize temperature distribution within each room.

Partitioning the dwelling into closable cells restricts air movement through the interior, and therefore helps resist infiltration. This applies to both horizontal air movement (between cracks in windward and leeward walls) and to vertical movement—since tall interior spaces create pressure differentials that propel air exchange.

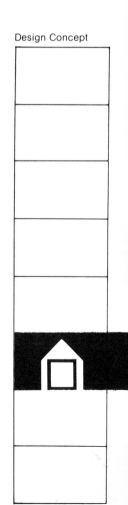

Design Concept

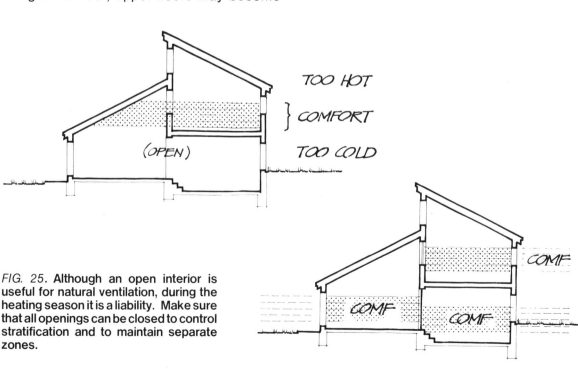

FIG. 25. Although an open interior is useful for natural ventilation, during the heating season it is a liability. Make sure that all openings can be closed to control stratification and to maintain separate zones.

Use "open plan" interior to promote interior air flow.

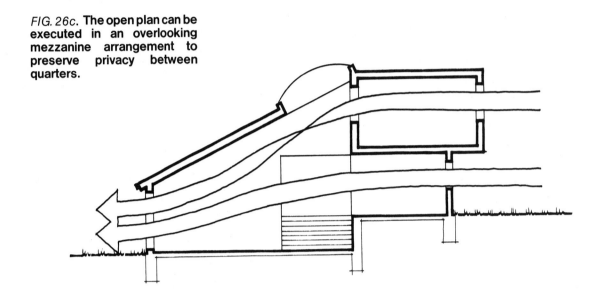

FIG. 26c. **The open plan can be executed in an overlooking mezzanine arrangement to preserve privacy between quarters.**

Design Concept

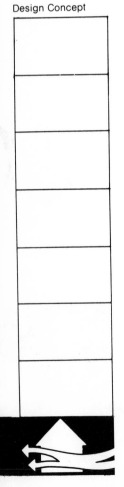

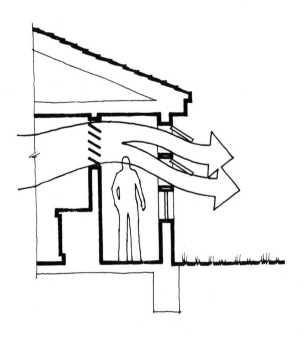

FIG. 26d. **Louvered wall panel affords visual privacy between rooms on the same level and a sensible semi-direct link to outside. This idea is especially appropriate to corridors and across master bath, dressing room, etc.**

Use "open plan" to promote interior air flow.

Any barrier or partition located in the internal air flow path between ventilating inlets and outlets will impede air circulation and the ventilation of the interior. In order to promote unrestricted air movement, therefore, partitioning should be adjustable and located so as to offer least resistance to airflow when it is desired for natural ventilation cooling. The open partitionless interior is the surest way of achieving good air movement, but this design strategy is usually only applicable in small apartments and portions of the house where privacy is not necessary.

Devices that permit air movement between rooms and that provide visual privacy include louvered doors, transom windows, and louvered walls. Folding and removable partitions are useful where the use of rooms and

associated need for privacy may vary from time to time (See FIGS. 26a through 26d).

Although the effect of any partition will usually be to slow down ventilating breezes, interior walls can sometimes be used to channel air movement through spaces where it is most needed. Conversely, poorly placed partitions can cut off air flow entirely, creating stagnant pools of still air. FIG. 26e illustrates the effects of interior walls on breezes flowing through a test model enclosure. The basic rule to follow in designing interiors for ventilation, especially in hot humid zones where air movement is desirable, is to minimize walls, doors, and other baffles and dampers to air flow. Where partitions are necessary, appropriate openings should be included to permit full house ventilation.

Promote Ventilation

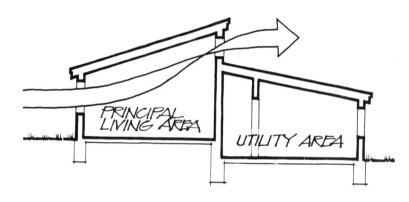

FIG. 26a. "True" cross ventilation requires both exterior inlets and outlets in same room. A popular way of achieving this in the 1950's and '60's was with split-shed roof and louvers or operable clerestory windows.

Design Concept

FIG. 26b. The best cross ventilation is obtained with single-room-deep house plan. House at right was designed by architect Albert Hill in the 1940's. Large terrace and carport funnel in wind flow. A second bedroom could be located at the opposite end of the house to preserve the single-loaded corridor scheme.

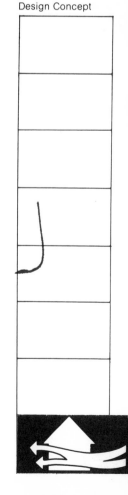

Use "open plan" interior to promote interior air flow.

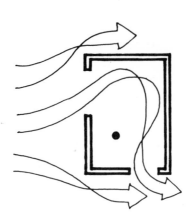

UNOBSTRUCTED AIR FLOW PATH
WILL BE DETERMINED BY LOCATION
OF INTAKE VENT IN FACADE.
NOTE: STATIC AREA "●".

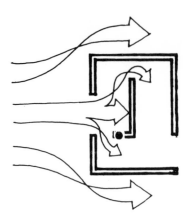

DIRECT INCOMING AIR FLOW IS IMMED-
IATELY BLOCKED BY PARTITION. LITTLE
FLOW AROUND OBSTRUCTION RESULT:
IN MEAGER COOLING EFFECT.

Design Concept

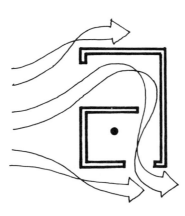

PLACING PARTITION IN STATIC AREA
WILL HAVE LITTLE EFFECT ON AIR
FLOW PATTERN

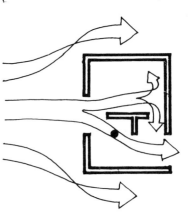

PARTITION PLACED TO "SPLIT" IN-
COMING FLOW DISSIPATES LITTLE
ENERGY. RESULT: OVERALL
ADEQUATE VENTILATION

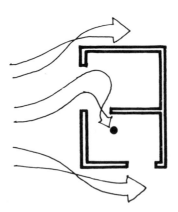

PARTITION PLACED IN FLOW ZONE
ABSORBS DYNAMIC FORCE.
NEITHER ROOM RECEIVES
ADEQUATE VENTILATION

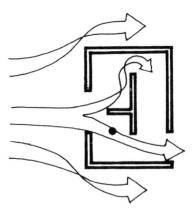

DIVIDER PARTITION SPLITS AIR
FLOW; LOWER ROOM IS WELL
VENTILATED, BACK ROOM RE-
CEIVES LITTLE AIR MOVEMENT

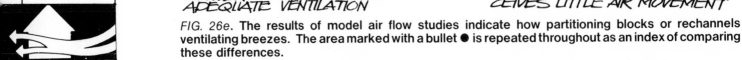

FIG. 26e. The results of model air flow studies indicate how partitioning blocks or rechannels
ventilating breezes. The area marked with a bullet ● is repeated throughout as an index of comparing
these differences.

Provide vertical air shafts to promote interior air flow.

Hot air will rise through the interior of a house, creating the potential for internal air flow even under breezeless outdoor conditions, although unassisted stratification is somewhat limited. The ability of this "stack effect" to promote ventilation depends in large part upon how freely air may rise within the house. A well-partitioned interior will restrict the air flow, whereas tall studio spaces and open-walled balcony rooms permit free flow of air.

Conventional interiors can also be adapted to promote unrestricted rising air flow by locating open stairways in a central area to serve as a vertical ventilation conduit.

In-floor metal grates, commonly used in old-fashioned gravity heating systems, can usefully serve their former purpose in hallways and corridors (*FIG. 27b*). These "flow through" floor panels can be fitted with lay-in damper panels to prevent excessive upward heat flow in winter. (A simpler solution would be to cover such grates with a dense throw rug for the winter season.)

Another approach is to organize the entire design around an interior shaft, terminating in a skylight/ventilating monitor combination. This could be combined with a stair or fireplace, or other vertical elements (*FIG. 27a*).

Promote Ventilation

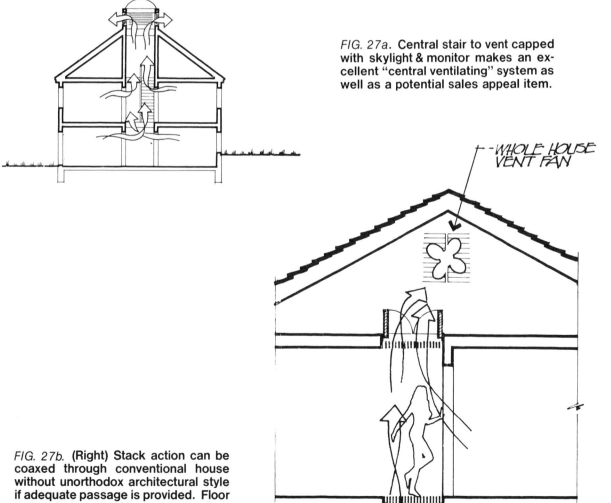

FIG. 27a. **Central stair to vent capped with skylight & monitor makes an excellent "central ventilating" system as well as a potential sales appeal item.**

Design Concept

FIG. 27b. **(Right) Stack action can be coaxed through conventional house without unorthodox architectural style if adequate passage is provided. Floor grates, louvered ceilings, gable ends or ridge vents can provide the route.**

Select insulating materials for resistance to heat flow through building envelope.

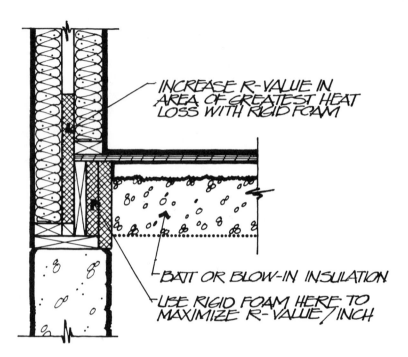

INCREASE R-VALUE IN AREA OF GREATEST HEAT LOSS WITH RIGID FOAM

BATT OR BLOW-IN INSULATION.

USE RIGID FOAM HERE TO MAXIMIZE R-VALUE/INCH

Design Concept

FIG. 28b. Thermal bridging at the band joist can be almost completely eliminated with twin stud framing, resting on staggered sill plates.

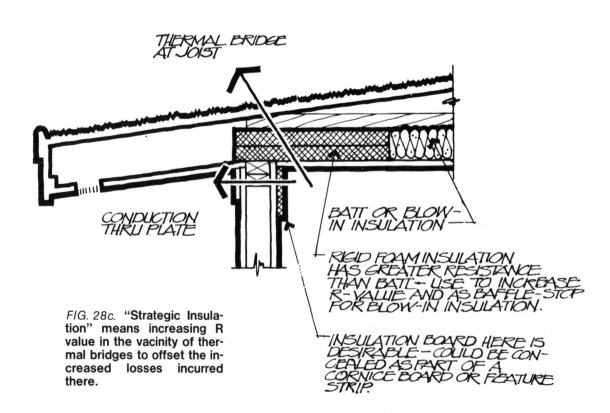

THERMAL BRIDGE AT JOIST

CONDUCTION THRU PLATE

BATT OR BLOW-IN INSULATION

RIGID FOAM INSULATION HAS GREATER RESISTANCE THAN BATT - USE TO INCREASE R-VALUE AND AS BAFFLE-STOP FOR BLOW-IN INSULATION.

INSULATION BOARD HERE IS DESIRABLE - COULD BE CONCEALED AS PART OF A CORNICE BOARD OR FEATURE STRIP.

FIG. 28c. "Strategic Insulation" means increasing R value in the vacinity of thermal bridges to offset the increased losses incurred there.

Select insulating materials for resistance to heat flow through building envelope.

All building materials have insulating value since all provide some amount of resistance to the flow of heat. Most construction systems have insufficient resistance in themselves to ensure economical heating, so insulation in the form of infill or applied materials is usually added to the building envelope.

Envelope strategies

Insulation will generally always be of value wherever and whenever average daily temperatures on the outside of the building envelope fall below the comfort zone (i.e., when indoor heating is necessary). Insulation will also be useful when mechanical cooling techniques are necessary to provide summer comfort. A third, related use of insulation is to keep interior wall surfaces at a higher temperature in the winter. This is an aid to thermal comfort, as well as being a means to prevent condensation on interior walls. It is also applicable to basement walls under summer conditions. Insulation can have an appreciable effect on the timing of the transmittance of solar heat gain when used in conjunction with materials of high thermal mass. This is not, however, a commonly employed use of insulating materials, and depends on where the insulation is placed in the wall or roof section.

Placement

In terms of placement, insulating materials should be located wherever there is a path of conductive heat flow; the major conductive paths are shown in *FIG. 28a*. Although conductive transfer through the ground is usually outward (heat loss), in overheated climates, perimeter and subsurface transfer may involve heat gains as well.

How much insulation to use

FIG. 28a shows that while heat passes through the building envelope by conduction, the processes that supply heat or remove it from the outside skin of the building envelope vary with exposure to sun, wind, and contact with the earth. For this reason, recommended insulation values differ for walls, roofs, and floor slabs, with respect to climate (geographic location). *FIG. 28d* presents a range of R values for energy efficiency. This map is based on an analysis of material and installation and fuel costs, with respect to regional climatic conditions, to determine "economically justifiable" levels of insulation. The amount of insulation that is cost effective increases proportionately to the cost of fuel. The basic procedure for determining optimum insulation levels is a cost/benefit analysis which will be heavily influenced by the selected payback period. In principle, such analysis plots the relationships of fuel cost vs. insulation cost over time, the most efficient combination of these being the corresponding insulation thickness at which the additive costs are lowest.

Recently, "superinsulation" as an energy conserving strategy has received increasing interest. The primary objective is to reduce conductive losses to such a level that much of the annual heating demand is supplied by internal heat gains from lights and appliances (3.4 Btu/hour per watt), occupants 400-500 Btu/hour for a sedentary adult), and other household activities such as cooking. Superinsulation may require R40 walls and R60 ceilings. Special attention must be given to detailing, to minimize thermal bridging (see *FIGS. 28c-28e*).

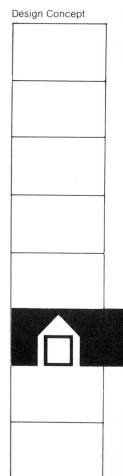

Design Concept

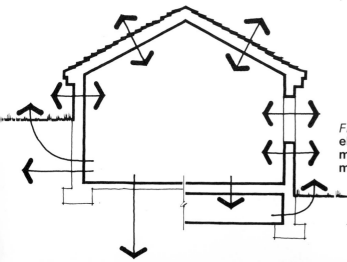

FIG. 28a. **Conductive heat transfer occurs thru all elements of the building shell—even though heat may ultimately be gained or lost at the surface by means of convection, radiation or evaporation.**

Select insulating materials for resistance to heat flow through building envelope.

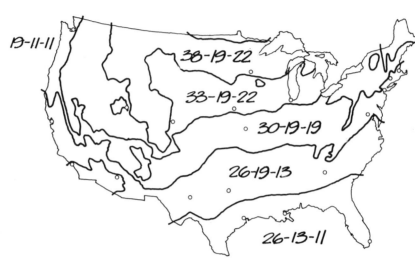

FIG. 28 d. Recommended ▸imum insulation (R) levels spectively for ceiling/wall/fl▸

FIG. 28e. Thermal brid▸ can be reduced by usin▸ inch stud spacing and sp▸ fying insulating sheath▸ outside of studs.

Design Concept

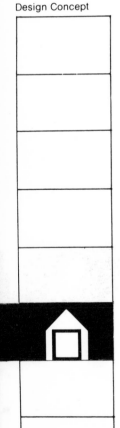

16" O.C.

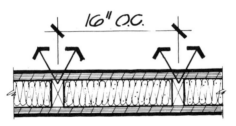

ALMOST 10% OF NORMAL 16" O.C. SECTION CONSISTS OF UNINSULATED "THERMAL BRIDGES"

INSULATING SHEATHING

INSULATING SHEATHING STO▸ SOME HEAT OUTFLOW AT EN▸ OF BRIDGE

24" O.C.

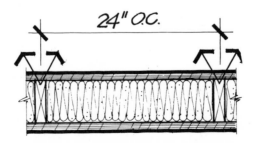

2×6 FRAMING AT 24" O.C. REDUCES THERMAL BRIDG▸ AREA TO 6.2% OF WALL LENGTH, PLUS PROVIDES LARGER CAVITY FOR INSUL▸

Use solar wall and roof collectors on south-oriented surfaces

Promote Solar Gain

The advantages of orienting a house so that its long facades and principal glazed areas face south have been discussed above. Another reason for southerly orientation is to increase wall and/or roof surfaces that are properly positioned for solar collector assemblies. A small amount of pre-planning in the design and siting of the structure can give home-buyers the solar option to begin with or for possible later retrofitting.

Maximum solar irradiation is received on a surface facing true south. However, since most collectors operate at higher efficiencies with higher (outdoor) air temperatures, there is some advantage in positioning the collector to maximize receipt of afternoon, rather than noonday, sun. An orientation between due south and 10° to 15° west of south is therefore acceptable. FIG. 29a shows that even with deviation as much as 25° from south, over 90% of the average daily irradiation will still be received.

The general rule of thumb for the optimum tilt angle of the collector (for solar heating only) is that it should equal the location latitude + 15° (FIG. 29b). This requires roof pitches of about 45° (12 in 12) along the Gulf Coast to 65° (25 in 12) near the Canadian border. These tilts may be compatible with conventional house styles —gambrel and mansard roofs for example; otherwise collectors can be affixed to vertical walls, instead of the roof. Again, departure from the rule of thumb standard for collector tilt angle may decrease collector efficiency somewhat, but deviations of plus or minus 10° from the optimum are considered acceptable. The increased heat input of reflected radiation from the ground and other surfaces (such as a flat roof) should also be considered and justifies the location of collectors on vertical surfaces if there is likely to be snow on the ground or roof in winter.

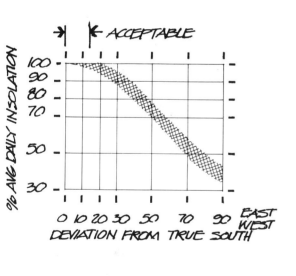

FIG. 29a. **Solar receipt as a function of orientation is represented in this graph. Note that even if collectors are oriented 30° off of south, they will still receive 90%± of the insolation as if facing due south.**

FIG. 29b. **Even if a solar system is not presently being considered, it is wise to plan ahead.**

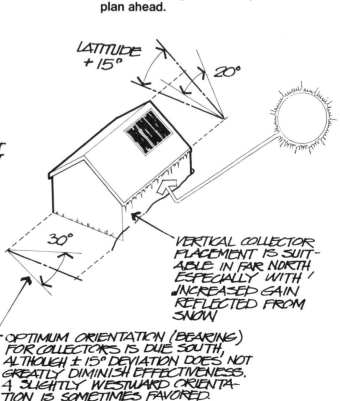

VERTICAL COLLECTOR PLACEMENT IS SUITABLE IN FAR NORTH ESPECIALLY WITH INCREASED GAIN REFLECTED FROM SNOW

OPTIMUM ORIENTATION (BEARING) FOR COLLECTORS IS DUE SOUTH, ALTHOUGH ± 15° DEVIATION DOES NOT GREATLY DIMINISH EFFECTIVENESS. A SLIGHTLY WESTWARD ORIENTATION IS SOMETIMES FAVORED.

Design Concept

Apply vapor barriers to control moisture migration.

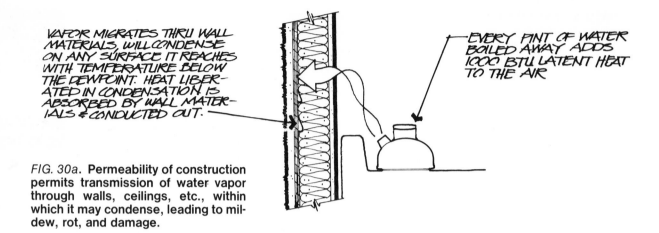

VAPOR MIGRATES THRU WALL MATERIALS, WILL CONDENSE ON ANY SURFACE IT REACHES WITH TEMPERATURE BELOW THE DEWPOINT. HEAT LIBER-ATED IN CONDENSATION IS ABSORBED BY WALL MATER-IALS & CONDUCTED OUT.

EVERY PINT OF WATER BOILED AWAY ADDS 1000 BTU LATENT HEAT TO THE AIR

FIG. 30a. **Permeability of construction permits transmission of water vapor through walls, ceilings, etc., within which it may condense, leading to mildew, rot, and damage.**

Design Concept

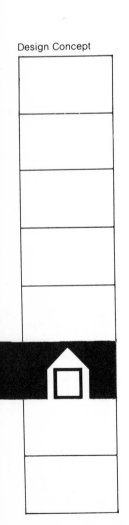

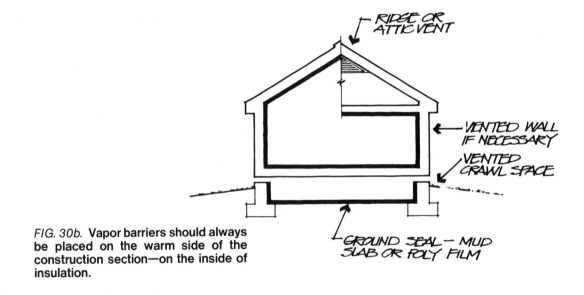

RIDGE OR ATTIC VENT

VENTED WALL IF NECESSARY

VENTED CRAWL SPACE

GROUND SEAL — MUD SLAB OR POLY FILM

FIG. 30b. **Vapor barriers should always be placed on the warm side of the construction section—on the inside of insulation.**

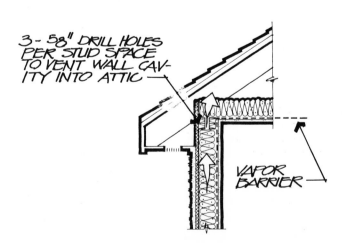

3 - 5⁄8" DRILL HOLES PER STUD SPACE TO VENT WALL CAV-ITY INTO ATTIC

VAPOR BARRIER

Apply vapor barriers to control moisture migration.

In the opposite way that evaporation produces a cooling effect by absorbing heat from building surfaces, condensation gives up latent heat contained in the air to the surfaces upon which it condenses. Condensation in buildings can be damaging and energy expensive.

The likelihood of condensation problems increases as buildings are made more weathertight and sealed against infiltration. Infiltration controls prevent the escape of airborne moisture that is generated through cooking, laundering, bathing and other activities; the resulting higher indoor humidity levels (actually *vapor pressure*) increase the propensity of water vapor to penetrate the building shell and where it meets a surface at or below the dew point temperature, it condenses (see *FIG. 30a*). Condensation has obvious harmful effects on wood, wood products, and steel, and where exposed to freeze-thaw potential, it can be destructive to masonry as well. Inasmuch as the conductivity of porous materials is usually related to moisture content, condensation within the wall can also reduce its resistance to heat flow.

The usual solution to condensation problems is to block the migration of moisture near the interior of the wall before the dew point is reached. This is achieved by means of a vapor barrier, a membrane that resists the penetration of water vapor. Vapor barriers come in a variety of materials. Some are integral with other building components, such as insulation or interior finish materials. Others are independently obtained and applied.

The effectiveness of a construction material or assembly in resisting vapor transmission is measured in *perm* units (1 perm = 1 grain moisture/ft^2 (hour) per inch of mercury pressure difference). The resistance of individual materials to vapor transmission is rated in terms of *perm inches,* which indicates the rate of transmission through a one-inch thickness of the material. Perms and perm-inches are measures of permeance and permeability, respectively.

The in-place effectiveness of barriers with even the lowest of perm ratings can be greatly diminished if they are poorly sealed at the joints, ruptured in construction, or frequently punctured for outlet boxes air registers, and other fixtures. If the wall is subject to a great deal of air leakage (or infiltration and exfiltration), the provision of a vapor barrier may be pointless. A recent report from the National Research Council of Canada states:

> Air leakage is now considered to be the prime cause of most condensation problems in walls and roof spaces. If, therefore, a building can be made tight against air leakage it may not need a vapor barrier, as defined. On the other hand, if there are openings that permit air to leak from the warm side to the cold side of the insulation, adding a vapor barrier (even of zero permeance) that does not seal off the openings will be useless.

Care should be exercised in handling and installation of the vapor membrane. Caution must also be given to the selection of sheathing and surfaces on the outside of the building: if a second vapor barrier is inadvertently created there, any condensation that does occur within the wall or roof will be trapped. In these instances (such as a bituminous insulated roof), venting of the cavity or interior of the shell is necessary to provide moisture escape. The rule is always to place the vapor barrier on the warm side of the insulation. Some typical installations are illustrated in *FIG. 30b*.

Use of a vapor barrier beneath slab-on-grade construction and on the "floor" of crawl spaces is usually recommended. This application is intended to restrict moisture inducement out of the ground and into the building. Studies undertaken by the U.S. Housing and Home Finance Agency indicated that the contribution of water vapor from unvented or untreated crawl spaces to the interior may be as high as 100 lbs. of water per day per 1000 square foot of enclosed crawl space. Much of this can be stopped simply by laying a plastic film on the crawl space floor.

Minimize Conductive
Heat Flow

Minimize Infiltration

Design Concept

149

31 Develop construction details to minimize air infiltration and exfiltration.

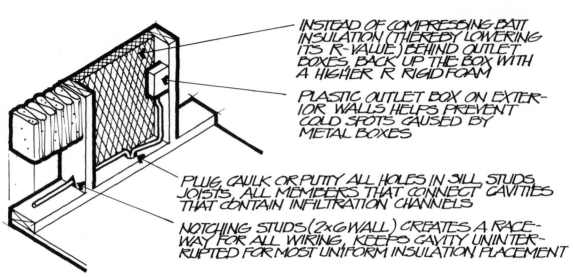

INSTEAD OF COMPRESSING BATT INSULATION (THEREBY LOWERING ITS R-VALUE) BEHIND OUTLET BOXES, BACK UP THE BOX WITH A HIGHER R RIGID FOAM

PLASTIC OUTLET BOX ON EXTERIOR WALLS HELPS PREVENT COLD SPOTS CAUSED BY METAL BOXES

PLUG, CAULK OR PUTTY ALL HOLES IN SILL, STUDS, JOISTS, ALL MEMBERS THAT CONNECT CAVITIES THAT CONTAIN INFILTRATION CHANNELS

NOTCHING STUDS (2x6 WALL) CREATES A RACEWAY FOR ALL WIRING, KEEPS CAVITY UNINTERRUPTED FOR MOST UNIFORM INSULATION PLACEMENT

Design Concept

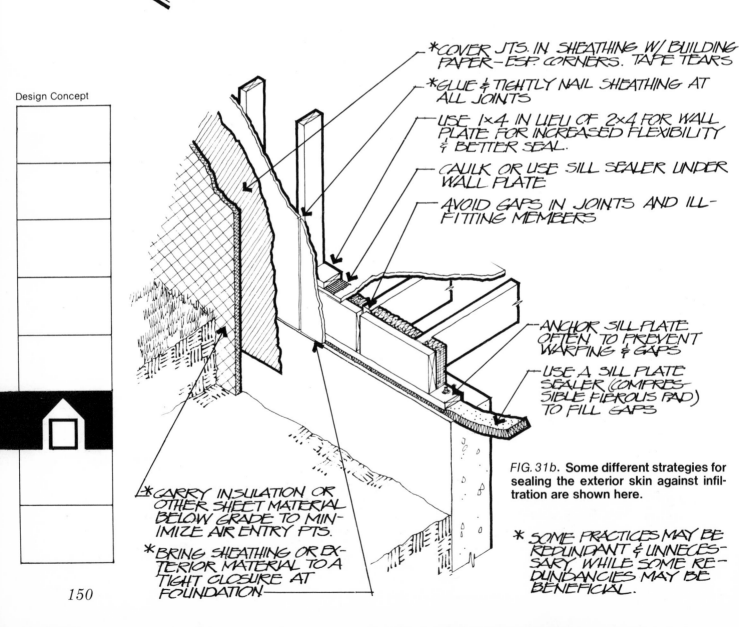

*COVER JTS. IN SHEATHING W/ BUILDING PAPER - ESP. CORNERS. TAPE TEARS

*GLUE & TIGHTLY NAIL SHEATHING AT ALL JOINTS

USE 1x4 IN LIEU OF 2x4 FOR WALL PLATE FOR INCREASED FLEXIBILITY & BETTER SEAL.

CAULK OR USE SILL SEALER UNDER WALL PLATE

AVOID GAPS IN JOINTS AND ILL-FITTING MEMBERS

ANCHOR SILL PLATE OFTEN TO PREVENT WARPING & GAPS

USE A SILL PLATE SEALER (COMPRESSIBLE FIBROUS PAD) TO FILL GAPS

FIG. 31b. Some different strategies for sealing the exterior skin against infiltration are shown here.

*CARRY INSULATION OR OTHER SHEET MATERIAL BELOW GRADE TO MINIMIZE AIR ENTRY PTS.

*BRING SHEATHING OR EXTERIOR MATERIAL TO A TIGHT CLOSURE AT FOUNDATION

* SOME PRACTICES MAY BE REDUNDANT & UNNECESSARY, WHILE SOME REDUNDANCIES MAY BE BENEFICIAL.

Develop construction details to minimize air infiltration and exfiltration.

Infiltration can occur anywhere that framing and finish members come together in construction. Details of blocking airflow and identifying vulnerable spots will vary with the construction type, but most often the best measures are to caulk cracks and construction joints. Some recommended practices are illustrated in FIGS. 31a and 31b.

Infiltration-prone areas can be grouped into three categories: (1) direct penetrations through the building skin, some of which can be eliminated through subsurface routing (electrical service, for example), others (such as vent stacks) must simply be carefully sealed; (2) cracks and gaps in the exterior and interior skins (where sheathing and foundation walls and floors meet, for example), and (3) cavities, holes, raceways, and other passages within walls or assemblies. Measures should be taken in each area so that if new cracks open, there will still be sufficient resistance to block flow.

Influence of construction types and finishes
Because of the varying nature of building

materials, some construction systems are more resistant to air leakage than others. *Table 31a* indicates that the infiltration rate through a brick wall can be reduced by as much as a factor of 100 by plastering the outside against air penetration. ASHRAE Fundamentals [1981] indicates that a heavy coat of certain latex paints can reduce air leakage by 50%, and three coats of oil paint, carefully applied, by 28%. As shown in *Table 31a,* a well plastered wall reduces infiltration to a negligible value.

A well secured and continuous vapor barrier (taped breaks and tears) serves a similar purpose at the interior side of the wall. Under shingles and siding, a blanket of building paper helps cover joints in sheathing (*FIG. 31a*), all of which should be carefully fitted and inspected for "daylight" before the final coverings are applied. A great deal of heat loss by air leakage occurs through ceilings, a seldom considered problem area for infiltration. This suggests detailing the construction of the ceiling to resist air flow, such as by careful installation of a continuous vapor barrier.

FIG. 31a. **Look for gaps and potentially leaky assemblies & caulk or gasket between members.**

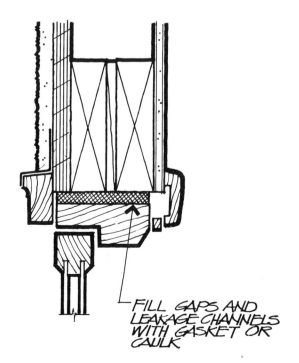

FILL GAPS AND LEAKAGE CHANNELS WITH GASKET OR CAULK

TABLE 31a

Infiltration characteristics of different walls (CFH/sq.ft.)

Type of wall	pressure difference (as wind velocity, mph)				
	10.2	14.4	20.4	24.9	28.8
8½″ brick wall a					
plain	5	9	16	24	28
plastered b	.05	.08	.14	.20	.27
13″ brick wall	5	8	14	20	24
plastered b	.01	.04	.05	.09	.11
plastered c	.03	.25	.46	.66	.84
Frame wall, lath and plaster d	.09	.15	.22	.29	.32

a constructed of porous brick and lime mortar—poor workmanship
b two coats of prepared gypsum plaster on brick
c furring, lath, and two coats of prepared gypsum plaster on brick
d painted bevel siding or cedar shingles, sheathing, building paper, wood lath, and three coats of gypsum plaster.

TABLE 31b

Equivalent orifice areas (sq. in.)

13″ porous brick wall, no plaster, 100 sq. ft.	3.1
13″ porous brick wall, 3 coats plaster, 100 sq.ft.	.054
frame wall, wood siding, 3 coats plaster, 100 sq. ft.	.33
door, tight fitting, 3′ x 7′	7.6
double hung window, loose fitting, 3′ x 4′	4.7
double hung window, tight fitting, 3′ x 4′	0.93

[Wilson 1961]

Minimize Infiltration

Design Concept

Select high-capacitance materials for controlled heat flow through the building envelope.

TABLE 32a.

Time lag is related to thermal conductivity, heat capacity, and thickness of material. For homogeneous building materials, time lag can be calculated from the formula:

$$\text{TIME LAG (in hours)} = 1.38 \text{ THICKNESS } \sqrt{\text{HEAT CAPACITY/CONDUCTIVITY}}$$

Computed lag values for some different materials are given below, per foot thickness. For thinner wall, simply multiply by corresponding fraction of one foot.

Material Description	Heat Capacity BTU/cu.ft.°F	Conductivity BTU-FT./hr.ft.°F	Time Lag/Foot (Hours)
Adobe	19.6	0.37	10
Brick, common (120 pcf)	24	0.42	10.4
Brick, face (130 pcf)	26	0.75	6.1
Concrete (140 pcf)	29.4	1.0	7.5
Gypsum	20.3	.25	12.4
Iron, cast	54	27.6	1.9
Limestone	22.7	.54	8.9
Marble	34	1.5	6.6
Particleboard (60.2 pcf)	18.7	0.1	18.9
Plasterboard	22.4	.43	10.0
Plywood	9.9	.067	16.8
Polystyrene, beadboard	0.3	.023	4.9
Rubber, soft vulcanized	68.6	0.08	40
Paraffin	38.6	0.075	31.3
Sand	18	0.19	13.4
Soil, light dry (80 psf)	18	0.2	13.1
Soil, average (damp, 131 psf)	30.1	.75	8.75
Soil, wet (117 psf)	35.1	1.4	6.9
Wood, hardwood	18.7	0.091	19.8
Wood, white oak	26.8*	0.1	22.6
Wood, softwood	10.6	0.067	17.4
Wood, white pine	18.1*	.063	23.4

*These relatively high values may be attributed to higher moisture content, compared to "softwood" and "hardwood" in general.

Design Concept

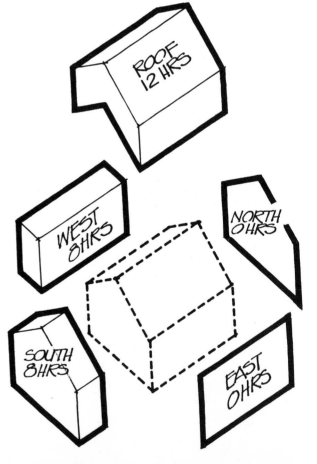

FIG 32b. Different orientations demand different lag times. These suggestions are made for a California climate. [From Burt, Hill, Kosar, Rittelmann 1977]. Care must be taken not to rotate the plan out of its intended orientation.

Select high-capacitance materials for controlled heat flow through the building envelope.

Lag time design

The heat capacities of different building materials and their assorted lag times provide a useful tool for the climatic designer, allowing the thermal response of house walls and roofs to be selected according to their respective solar orientations.

The amount of solar energy received at building surfaces facing the four cardinal directions is diagrammed in *FIG. 32a.* Assuming that all surfaces have essentially the same solar absorptivity and thermal characteristics, the graphs in *FIG. 32a* also represent the relative increases in temperature at the exterior of these surfaces (neglecting the effect of outdoor air temperature). If the walls are constructed of materials having negligible thermal capacity, the heat gain of the exterior will be transmitted almost instantaneously to the interior. The interior temperature curve, therefore, will closely follow the summary curve, reaching peak temperatures just before and after solar noon. (For simplicity of illustration, the effect of air temperature increase throughout the day is ignored. In reality, air temperature usually peaks about 3 hours following solar noon, thereby shifting the internal maximum temperature to later in the day.) A building simulating these conditions would be a cubical summer cabin, fully exposed to the sun.

Just as knowledgeable specification of thermal mass in the building shell can help reduce peaks in cooling loads, unwitting distribution of mass materials can concentrate peak loads. Similarly, reorientation of a house that is tuned to specific compass points can undo benefits built in to a well conceived design. Placing high capacitance in the wrong places can be a costly mistake for future homeowners. *(FIG. 32b).*

Lag time tables

Table 32a gives representative lag times for various homogeneous wall materials. The lag time of composite walls can be manipulated considerably through the choice of materials and the location of insulation within the wall section. Some values for composite wall sections are reproduced in *Table 32b.*

The use of thermal mass materials in house construction has largely been ignored, in spite of its being a low cost, low technology approach to energy conservation. Manufacturers of masonry products have recognized this and are a source of some literature on the subject.

FIG. 32a. **Relative solar radiation intensities for roof and 4 wall orientations for 35° N. latitude (idealized values, discounting weather conditions), plotted as a function of time.**

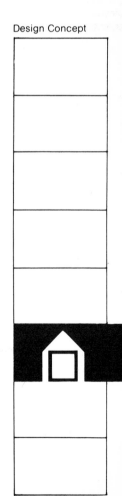

Design Concept

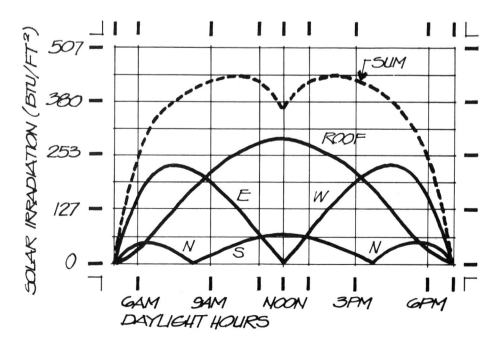

Select high-capacitance materials for controlled heat flow through the building envelope.

TABLE 32b

Time lag values for 14 different composite wall sections [from Ward, 1977].

WALL DESCRIPTION	WEIGHT LBS. PER SQUARE FOOT	U-VALUE (WINTER)	HEAT GAIN Btu/hr/sq.ft (DARK COLOR)		TIME-LAG (HOURS)
			AVG. ORIEN-TATION	WEST ORIEN-TATION	
6″ BRICK WALL (SOLID UNITS), 2″ POLYSTYRENE IN-SULATION BOARD 1/2″ GYPSUMBOARD	63.5	0.080	2.25	1.91	4
6″ BRICK WALL (HOL-LOW UNITS), 1″ x 2″ FURRING 1/2″ GYP-SUMBOARD	47.5	0.325	9.95	8.44	4
6″ BRICK WALL (HOL-LOW UNITS, 2″ POLY-STYRENE INSULA-TION BOARD 1/2″ GYPSUMBOARD	46.5	0.078	2.29	1.94	4
8″ BRICK AND LIGHT-WEIGHT CONCRETE (100-POUND DEN-SITY) BLOCK. 1″ x 2″ FURRING. 1/2 ″ GYP-SUMBOARD	65.0	0.263	8.51	7.98	4
8″ BRICK AND LIGHT-WEIGHT CONCRETE (100 POUND DEN-SITY) BLOCK. 2″ POLY-STYRENE INSULA-TION BOARD. 1/2″ GYPSUMBOARD	64.0	0.073	2.06	1.75	4
4″ BRICK CURTAIN WALL (PARTIALLY REINF. OR WITH HIGHBOND MORTAR) 2″ POLYSTYRENE INSULATION BOARD, 1/2″ GYPSUMBOARD	42.0	0.082	2.77	2.84	3
6″ PRECAST CON-CRETE (140-POUND DENSITY) SANDWICH PANEL, 2″ POLY-URETHANE CORE	47.5	0.065	1.82	1.55	4

Design Concept

Select high-capacitance materials for controlled heat flow through the building envelope.

TABLE 32c
Time lag values for 14 different composit wall sections

WALL DESCRIPTION	WEIGHT LBS. PER SQUARE FOOT	U-VALUE (WINTER)	HEAT GAIN Btu/hr/sq.ft (DARK COLOR)		TIME-LAG (HOURS)
			AVG. ORIEN-TATION	WEST ORIEN-TATION	
1/2″ PLYWOOD SIDING, 1/2″ INSULATION BOARD SHEATHING, WOOD STUDS, FULL BATT (R-11) INSULATION. 1/2″ GYPSUMBOARD	5.0	0.076	3.05	4.60	2
10″ BRICK AND BRICK CAVITY WALL WITH 2″ POLYURETHANE INSULATION BOARD IN CAVITY.	79.0	0.058	1.17	1.06	8
10″ BRICK AND BRICK CAVITY WALL	78.5	0.370	7.73	8.29	8
10″ BRICK AND LIGHTWEIGHT CONCRETE (100 POUND DENSITY) BLOCK CAVITY WALL WITH 2″ POLYSTYRENE INSULATION BOARD IN CAVITY	62.0	0.071	1.72	1.35	6
10″ BRICK AND LIGHT WEIGHT (100 POUND DENSITY) BLOCK CAVITY WALL	61.0	0.273	6.09	6.09	6
4″ BRICK VENEER 1/2″ INSULATION BOARD SHEATHING, WOOD STUDS, FULL BATT (R-11) INSULATION. 1/2″ GYPSUMBOARD	43.0	0.077	2.18	1.95	4
8″ BRICK WALL (HOLLOW UNITS), 1″ x 2″ FURRING, 1/2″ GYPSUMBOARD	56.0	0.316	7.37	5.90	6

Design Concept

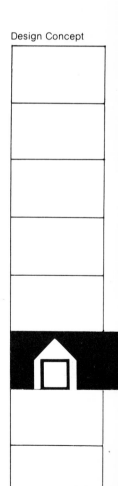

Use sod roofs.

Sod is a complex thermal system that provides both summer and winter climatic benefits. The physics of heat transfer in and out of the soil is difficult to describe accurately, yet some of its beneficial influences can be generalized:

1. Grass and other leafy ground covers intercept most radiation received at the ground, reflecting 20 to 30 percent and absorbing the remainder at leaf-level. The ground surface, therefore, is shaded and receives much less heat gain than a conventional roof.

2. A well-irrigated low grass dissipates through evaporation about 1000 to 1200 Btu/ft^2 of solar heat received per day during summer months — often 80 percent of incident radiation. The soil, therefore, experiences very little net heat gain as a result of this cooling effect.

3. The thermal mass of a soil cover damps out temperature variations so that only 30 percent of the daily fluctuation is felt at a depth of 18 inches. Also as a result of its mass, a typical soil possesses a timelag of about 9 hours per foot (this varies with moisture content; see *Table 32a*). Since heat stored in the soil during daylight hours is reradiated as thermal energy at night, the temperature of the soil at a depth of about 18 inches approximates the daily average surface temperature, which roughly equals the average air temperature. (*FIG. 33*)

4. Because of the thermal mass of 12 to 18 inches of soil damps out daily temperature fluctuations in both winter and summer, the exterior of the structure is subjected to much smaller heating and cooling loads. As a result, insulation requirements can be substantially reduced as compared to normal recommended roof and ceiling R values.

5. The design and accessibility of sod roofs allows these to be mulched with leaves, straw, or other compost materials in the fall, adding to winter insulation. Snow, too, can be exploited for its thermal resistance, which ranges from R4 to R28 per foot, depending on density.

Earth weighs as much as 120 pounds per cubic foot or more when saturated, so supporting soil cover adds to the cost not only of the roof, but of structural support and foundations as well. A dependable water proofing system also is necessary, so sod roofs are not common, except on houses which are largely recessed into the earth or bermed. Sod-roofed underground construction has been used in tornado zones of the midwest, where the roof mass contributes to storm resistance.

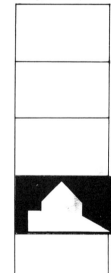

Minimize Conductive
Heat Flow

Minimize Solar Gain

Promote
Evaporative Cooling

Promote
Radiant Cooling

Design Concept

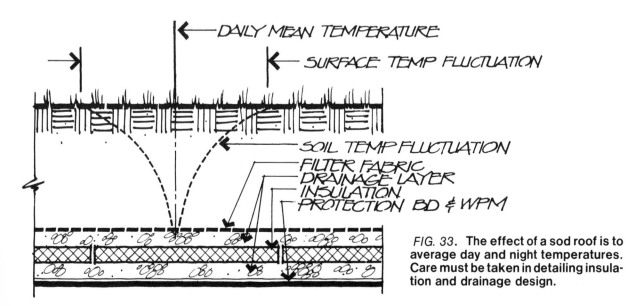

DAILY MEAN TEMPERATURE

SURFACE TEMP FLUCTUATION

SOIL TEMP FLUCTUATION
FILTER FABRIC
DRAINAGE LAYER
INSULATION
PROTECTION BD & WPM

FIG. 33. The effect of a sod roof is to average day and night temperatures. Care must be taken in detailing insulation and drainage design.

Provide shading for walls exposed to summer sun.

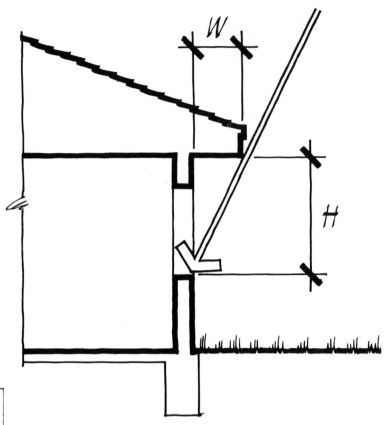

FIG. 34b. Roughly appropriate overhang dimension W can be calculated by selecting the shade line factor (slf) from TABLE 34a and inserting in the formula:

Desirable overhang $W = \dfrac{H}{SLF}$

Design Concept

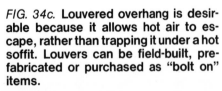

TABLE 34a

Shade Line Factors — Latitude in degrees

Window faces	25	30	35	40	45	50	55
East	.8	.8	.8	.8	.8	.8	.8
Southeast	1.9	1.6	1.4	1.3	1.1	1.0	0.9
South	10.1	5.4	3.6	2.6	2.0	1.7	1.4
Southwest	1.9	1.6	1.4	1.3	1.1	1.0	0.8
West	0.8	0.8	0.8	0.8	0.8	0.8	0.8

FIG. 34c. Louvered overhang is desirable because it allows hot air to escape, rather than trapping it under a hot soffit. Louvers can be field-built, prefabricated or purchased as "bolt on" items.

Support structure can be cantilevered ends of joists. Rafters or outriggers framework can also be supported on struts from below, or suspended on cable ties anchored above.

Louvers themselves may be wood or metal; fabrics are also a possibility.

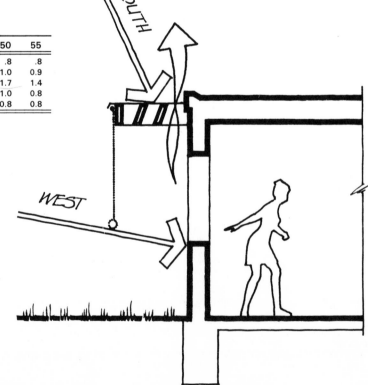

Provide shading for walls exposed to summer sun.

Reductions in solar gain through building walls can be achieved by shading devices. For a dark gray wall receiving 160 Btu per hour of solar irradiation per square foot (typical of a southwest-facing wall in a mid-afternoon for north central U.S. latitudes), shading the surface will have the equivalent effect of lowering the outside air temperature by 37°F. This in turn will reduce heat flow from 4.0 to 1.15 Btu per hour through a low mass wall of overall resistance R13 when air temperature is 87°F.

A very effective means of shading walls (and window openings) is to extend or project the roof to provide a useful overhang (*FIG. 34a*). The Small Homes Council of the University of Illinois reports:

> A study of the weather conditions and the sun angles at various locations between 30° and 50° latitude indicates that a standard 30/16 overhang (horizontal projection of 30″, located 16″ above the top of the window) will provide good sun control on south windows for this range of latitudes . . . when glass doors or tall windows are used, it may be desirable to increase the overhang to provide more shade.

Overhangs can be tailored to specific latitudes, local heating and cooling seasons, and varying orientations. The method for sizing these is described in Olgyay and Olgyay [1957]. A simplified assessment of appropriate overhang design for different latitudes is given in *FIG. 34b.* Although projecting the eaves is probably the simplest technique for shading walls and windows of single-storied buildings, numerous other devices have been utilized over the years. Many of these associated with specific periods or architectural styles, but most can readily be adapted to contemporary designs.

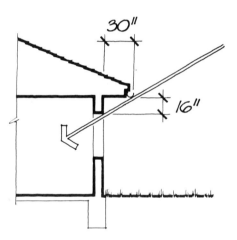

FIG. 34a. **The Small Homes Council of the University of Illinois suggests that a 30/16 overhang is a good solution for south elevations from 30° to 50° latitude—for conventional window sill and ceiling heights or similar proportions.**

A standard overhang may be justified on the basis that as latitude increases, the duration of the overheated season decreases. So, although solar altitude decreases, the overhang need not increase.

Minimize Conductive Heat Flow

Minimize Solar Gain

Promote Radiant Cooling

Design Concept

Use heat reflective materials on surfaces oriented to summer sun.

When solar radiation strikes an opaque building surface, a portion of the heat is reflected away and the remainder is absorbed. The percentage of the incident radiation that is reflected by a given material is designated as the *reflectance* value of the surface. The percentage of radiation that is absorbed accordingly, is called the *absorptance* of the material. Since all the radiation striking a surface is either absorbed or reflected, the sum of these values must always equal 100% (radiation transmitting materials are an exception). A list of surface characteristics of common building materials is presented in *Table 35 a.*

It is commonly acknowledged that painting a house a light color will keep it cooler in summer (and winter as well). The magnitude of effect and subtleties of application of the idea is shown in *FIG. 35a.* Surface reflectivity will be of greatest significance to light-weight buildings of low insulation levels. Maximization of surface reflectance will therefore apply most advantageously to the southern regions where high R values for walls are not demanded. A simple analysis of summer and winter considerations is presented in *FIG. 35b.*

Minimize Conductive
Heat Flow

Minimize Solar Gain

Promote
Radiant Cooling

Design Concept

FIG. 35a. **Whitewashing a dark wall can considerably reduce solar heat gain. The benefit will be greatest on poorly insulated sections.**

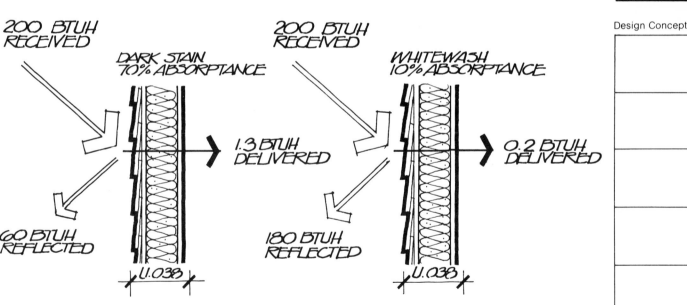

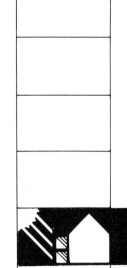

SOLAR EXCESS HEAT FLOW:

$$\frac{200 \times 70\% \times .038}{4^*} = 1.3 \; BTUH$$

SOL-AIR EXCESS TEMPERATURE:

$$\frac{200 \times 70\%}{4^*} = 35°F$$

SOLAR EXCESS HEAT FLOW:

$$\frac{200 \times 10\% \times .038}{4^*} = 0.19 \; BTUH$$

SOL-AIR EXCESS TEMPERATURE:

$$\frac{200 \times 10\%}{4^*} = 5°F$$

**4 BTUH/FT² SURFACE CONDUCTANCE
COEFFICIENT FOR SUMMER (7.5 MPH WIND)*

35 Use heat reflective materials on surfaces oriented to summer sun.

Absorptivity vs. emissivity

While some different surfaces may show the same behavior in reflecting solar radiation, they do not necessarily dissipate the heat they absorb in the same manner, or to the same degree of effectiveness. Within a narrow band of radiation wave lengths, the emissivity value of a surface will approximate its absorbtance, hence the aphorism "a good absorber is a good emitter." This rule does not generally apply to radiation receipt at one frequency (solar) and emittance at another (thermal). The result is that some surfaces have a heat trapping effect, at least relative to other surfaces of similar absorbance and solar reflectivity. The significance of this in terms of building climate is illustrated in *FIG. 35c.*

The heat trapping effect of polished metal and foil surfaces can be used to advantage in double-wall construction. Facing the shiny surface of a metal exterior panel or foil liner across a cavity toward the inside will help preclude radiant transfer of heat through the air space to the interior wall. This arrangement should be familiar to builders, inasmuch as it is commercially available in the form of reflective foil insulation and is effective in blocking radiant heat transfer if next to an air space. Some new "insulating shades" for windows use reflective films with spacers and thus are effective in winter and summer.

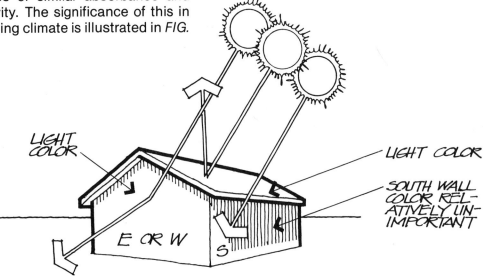

FIG. 35b. **Above: High summer sun delivers greatest heat load to roof, east and west walls (the latter receive about half of a flat roof). South facade receives relatively little radiation—especially if there are even small overhangs.**

Below: Low and shorter winter sun path makes roof solar receipt small— only about one-half of south exposure. North, east and west gains are all very small, provide no significant useful gain. Dark colors on the south wall provide some benefit for reducing heat loss by elevating exterior surface temperatures.

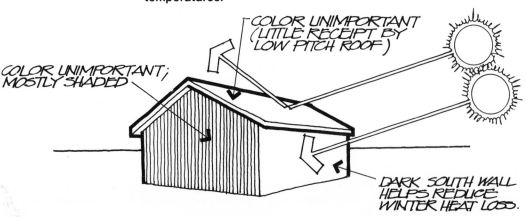

Design Concept

Use heat reflective materials on surfaces oriented to summer sun.

TABLE 35a

Surface properties of materials

	Solar Reflectance(%)	Thermal Reflectance(%)	Thermal Emittance
aluminum foil, bright	95		.5
white plaster	93		.91
fresh snow	87		.82
aluminum foil, oxidized	85		.12
aluminum sheet, polished	85	92	.08
whitewash, new	80		.90
white painted aluminum	80		.91
white paint	70-75	5-10	.9-.95
chromium plate	72	80	.20
polished copper	75	85	.15
snow, re granules	67		.89
light gray paint	60	5	.9-.95
white powdered sand	55		.90
aluminum, weathered	47		
aluminum, paint	45-50	45	.33-.73
polished marble	40-50		.90
granite	45		.44
Indiana limestone	43	5	.95
concrete	40		.88
wood, pine	40	5	.95
brick (light-dark)	23-48	5	.95
dark gray paint	30	5	.95
asbestos, slate	19		.96
galvanized iron, aged	10-20	72	.28
black gloss paint	10		.90
black tar paper	7		.93
lamp black	2		.95

Design Concept

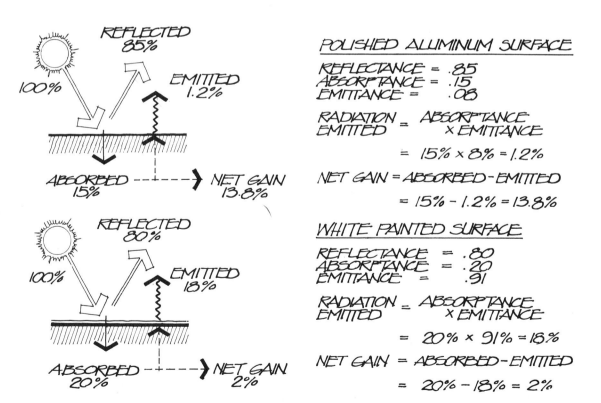

POLISHED ALUMINUM SURFACE

REFLECTANCE = .85
ABSORPTANCE = .15
EMITTANCE = .08

$$\frac{RADIATION}{EMITTED} = ABSORPTANCE \times EMITTANCE$$

$$= 15\% \times 8\% = 1.2\%$$

$$NET\ GAIN = ABSORBED - EMITTED$$

$$= 15\% - 1.2\% = 13.8\%$$

WHITE PAINTED SURFACE

REFLECTANCE = .80
ABSORPTANCE = .20
EMITTANCE = .91

$$\frac{RADIATION}{EMITTED} = ABSORPTANCE \times EMITTANCE$$

$$= 20\% \times 91\% = 18\%$$

$$NET\ GAIN = ABSORBED - EMITTED$$

$$= 20\% - 18\% = 2\%$$

FIG. 35c. A polished aluminum surface will reflect most of the sun's radiation it receives. Being a poor emitter of its own heat, however, it will become hotter than if it were painted and less reflective—although the painted surface absorbs more radiation, it also is capable of emitting more, so its net gain is lower.

163

Use planting next to building skin.

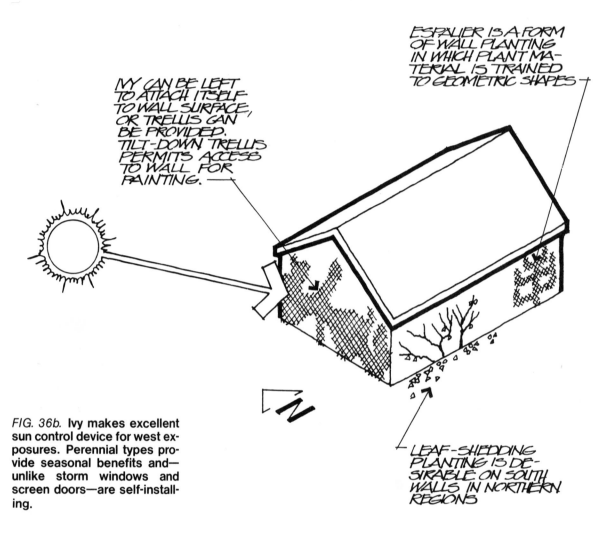

IVY CAN BE LEFT TO ATTACH ITSELF TO WALL SURFACE, OR TRELLIS CAN BE PROVIDED. TILT-DOWN TRELLIS PERMITS ACCESS TO WALL FOR PAINTING.

ESPALIER IS A FORM OF WALL PLANTING IN WHICH PLANT MATERIAL IS TRAINED TO GEOMETRIC SHAPES

LEAF-SHEDDING PLANTING IS DESIRABLE ON SOUTH WALLS IN NORTHERN REGIONS

Design Concept

FIG. 36b. Ivy makes excellent sun control device for west exposures. Perennial types provide seasonal benefits and—unlike storm windows and screen doors—are self-installing.

Use planting next to building skin.

Planting, such as climbing ivy, is a useful climatic device when located next to the building skin because it performs several functions simultaneously. The most obvious benefit of vegetation is its summer shading ability. A dense cover of planting will intercept the sun's radiation before it reaches the building skin, thereby reducing the exterior surface temperature and the amount of heat conducted into the interior. (*FIGS. 36a & b*).

If, for example, an ivy cover on a wall trellis transmits only half of the sunlight striking it, it will cut the solar gain at the surface by one-half. A 50% reduction in solar irradiation can easily mean a reduction in heat gain at the surface of a west wall 57 Btu per square foot on a summer day.

A (summer) disadvantage of planting is that it traps air near the surface of the building, thereby decreasing the effect of breezes in washing away the boundary layer of heated air. Any breeze that is strong enough to ripple the leaves certainly will override this disadvantage, and the positive cooling effect of evaporation of water from the leaf surfaces will also counteract against overheating of this air layer.

In winter, the boundary layer created by an evergreen leaf blanket serves an insulative function that restricts connective heat loss. A planting strategy for many climates, therefore, would include evergreen planting along north, east and west walls, and a perennial ivy (that sheds its leaves) on southern exposures.

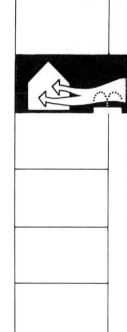

Minimize Conductive Heat Flow

Minimize Solar Gain

Promote Evaporative Cooling

Promote Radiant Cooling

Design Concept

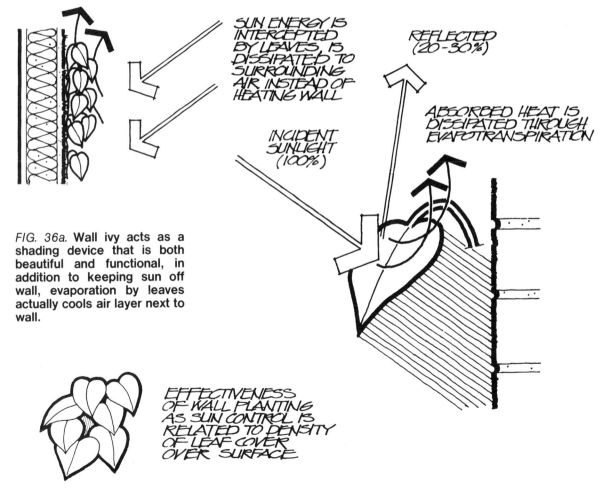

SUN ENERGY IS INTERCEPTED BY LEAVES, IS DISSIPATED TO SURROUNDING AIR INSTEAD OF HEATING WALL

REFLECTED (20-30%)

INCIDENT SUNLIGHT (100%)

ABSORBED HEAT IS DISSIPATED THROUGH EVAPOTRANSPIRATION

FIG. 36a. Wall ivy acts as a shading device that is both beautiful and functional, in addition to keeping sun off wall, evaporation by leaves actually cools air layer next to wall.

EFFECTIVENESS OF WALL PLANTING AS SUN CONTROL IS RELATED TO DENSITY OF LEAF COVER OVER SURFACE

Use double roof and wall construction for ventilation within the building shell.

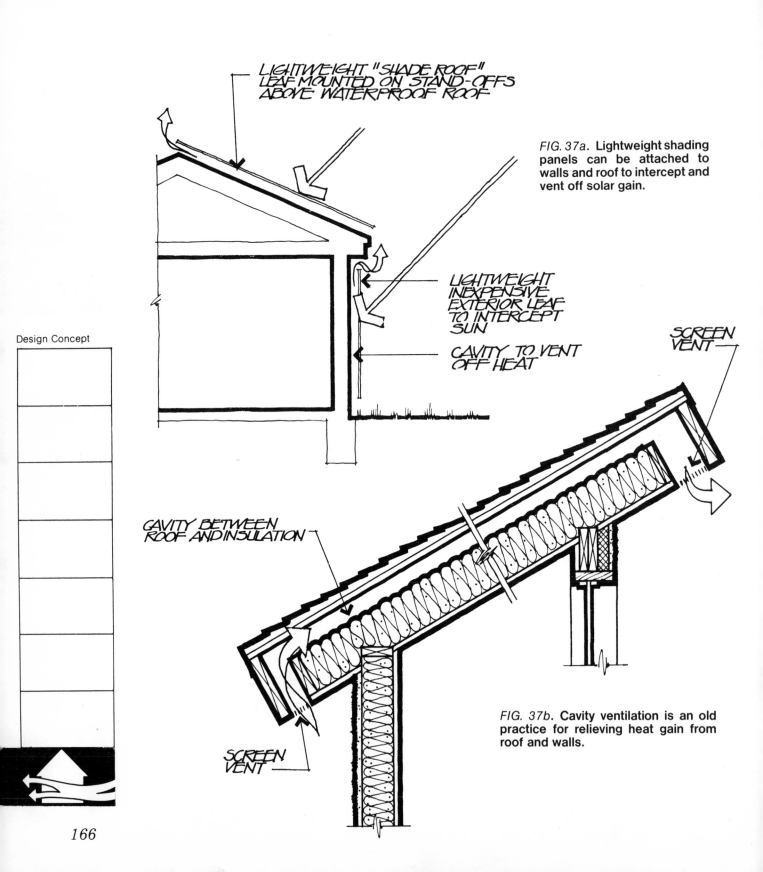

LIGHTWEIGHT "SHADE ROOF"
LEAF MOUNTED ON STAND-OFFS
ABOVE WATERPROOF ROOF

FIG. 37a. Lightweight shading panels can be attached to walls and roof to intercept and vent off solar gain.

LIGHTWEIGHT
INEXPENSIVE
EXTERIOR LEAF
TO INTERCEPT
SUN

CAVITY TO VENT
OFF HEAT

Design Concept

CAVITY BETWEEN
ROOF AND INSULATION

SCREEN
VENT

SCREEN
VENT

FIG. 37b. Cavity ventilation is an old practice for relieving heat gain from roof and walls.

166

Use double roof and wall construction for ventilation within the building shell.

Three fundamental approaches to building shell design can be identified: (1) design for isolation from external conditions through thermal resistance (insulation); (2) design for controlled timing of diurnal interaction with external conditions (thermal mass); and (3) design of the shell as a ducting system to collect, transport, and distribute energy. The third approach is unlike the first two, in that the shell itself plays a role in promoting the transfer of heating energy throughout the structure. Double-shell (cavity) walls and roofs can be used as independent devices designed to perform specific tasks. They may also be designed as component parts of a solar/convective system, sometimes referred to as an "energy envelope" design.

Cavity shell convector

When a lightweight cavity wall or roof is exposed to sunlight, the heated air column inside will attempt to rise by natural convection. If openings are provided at both top and bottom, the wall will ventilate itself, effectively washing away the excess solar heat gain and maintaining an internal temperature near that of outdoor air. This technique has long been applied to roof ventilation and is applicable to walls as well (*FIGS. 37a* and *37b*).

Cavity shell collector

If the exterior skin of a south-facing cavity wall is a glazing material, the air column will heat faster and the air flow rate of the system will be increased. This configuration describes a basic solar heating system, early versions of which were devised by Edward Morse in the 1880s and now popularly referred to as "Trombe" walls, named after a French engineer who has investigated this approach in the last ten years.

Depending on the arrangement of dampers and vent openings, such systems can be used selectively for heating or for ventilation, or a combination of the two. The inner skin of the cavity which receives the irradiation can be designed to serve in itself as a storage medium such as the Trombe wall, or the heated air can be ducted mechanically or by "thermosiphoning" into a separate storage location. The cavity itself may be only a few inches in thickness or may consist of an entire room-sized gallery or greenhouse.

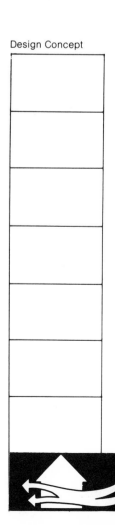

Promote Ventilation

Design Concept

Use roof spray or roof ponds for evaporative cooling.

Minimize Conductive
Heat Flow

Exterior building surfaces exposed to the sun can benefit from the cooling effects of surface evaporation through a variety of techniques. The two major types are roof ponds, which must be planned as an integral part of the structure, and roof sprays or washes, which can be built-in or retrofitted to most conventional residential roofs.

Effect of roof spray

The primary function of a roof spray system is to relieve the roof surface of the solar load by dissipating solar heat as it arrives at the roof surface. Roof spray will be of greatest benefit on roofs, (1) of light-weight construction and (2) which are exposed on the underside to an inhabited space (as opposed to an attic space).

Generally speaking, houses constructed with well ventilated attics are not likely to obtain much benefit from roof spraying, since heat that is conducted into the attic space can be dispelled by ventilation. Similarly, if the attic floor is well insulated, heat built up under the roof will be effectively isolated from the living space, thereby making dissipation at the exterior surface less necessary. Houses without attics or with poorly insulated overhead assemblies, therefore, are the most

likely beneficiaries of roof spraying techniques.

The foremost effect of roof spraying is to lower the temperature of the roof surface. In addition to the possible interior cooling benefits, it washes the exterior walls of the structure with a curtain of falling cool air. See *FIG. 38b*.

The lower average surface temperature maintained by roof spraying reduces blistering and protects the vitality of roofing asphalt by inhibiting the evaporation of volatile oils. Roof spray will be particularly valuable in protecting roofs which are insulated at the roof deck, rather than in the attic ceiling.

Promote
Evaporative Cooling

Promote
Radiant Cooling

Design Concept

FIG. 38a. **Only two builder-provided features are necessary for installation of an evaporative spray system. 1) a conveniently located exterior hose bib 2) a relatively low roof pitch.**

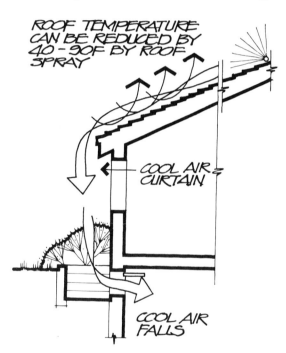

FIG. 38b. **Spraying roof keeps surface temperature constant, and prevents rapid expansion and contraction that ages roof quickly. Roof spray has advantage of being operable only when needed.**

38   Use roof spray or roof ponds for evaporative cooling.

Effectiveness of roof spraying

In a fairly efficient operation, over 90 percent of the solar load upon the roof can be alleviated. Roof surface temperatures commonly in the range of 140-160°F can be reduced to 80-100°F under summer sun conditions. In relatively dry climates, such as the arid southwest, selective roof spraying can maintain the average daily temperature of the roof below the average of the air during the diurnal cycle. John Yellott reports in an article entitled "Roof Cooling with Intermittent Water Sprays"[ASHRAE *Transactions* 1966]:

> From these tests, conducted under maximum insolation and very low relative humidity, it may be concluded that the intermittent sprays can offset the absorbed solar radiation almost completely, and that the average temperature differential between the ambient air and the sprayed deck can be reduced to virtually zero.

Roof spray systems

The simplest form of roof spray—a perforated garden hose or lawn sprinkler on a roof—can produce significant reductions in roof temperature (and indeed have been used for this purpose). This installation requires no more from a builder than a conveniently located outdoor hose bib and a relatively low pitched roof.

Simple, improvised systems are likely to waste water, however, so temperature control systems are desirable. Several companies supply package systems consisting of a roof surface temperature sensor, timer for intermittent spray (to prevent flooding and runoff water waste), valves, and a patented self-cleaning perforated spray pipe. Such systems typically mist the roof surface for 10 to 15 seconds every 10 minutes when the surface temperature exceeds 90°F. This is claimed to consume less than 25¢ worth of water per day, and less electricity than an electric clock. *FIG. 38a* diagrams a typical installation.

The value of spray cooling lies in its simplicity and multiple climate control benefits. Readily utilized as a retrofit system as well as built-in, it is an effective cooling strategy in a variety of climatic regions. However, its applicability is limited primarily by the availability of water. The nontechnological basis of the idea also lends itself to test installations, to determine its effectiveness on stock house designs and in local climates. There are few guidelines to specify the best applications of roof spray, but it has been recommended for use in regions where sol-air temperatures exceed 100°F for 250 hours or more, where the roof has a U value larger than 0.2, and the roof has a mass less than 50 psf.

Roof ponds

Ponded roofs perform essentially the same cooling function as do spray systems, but with the added component of thermal mass (heat storage). Open roof ponds have seen limited application over the past several decades, largely due to the extra costs associated with supporting the weight of water (5.2 psf per inch depth) and the problem of ensuring waterproofing. Needless to say, roof ponds must be designed as an integral part of the structure and require flat roof design. Tests conducted by Yellott in Arizona indicate little difference in performance between ponded and sprayed roofs, concluding that intermittent spray provided as effective cooling as flooding with two inches of water.

Design Concept

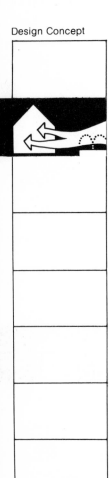

Provide insulating controls at glazing.

Of all the elements in the building envelope, windows and other glazed areas are the most vulnerable to unwanted heat transfer. This is evident in comparing the overall thermal resistance for a pane of ordinary window glass (R = .9) to typical insulated wall values (R13 to R30).

The insulative value of a window assembly can be improved by increasing the R value of the glazing unit itself, as well as by adding insulative controls to the inside or outside of the assembly, as discussed below.

Insulativeness of glazing unit

The insulation value of the glazing unit is dependent on the resistance of the glazing material, the number of glazing panels and air spaces between, and the detailing of the frame with respect to conductive heat flow (*FIG. 39a*). *Table 39a* compares the overall heat flow resistance of several different glazing materials. The total insulating value of the window assembly must be considered and is a function of the window glazing the frame design and its installation detailing.

Interior insulating aids

Interior insulating aids include all devices that cover or can be fitted into the inside of the window opening. Generic types comprise overhead roller shades and unfolding accordion shades, sliding, hinged, or detachable panels and shutters, plus drapes and interior storm windows. Some examples are illustrated in *FIGS. 39b & c*; the thermal resistance of several devices are given in Table 39b.

Exterior insulating aids

The storm window is the most familiar external insulating device. Its popularity can be attributed to being transparent and non-operable, requiring only semi-annual placement and can be "retrofitted" to older windows. Several other exterior devices also offer additional thermal resistance to window areas. These can be assumed to be useful only if they are controllable from the interior and are weather and decay-resistant. A list of thermal performance characteristics of different products is given in *Table 39c*.

FIG. 39a. Thermal benefit of double and triple glazing is due to the fact that air has a conductivity of about 1/40 that of glass. In principle, the thicker the air space, the greater the resistance; for air spaces thicker than 5/8″, the increased air resistance is offset by greater facility for internal convection. For air spaces in excess of 1¼″, the air space conductance is practically independent of thickness.

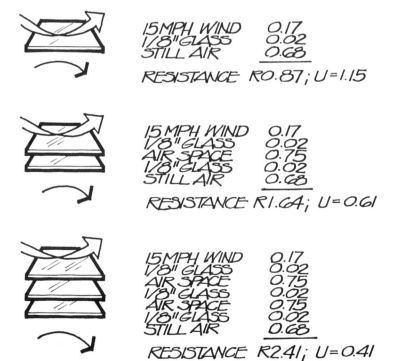

15 MPH WIND	0.17
1/8″ GLASS	0.02
STILL AIR	0.68
RESISTANCE R0.87; U=1.15	

15 MPH WIND	0.17
1/8″ GLASS	0.02
AIR SPACE	0.75
1/8″ GLASS	0.02
STILL AIR	0.68
RESISTANCE R1.64; U=0.61	

15 MPH WIND	0.17
1/8″ GLASS	0.02
AIR SPACE	0.75
1/8″ GLASS	0.02
AIR SPACE	0.75
1/8″ GLASS	0.02
STILL AIR	0.68
RESISTANCE R2.41; U=0.41	

Design Concept

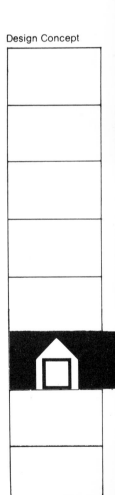

Provide insulating controls at glazing.

TABLE 39a
Comparative thermal transmission of glazing materials, vertical*

Glazing type	Winter U**	Overall R**
Single sheet glass, 1/4″ (e = .84)	1.14	.88
Insulating glass, double glazed (e = .84)		
1/8″ with 3/16″ air space	.62	1.6
1/8″ with 1/4″ air space	.58	1.7
1/4″ with 1/2″ air space	.49	2.0
1/2″ air space with low emittance coating†		
e = .60	.43	2.3
e = .40	.38	2.6
e = .20	.32	3.1
Insulating glass, triple glazed (e = .84)		
1/4″ air spaces	.39	2.6
1/2″ air spaces	.31	3.2
Storm window (conventional), 1″ to 4″ air space	.50	2.0
Single sheet plastic, 1/8″	1.06	.94
1/4″	.96	1.0
1/2″	.81	1.2
Insulating plastic, double glazed		
1/8″ with 1/4″ air space	.55	1.8
1/4″ with 1/2″ air space	.43	2.3
Glass block		
6 x 6 x 4″ thick	.60	1.7
8 x 8 x 4″	.56	1.8
8 x 8 x 4″, double cavity	.48	2.1
12 x 12 x 4″	.52	1.9
12 x 12 x 4″, double cavity	.44	2.3
Double skinned acrylic sheet	.6	1.7
Corrulux (8 oz. corrugated fiberglass) ‡	1.09	.92
Sunwall I‡	.45	2.2
Sunwall II‡	.38	2.6
Sunwall III‡	.30	3.3
Sunwall IV‡	.25	3.9
Sunwall V‡	.22	4.5

*Values from ASHRAE Fundamentals [1981] except as noted
**U values from air-to-air, Btu/hr(ft^2)F; R=1/U
†Reflective coating on one sheet facing air space; other faces uncoated
‡From manufacturer's literature

TABLE 39b
Comparable thermal transmission through operable insulation devices, interior

Insulating device	Unit R	Winter U*	Overall R*
Drapery			
tight fitting, tight weave closed†			
loose fitting, closed†			
Window Blanket	2.0	.34	2.9
Panel type			
cellular glass†	2.5/in.		
rigid fiberglas board†	4.5/in.		
expanded polyurethane†	6.25/in.		
expanded polystyrene, extruded†	5.3/in.		
expanded polystyrene, beadboard†	3.6/in.		
Nightwall		.27	3.7
Insul Shutter	6.25	.14	7.14
Skylid		.26	3.8
Roll down shades			
ordinary roller shade, drawn†		.88	1.1
ordinary shade, with metallized film		.52	1.9
ordinary shade, with edge tracks			
or seals		.50	2.0
Insealshaid	4.0		
Curtain Wall	9.1	0.1	10.0
Window Quilt	3.2	.24	4.2
Thermo Shade		.40	2.5
High R Shade	.08	.07	14.3
Integral			
Slimshade, interior louvers,		.41	2.4
closed‡			
Beadwall	3.3/in.		
Interior storm window		.5	2.0

*for device in combination with single glazing
† from ASHRAE Fundamentals [1981]
‡ from manufacturer's literature

TABLE 39c
Comparative thermal transmission through exterior insulating devices

Insulating device	Unit R	Winter U*	Overall R*
Roll down shutters			
Rolladen	1.1	.45-.50	
Roll-Awn	1.6	.40	
Rolsekur	1.1	.45-.50	
Everstrait	1.1	.45-.50	
Storm window, conventional		.56	1.78
Shading devices			
Koolshade solarscreen (reduces solar gain)†		.96	1.04

*for device in combination with single glazing
† from manufacturer's literature

Design Concepts

Provide insulating controls at glazing.

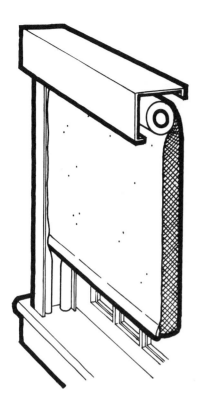

FIG. 39b. An insulating shade is designed to minimize conductive & radiative heat transmission, as well as a seal against infiltration.

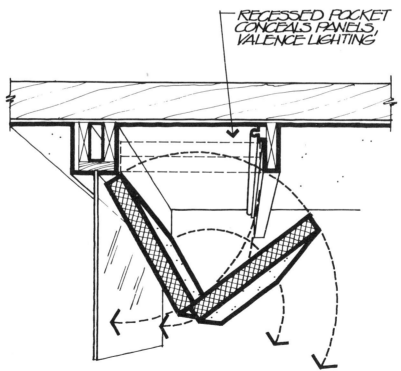

RECESSED POCKET CONCEALS PANELS, VALENCE LIGHTING'

Design Concept

FIG. 39c. Hinged and fold-down panels offer large additional resistances to window units, can be made from aluminum-faced urethane or polystyrene sandwich panels, or insulation board w/laminated facings.

Provide insulating controls at glazing.

Applications

The effectiveness of operable insulation devices will depend often on their ease of use. Insulating shades or shutters are of no value if they are not closed at night. In view of this, triple glazing–even with its reduced solar transmission and comparatively poor R Value–may be preferred. Glazing and insulating devices should, therefore, be considered together. Without operable insulation, for example, triple glazing should be utilized in regions of more than 4500 heating degree days (average winter temperature below 30°F). If operable insulating devices are used, only double or even single glazing need be used on solar-oriented windows, so as to take advantage of their higher solar transmittance.

Design Concept

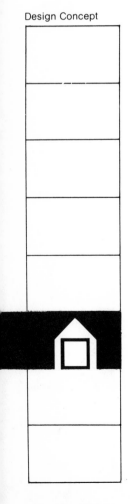

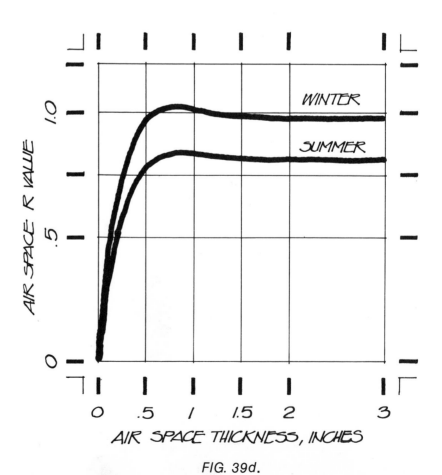

FIG. 39d.

Minimize window and door openings on north, east, and/or west walls.

The number of doors and windows of a house are related to winter heat loss in two major ways: (1) doors and windows both are inferior in insulative value to opaque walls, thereby lowering the overall average resistance of the envelope; (2) infiltration increases with the number of windows and doors.

Minimizing the door and window openings will reduce conductive and infiltrative heat transfer, regardless of their location. Placement in north, east and west elevations, however, is particularly disadvantageous due to the absence of compensatory winter sun at these exposures and the high solar gain (in east and west) in summer. Furthermore, winter winds frequently come from a northerly direction. (*FIG. 40a*).

Window area can be reduced by decreasing the area of individual windows and/or by cutting down on the number of windows. (*FIG.*

40b). The shape of windows should be designed to minimize perimeter infiltration. Square shapes should be favored over elongated ones and the unit should be tightly sealed and weatherstripped. An equally effective approach is to use large "fixed" glazing where it is needed for view or illumination and to specify smaller operable sash as required for ventilation.

Because doors are vulnerable areas within building's insulative envelope, they should be selected for their thermal resistance. *Table 40* compares the heat transmission values for different types of commonly specified doors. As can be seen, the overall resistance of typical doors is far less than the walls in which they are placed.

Minimize Conductive Heat Flow

Minimize Infiltration

Design Concept

TABLE 40
Conductivity coefficients for various door types, winter conditions.

Door Type	With Wood storm door		With metal storm door		With no storm door	
	U	R	U	R	U	R
* 1″ solid wood	.30	3.3	.39	2.6	.64	1.6
* 1¼″ solid wood	.28	3.6	.34	2.9	.55	1.8
* 1½″ solid wood	.27	3.7	.33	3.0	.49	2.0
* 2″ solid wood	.24	4.2	.29	3.4	.43	2.3
** 1¾″ steel, mineral fiber core b					.14-.24	4-7
** 1¾″ steel, solid urethane foam core					.09-.075	11-13
** 1¾″ steel, solid polystyrene					.13-.24	4-7.5
** 1¾″ steel, honeycomb core					.41-.45	2.2-2.4
** 1¾″ steel, fiberboard core					.28	3.6

* from ASHRAE Fundamentals [1981]
** from various manufacturers

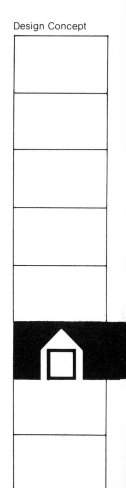

Minimize window and door openings on north, east, and/or west walls.

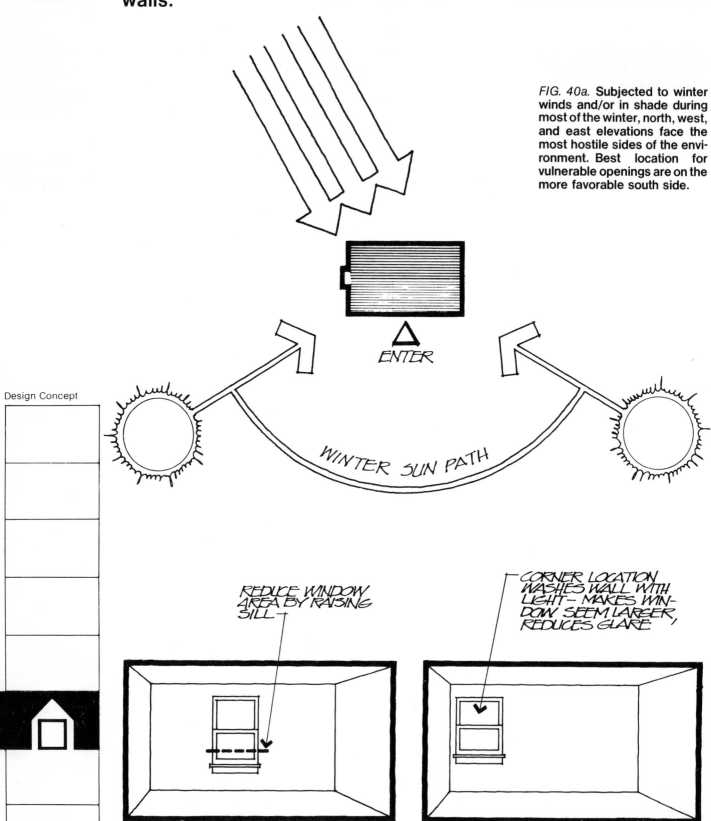

FIG. 40a. Subjected to winter winds and/or in shade during most of the winter, north, west, and east elevations face the most hostile sides of the environment. Best location for vulnerable openings are on the more favorable south side.

ENTER

WINTER SUN PATH

Design Concept

REDUCE WINDOW AREA BY RAISING SILL

CORNER LOCATION WASHES WALL WITH LIGHT — MAKES WINDOW SEEM LARGER, REDUCES GLARE.

FIG. 40b. When windows must be placed in north, west and east walls, keep the opening small, and use internal placement to best advantage.

Maximize south-facing glazing.

The winter heating load of houses in most climates can be appreciably reduced by maximizing south glazing to capture solar heat. January solar receipt on a south surface ranges from about 560 Btu/ft^2/day in cloudy, northern regions to almost 1500 Btu/ft^2/day in clear, southern zones (see *Table 41a*).

Conventional houses can be improved by organizing the plan so that major rooms occur on the south facade if the window area is then shifted to the south from other orientations. Picture windows can thus be used to advantage. Sliding patio doors can serve equally well. See also *FIG. 41b*.

Dormer windows are often used to admit light to upper floors; these can be used as solar-oriented windows in both upper story rooms and in two-story spaces. Although dormer windows admit a smaller solar beam per given window area than would a skylight (averaged over the year), the dormer favors winter penetration while accepting less summer solar gain (*FIG. 41a*).

Another method of admitting solar energy to the interior is to raise or lower planes of the roof to create additional south-facing walls with windows. Like gables, this approach maximizes the receipt of winter radiation while reducing the amount of window area exposed to the summer sun. Winter solar gains are also increased by reflection from nearby roof or terrace surfaces. South facing windows and also roof monitors can be designed with overhangs to exclude the worst of summer sun.

How much glazing?

The amount of desirable solar contribution with regard to both comfort and internal heating requirements is related to the overall insulation of the structure and its thermal storage capacity (its ability to dampen internal temperature fluctuations). Given certain assumptions about the conductivity of the shell and heating load, some preliminary rules of thumb are summarized [from Balcomb 1980] in *Table 41b*.

Less glazing is required for better insulated houses. For example, in the Small Homes Council's "Lo-Cal" House in Illinois (R-40 ceiling, R-33 wall, R-20 floor, triple glazing), the optimum area of glazing was determined to be only 7% to 10% of the gross floor area. [Shick 1979]. According to SHC calculations, the average *net* solar gain of a south-facing, triple-glazed window is 200-400 Btu per square foot per day in winter at 40°N latitude in the midwest—easily enough to provide one third of the annual heating requirements of a well insulated house. This gain could be significantly increased with use of operable window insulation as well as by increasing the amount of internal heat storage mass.

Promote Solar Gain

Design Concept

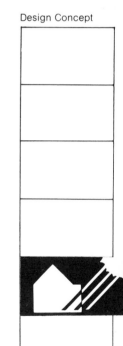

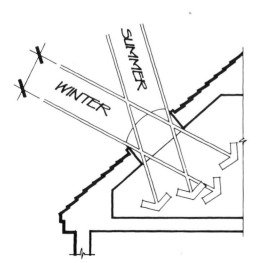 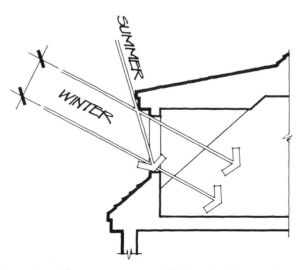

FIG. 41a. **Skylights on pitched roofs do not discriminate between summer and winter admittance. If the roof angle equals the latitude of the site, the skylight will admit the same area of solar beam at winter noon as at summer noon. Vertical glazings, such as dormer windows, favor winter sun to the exclusion of summer heat gain.**

Maximize south-facing glazing.

TABLE 41a
Solar radiation received on vertical south facing surface, monthly avg., BTU/ft^2(day)

	J	F	M	A	M	J	J	A	S	O	N	D
Atlanta	1041	1127	1122	1000	843	791	775	889	1052	1258	1205	1032
Boston	878	1052	1127	1030	944	927	940	1009	1212	1192	874	788
Chicago	921	1106	1207	1113	1023	1006	1026	1146	1280	1276	967	771
Dallas	1164	1218	1195	942	843	870	893	992	1146	1308	1250	1168
Denver	1440	1551	1572	1344	1147	1114	1130	1277	1535	1616	1424	1327
Hartford	869	1045	1070	1006	900	848	873	951	1094	1127	852	742
Houston	772	1034	1297	1522	1775	1898	1828	1686	1471	1276	924	730
Kansas City	1098	1218	1222	1113	994	975	1034	1149	1276	1343	1176	1005
Los Angeles	1353	1424	1405	1168	944	880	993	1091	1255	1392	1382	1309
Miami	1236	1224	1088	881	725	650	677	696	848	1088	1231	1260
Midland, Tx.	1496	1534	1504	1235	1062	1030	989	1096	1295	1520	1532	1462
Minneapolis	921	1212	1312	1208	1092	1057	1140	1237	1297	1233	895	742
New Orleans	1097	1169	1093	948	827	786	728	809	1003	1262	1203	1081
New York	884	1023	1100	1009	908	834	866	959	1115	1147	886	755
Omaha	1139	1286	1318	1174	1058	1052	1099	1223	1283	1368	1089	972
Phoenix	1472	1589	1552	1388	1211	1128	1059	1186	1482	1643	1561	1419
Salt Lake City	1129	1403	1542	1401	1311	1249	1328	1457	1692	1657	1310	1065
San Francisco	1145	1311	1408	1288	1126	1068	1124	1240	1455	1438	1250	1097
Seattle	559	841	1082	1166	1164	1066	1405	1278	1273	1007	674	478
D.C.	959	1097	1130	1019	902	882	884	986	1163	1221	1027	852

Source: Balcomb [1980] a

Design Concept

TABLE 41b
Rules of thumb for estimating design areas for south glazing.

KEY: A solar collection area of (R1)% to (R2)% of the floor area can be expected to reduce the annual heating load of a building in (location) by (S1)% to (S2)%, or, if R9 night insulation is used, by (S3)% to (S4)%.

CITY	R1	R2	S1	S2	S3	S4
Atlanta	08	17	22	36	34	58
Boston	15	29	17	25	40	64
Chicago	17	35	17	23	43	67
Dallas	08	17	27	44	38	64
Denver	12	23	27	43	47	74
Hartford	17	35	14	19	40	64
Houston	06	11	25	43	34	59
Kansas City	14	29	22	32	44	70
Los Angeles	05	09	36	58	44	72
Miami	01	02	27	48	31	54
Midland, Tx.	09	18	32	52	44	72
Minneapolis	25	50	NR	NR	55	76
New Orleans	05	11	27	46	35	61
New York	15	30	16	25	36	59
Omaha	20	40	21	29	51	75
Phoenix	06	12	37	60	48	75
Salt Lake City	13	26	27	39	48	72
San Francisco	06	13	34	54	45	71
Seattle	11	22	21	30	39	59
Washington	12	23	18	28	37	61

Source: Balcomb [1980] a

Provide reflective panels outside of glazing to increase winter irradiation.

The amount of direct solar irradiation received at a window opening is fixed by the season, weather conditions, and geometry of the building with respect to the sun. Although the intensity of direct solar radiation itself cannot be increased, irradiation received by surrounding surfaces can be reflected into the opening, thereby increasing the effective collecting area of the window.

Outside reflecting surfaces may be either fixed or operable. Fixed reflectors may easily lead to summertime overheating, however, so that in terms of year-round comfort, their potential effectiveness must often be compromised. One way to deal with this is to design overhangs to shade stationary horizontal reflecting panels; another method is to make the panels themselves demountable for seasonal installation (FIG. 42a).

A simple operable system would use hinged shutters with highly reflective surfaces that can be positioned to direct sun into the window. These will be of greatest value on walls that face east, southeast, southwest and west. In summertime, the opposite shutter can be stationed in the opposite position for shading (FIG. 42b).

Many operable and adjustable reflecting panel designs have been developed to augment the collecting ability of solar systems. Reflector geometry is diagrammed in FIG. 42c.

TABLE 42a

Reflectivity of various surfaces

Material	Reflectance (%)
perfect mirror (theoretical)	100
polished aluminum (Alzak)	75-95
snow cover, fresh	75-95
silver-backed glass	88-90
aluminized mylar (clean)	60-80
polished stainless steel	60-80
white porcelain enamel	70-77
white lead, zinc oxide paints (new)	68-76
mercury-backed glass	70-73
snow cover, old	40-70
concrete	30-50
snow, dirty, firm	20-50
dry grass	32±
water	8-10

Reflectance data taken from various sources.

Design Concept

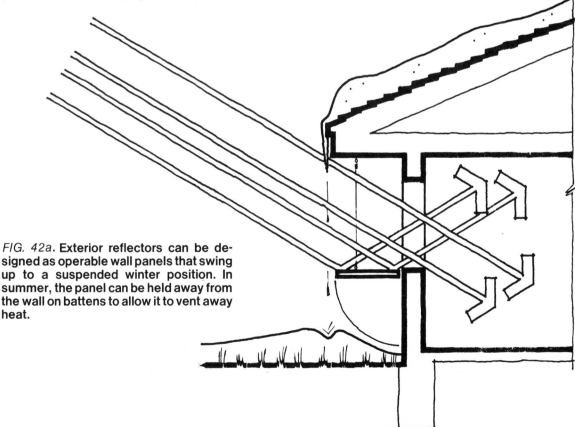

FIG. 42a. Exterior reflectors can be designed as operable wall panels that swing up to a suspended winter position. In summer, the panel can be held away from the wall on battens to allow it to vent away heat.

Provide reflective panels outside of glazing to increase winter irradiation.

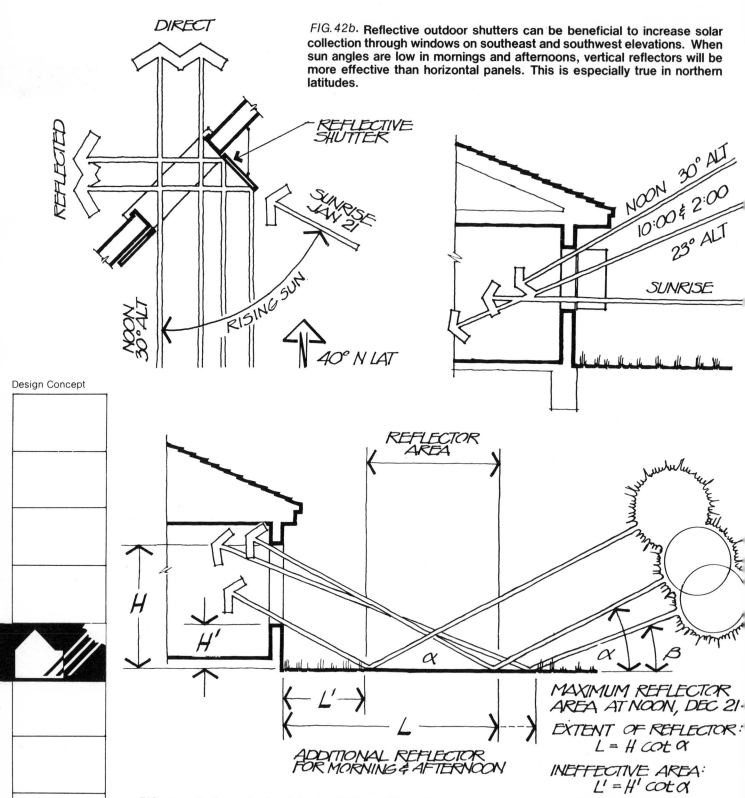

FIG. 42b. Reflective outdoor shutters can be beneficial to increase solar collection through windows on southeast and southwest elevations. When sun angles are low in mornings and afternoons, vertical reflectors will be more effective than horizontal panels. This is especially true in northern latitudes.

DIRECT

REFLECTED

REFLECTIVE SHUTTER

SUNRISE JAN 21

RISING SUN

NOON 30° ALT

40° N LAT

NOON 30° ALT

10:00 & 2:00

23° ALT

SUNRISE

Design Concept

REFLECTOR AREA

H

H'

α

α   β

L'

L

ADDITIONAL REFLECTOR FOR MORNING & AFTERNOON

MAXIMUM REFLECTOR AREA AT NOON, DEC 21·

EXTENT OF REFLECTOR:
$$L = H \cot \alpha$$

INEFFECTIVE AREA:
$$L' = H' \cot \alpha$$

FIG. 42c. Optimum horizontal extent of reflecting surface is easily calculated by simple geometry. Illustration shows worst condition, December 21, when sun is lowest. Coldest temperatures usually occur in late January, when noon sun altitude is 3½° higher. Higher sun requires less reflector area, and placement closer to the building. Precise angles are applicable only to specular reflection.

Use skylights for winter solar gain and natural illumination.

Skylights and roof windows are readily available and compatible with most house designs, making them a convenient way to bring solar heat and light into the dwelling. The benefits of a bright and sunny daylit interior are particularly appreciated by home-owners during winter days. Provisions must also be made to prevent summer solar gain. The solar orientation of skylight installations must also be carefully studied in terms of construction details. Roof pitch, thickness, curb height, and the size of the skylight opening are all important factors in determining winter solar penetration of the interior (*FIG. 43b* and *FIG. 43c*).

Some form of night insulation should be considered to prevent the skylight from becoming a net heat loser rather than a heat source. In addition the skylight assembly and supporting frame itself must be well insulated. If winter heat gain is to be maximized, one large skylight opening is preferable to a cluster of smaller ones, since the latter will produce an egg crate shading device effect.

A major disadvantage of skylighting is the admittance of solar heat during the summer. This problem will be most severe on low pitched or flat roofs, where the roof plane becomes almost perpendicular to the sun's rays. Methods of blocking solar gain include shading the skylight with louvers installed on the outside of the skylight during the summer, using operable skylights to vent overheated indoor air, and specifying roof window assemblies that are equipped with exterior awnings, shades, or internal blinds. Other sun controls include interior hinged or sliding panels that close off the entire opening (*FIG. 43a*) and interior operable louver systems.

Promote Solar Gain

Design Concept

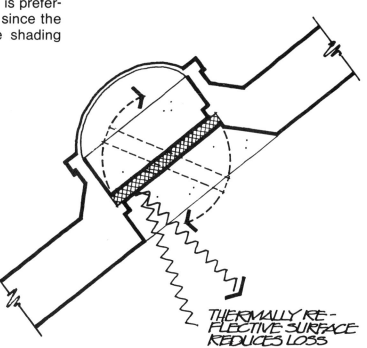

THERMALLY RE-FLECTIVE SURFACE REDUCES LOSS

FIG. 43a. **Interior insulating panels can be pivoted, hinged, or slide to close off skylight. Condensation is frequently a problem and must be anticipated.**

Use skylights for winter solar gain and natural illumination.

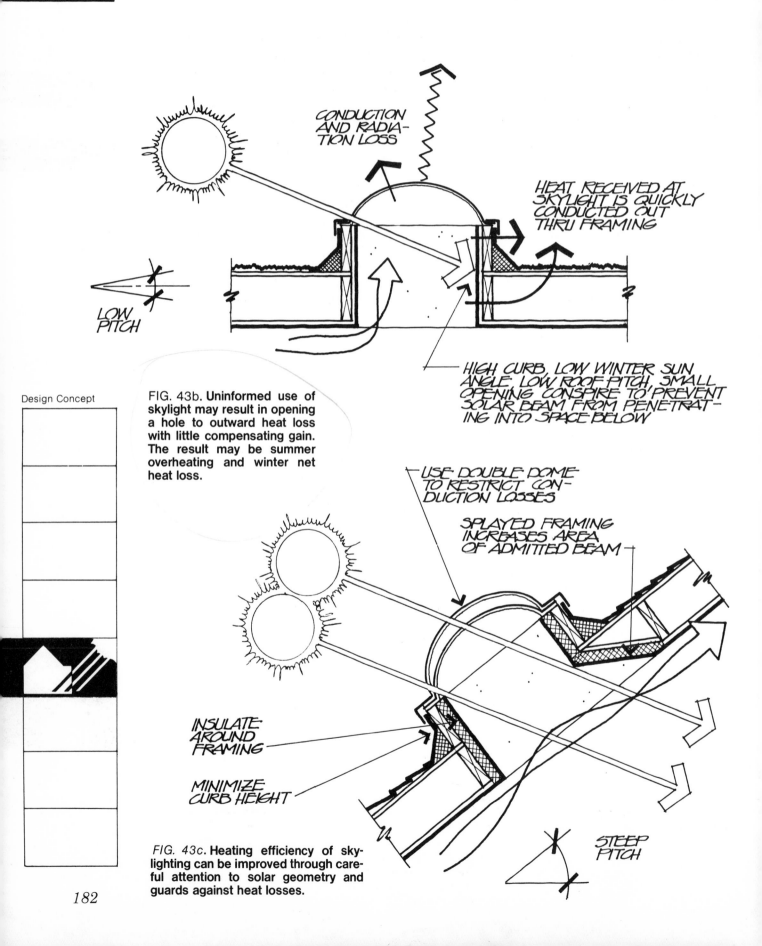

CONDUCTION
AND RADIA-
TION LOSS

HEAT RECEIVED AT
SKYLIGHT IS QUICKLY
CONDUCTED OUT
THRU FRAMING

LOW
PITCH

HIGH CURB, LOW WINTER SUN
ANGLE, LOW ROOF PITCH, SMALL
OPENING CONSPIRE TO PREVENT
SOLAR BEAM FROM PENETRAT-
ING INTO SPACE BELOW

Design Concept

FIG. 43b. Uninformed use of skylight may result in opening a hole to outward heat loss with little compensating gain. The result may be summer overheating and winter net heat loss.

USE DOUBLE DOME
TO RESTRICT CON-
DUCTION LOSSES

SPLAYED FRAMING
INCREASES AREA
OF ADMITTED BEAM

INSULATE
AROUND
FRAMING

MINIMIZE
CURB HEIGHT

STEEP
PITCH

FIG. 43c. **Heating efficiency of sky-lighting can be improved through careful attention to solar geometry and guards against heat losses.**

Detail window and door construction to prevent undesired air infiltration and exfiltration.

Infiltration around the sash and frames of operable windows and doors constitutes a major source of heat loss. While much of the potential for infiltration can be reduced by use of vestibules, storm windows, windbreaks, and orientation of these openings, infiltration rates at the openings can be improved further through careful detailing and selection of tight sealing door and window units.

Windows

Window type is closely related to air leakage, in that the geometry and proportion of crack length to window area of some window types makes them more difficult to seal than others. Window leakage is usually rated in terms of cubic feet of air transmitted in one hour (or one minute) through 1 linear foot of crack perimeter (CFH or CFM per foot of crack). Obviously, the lower the CFH value, the better the unit. Some typical leakage rates for different window types are presented in *FIG. 44b* and *FIG. 44d*.

The actual performance of any given window assembly will depend on the quality of its manufacture. Window units should be selected on the basis of published "third party" test data.

Doors

While windows can be rated as to air leakage as entire units or assemblies, doors are often purchased independently of the surrounding frame. Each door can be an individual problem, therefore, and must be customized upon installation or afterwards to seal against leakage. A solution is offered by prehung integrally weatherstripped door units, some with magnetic seals. Adjustable thresholds are important in creating a tight seal. Numerous weatherstripping products are readily available, the generic types of which are illustrated and compared to *Table 44a.*

Minimize Infiltration

Design Concept

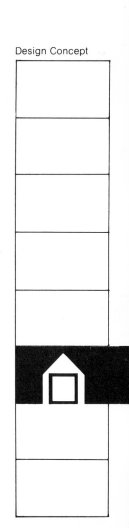

TABLE 44a Comparison of different types of weatherstripping for doors and windows. These are listed in order of estimated overall durability.

KEY: E—Excellent; VG—Very Good; G—Good; F—Fair; P—Poor					
TYPE	MATERIAL	Estimated Overall Durability	Effective Uses (1)	Suitable for Non-uniform gaps	Visibility When Installed
FLAT METAL STRIP	Brass or bronze	E	C/A	No	Very low
	Aluminum	VG to E	C/A	No	Very low
TUBULAR GASKET	Vinyl or rubber, foam-filled	VG	C/A	Yes	High
	Vinyl or rubber, hollow	VG	C/A	Yes	High
REINFORCED GASKET	Aluminium and vinyl	VG	C/A	Yes	High
REINFORCED FELT	Wool felt and aluminum	G	C	No	High
	Nonwool felt and aluminum	F to G	C	No	High
NONREINFORCED FELT	Wool	G	C	No	(2)
	Other	F to G	C	No	(2)
RIGID STRIP	Aluminum and vinyl	G	C	Yes	Low (3)
	Wood and foam	F	C	Yes	Low
FOAM STRIP	Neoprene or rubber	F	C	Yes	(2)
	Vinyl	F	C	Yes	(2)
	Polyurethane	P to F	C	Yes	(2)

(1) C—Where material will be subject to compression
A—Where material will be subject to abrasion
(2) Low if under sash or inside doorjamb. High if used along window frame or against door stop.
(3) On aluminum door, its primary use.

Detail window and door construction to prevent undesired air infiltration.

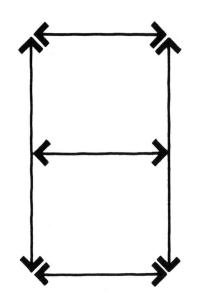

FIG. 44a. **Crack length for a double hung window is the distance around the perimeter of the operable sash.**

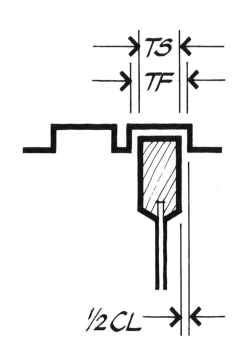

FIG. 44b. **Clearance is the combined gap area of the internal and external surfaces of the sash clearance. CL = TF - TS.**

Design Concept

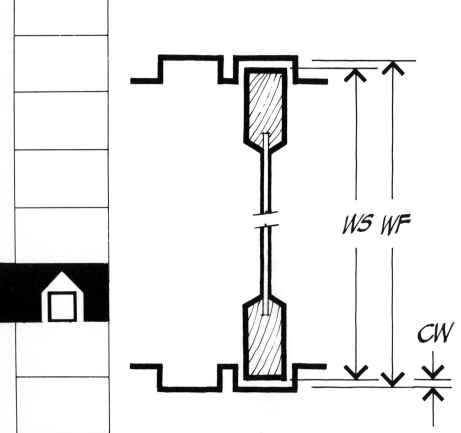

FIG. 44c. **Crack width is the distance between outside edge of the operable sash and the inside edge of the window frame. Crack Width CW = 1/2 (WF - WS)**

Detail window and door construction to prevent undesired air infiltration.

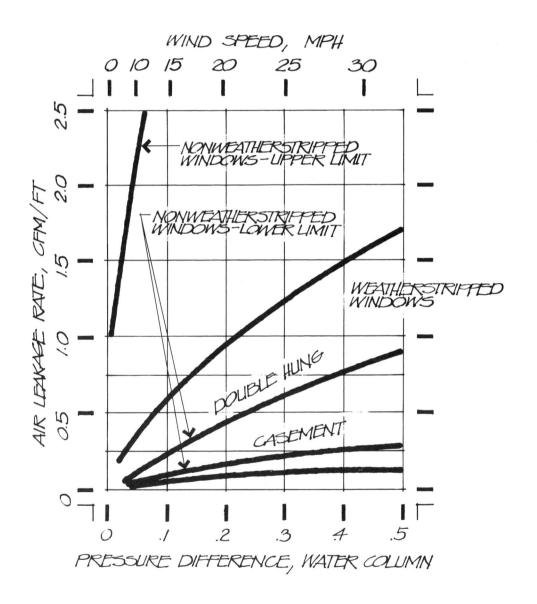

WIND SPEED, MPH

NONWEATHERSTRIPPED WINDOWS-UPPER LIMIT

NONWEATHERSTRIPPED WINDOWS-LOWER LIMIT

WEATHERSTRIPPED WINDOWS

DOUBLE HUNG

CASEMENT

AIR LEAKAGE RATE, CFM/FT

PRESSURE DIFFERENCE, WATER COLUMN

FIG. 44d. **This graph illustrates relationship between different window types, wind speed, and infiltration rates. (Redrawn from Canadian Building Digest, No. 35, "Window Air Leakage.")**

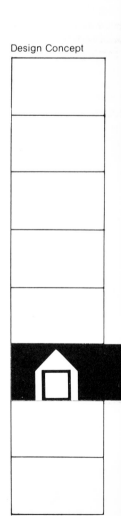

Design Concept

45

Provide ventilation openings for air flow to and from specific spaces and appliances.

FIG. 45a. Range hood with exterior exhaust removes undesirable heat from kitchen before it can cause discomfort. Venting to outside also guarantees positive removal of smoke and odors—not just filtration. Make sure unit has well sealing damper to prevent exfiltration in winter.

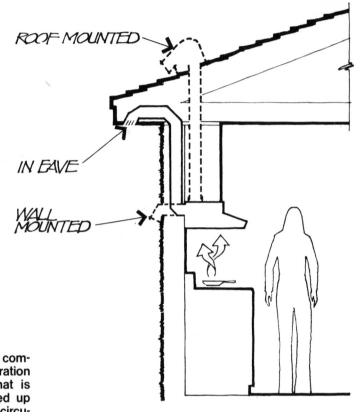

FIG. 45b. Fireplaces with outside combustion air intake reduce infiltration through doors and windows that is drawn in to replace air released up chimney. Shown here is a heat recirculating fireplace with outside intake.

Design Concept

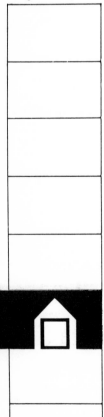

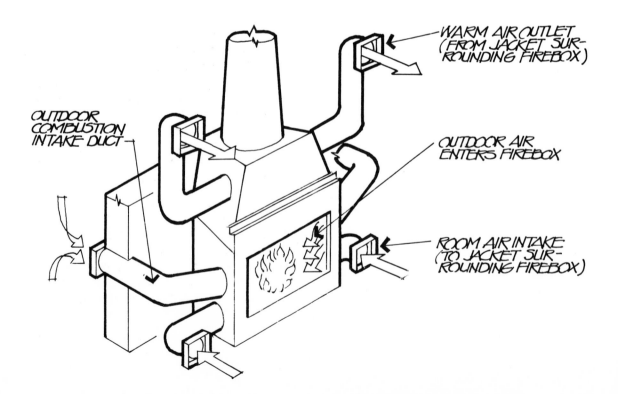

Provide ventilation openings for air flow to and from specific spaces and appliances.

Summer strategy

A simple way of reducing the cooling load of a house in summertime is to exhaust internally generated humid and overheated air directly to the outside before it reaches the living areas. Areas that deserve particular attention are kitchens, bathrooms, and laundry rooms. These should be provided with windows or other vents to the outside.

Fan-driven vents or exhaust stacks, especially from range hoods, can help reduce the house's need for air conditioning by reducing the cooling load before the thermostat calls the air conditioner on. Once the air conditioning system has been activated, however, outside venting should not normally be used, as this will promote infiltration of warm outdoor air (*FIG. 45a*).

Winter strategy

In a similar way, the heating requirement of the house in winter can be reduced by drawing upon outside air directly for combustion devices such as fireplaces and furnaces. If this is not done, combustion air is drawn from the interior of the house (or the basement mechanical room area), creating a reduced air pressure that induces infiltration through doors, windows. (*FIG. 45b*).

In winter heat generated through cooking, laundering and bathing should be conserved rather than exhausted to the outside. "Dual-exhaust" range hood fans are now available that exhaust to the outside in summer and recirculate through a filter in winter. Similarly, heat recovery methods and mechanisms should be investigated to "squeeze" the maximum amount of heat out of combustion exhaust from furnaces and fireplaces.

Houses built to very high standards of insulation will require some form of forced (controlled) air change in order to exhaust pollutants and to admit fresh air. This is best accomplished by a heat exchanger at the air inlet/outlet to recover heat from the exhaust and thus "preheat" the incoming air.

Minimize Infiltration

Promote Ventilation

Promote Evaporative Cooling

Design Concept

Provide shading for glazing exposed to summer sun.

TABLE 46a
Properties of shading devices.

	% Transmitted	% Reflected	% Absorbed	S.C. [a]
Venetian blinds				
light-colored horizontal	5	55	40	.55
med.-colored horizontal	5	35	60	.64
white (closed) vertical	0	77	23	.29
Roller shades				
white color, translucent	25	60	15	.39
white, opaque	0	80	20	.25
dark, opaque	0	12	88	.59

[a] shading coefficient given for device in combination with ⅛" clear float glass (87% transmittance): except for vertical blind, with 71-80% transmittance.

Design Concept

TABLE 46b
Solar and visible transmission of different glazing materials.

Material	%Daylight	%Solar	S.C.
Clear float glass, 1 sheet, ⅛"	90	86	1.00
2 sheets, ⅛"	82	71	.88
1 sheet, ¼"	88	77	.93
2 sheets, ¼"	78	60	.80
Sunadex "water white" 1 sheet, ⅛"		91.6	
Lo-iron 1 sheet, ⅛"		89.1	
heat absorbing ⅛"	84	65	.82
¼"	76	48	.68
Plexiglas 1 sheet, 3/16"	92	90	
Lexan polycarbonate sheet, ⅛"	86	89	—
¼"	82	86	—
Lascolite corrugated fiberglass			
crystal clear, 4 oz.	93	82	
clear, 5 oz.	87	81	
frost, 8 oz.	66	60	
solar white, 5 oz.	32	21	
Kalwall sunlite, 1 sheet		85-90	
sunwall 1 sheet	77		
acrylic double skinned sheet			
clear	83	83	.97
white	70	67	.81
white	20	19	.31
polycarbonate double skinned sheet	80	77	
polycarbonate double skinned sheet	73	74	.88
gray float glass, 1 sheet, ⅛" (LOF)	62	63	.82
¼"	42	44	.67
gray reflective, 1 sheet, (LOF)	8-34	11-37	.36-.62
Crystaflex 40-A, translucent, ⅛"	83	18	
Glasshade II	Varies	Varies	

Provide shading for glazing exposed to summer sun.

Window shading devices may be grouped into three categories:
(1) devices which are independent of the glazing material and applied at the *interior* of the opening;
(2) properties inherent to the glazing itself;
(3) devices that are applied within the window opening at the *exterior* of the window assembly.

The effectiveness of shading devices is expressed numerically as its *shading coefficient* (SC). The shading coefficient represents the ratio of total solar heat gain through the device as compared to the total gain through a single thickness of unshaded ordinary window glass; the latter is given the value of 1.0. All shading devices have values less than 1.0. Low shading coefficients indicate low transmittance and, therefore, good shading. For many devices (louvers, in particular), the value of the SC changes with the sun position and with the adjustment of the device itself.

1 – *Interior Shading Devices.* Interior shading mechanisms include roller shades, venetian blinds, curtains, and drapes. More recent additions include hinged and sliding panels (used primarily for insulation) and interior shutters.

Interior shading devices work primarily by reflecting solar radiation that arrives through the window. The major disadvantage of shades, drapes and the like is that regardless of how reflective they are made, they trap heat on the interior of the glass, so it remains indoors (See *FIG. 46* and *Table 46*).

In response to the rising energy cost of maintaining comfort, a number of manufacturers are examining methods of increasing the reflectivity of, and creating an air seal around, roller shades so that they can be used both for summer sun-shading and winter night insulation.

Several special window units circumvent some of the thermal disadvantages of internal shades or blinds by placing narrow louvers within the airspace between glazing sheets.

2 – *Solar Transmittance of Glazing Materials.* A number of glazing materials and manufactured window assemblies have been developed which reduce the amount of radiation that passes through the glazing itself. The heat admitting or rejecting characteristic of glazing materials is described as *solar transmittance*, although shading coefficients are often given as well. Instead of using regular sheet glass as an index, *transmittance* indicates the ratio of irradiation which passes through a glazing material in reference to the amount falling upon it (taken as 100%). Some typical values for different glazing materials are given in *Table 46b*.

As can be seen from *Table 46b* solar transmittance can be reduced considerably through selection of glazing materials (when the intent is to reduce solar gain year-round). Heat-absorbing clear and tinted glazing reduce solar transmission by absorbing heat within the glazing material itself. Absorption by glass will result in high glass temperatures (30°F or more above air temperatures is not uncommon) which, although a lesser evil than transmitted gain, adds heat to the interior by conduction and thermal radiation; this itself can be a source of discomfort to occupants. Reflective coatings on the exterior and solar control films applied to the interior of the glass offer another choice for keeping heat out of the dwelling. A major disadvantage of heat-reflecting and heat absorbing glazing types is their nonselective nature: they block needed solar gain in winter as well as blocking summer solar irradiation.

Minimize Conductive
Heat Flow

Minimize Solar Gain

Design Concept

46 Provide shading for glazing exposed to summer sun.

3 – *Exterior window shades.* The third category of window shades are devices affixed to the outside of the window jamb. Among the operable units are jalousies made of wood or metal, exterior venetian blinds and shutters.

Vertical baffles or louvers are sometimes applied to the wall outside windows for sun control. These cannot prevent the sun from fully entering the window on east, south, or west elevations unless they are pivoted to track the sun's movement. They can offer nearly full shading on eastern and western facades if pitched towards the north, but this will always result in some sacrifice of view.

Fixed horizontal louvers are effective in sun control, although some may find them to be visually disruptive. One response to the slatted view is to miniaturize the louver system, giving it a fabric-like scale and appearance.

A less selective group of solar screening devices are loose-weave fiberglass cloths that may be used in place of wire insect screens on windows, or draped on trellises—or used in lieu of opaque and translucent plastic sheets on roller shades. These fabrics permit vision and daylight passage, but can reduce solar transmission up to 75 percent.

The major important advantage of exterior-mounted sun shades is to block the heat on the outside of the building. There is a disadvantage to any fixed sunshade. If it shades a window opening from June to September (the typical "over-heated" months in United States climates), it also blocks the sun from March to June, when solar gain might be desired. This is because there is a time delay of about one and one-half months between the peak of the overheated season (late July through August) and the period of maximum insolation (June). The ideal solution is to have shades that can be moved, either seasonally or on a daily basis. A compromise is to use fixed devices for the later summer months.

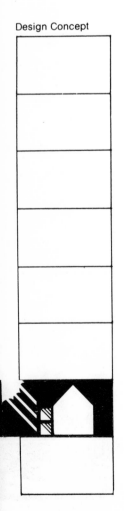

Design Concept

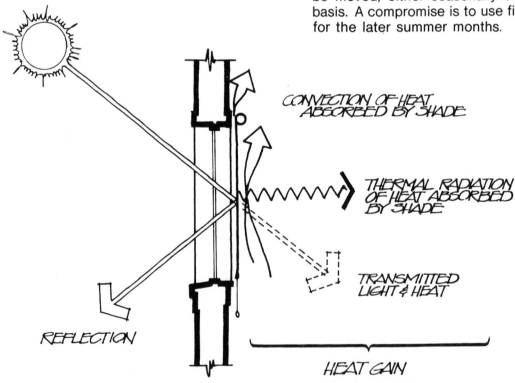

CONVECTION OF HEAT ABSORBED BY SHADE

THERMAL RADIATION OF HEAT ABSORBED BY SHADE

TRANSMITTED LIGHT & HEAT

REFLECTION

HEAT GAIN

FIG. 46. The color of a roller shade affects its overall heat-reducing performance. The best shade is one with high reflectance—not necessarily the highest shading coefficient; a good absorber is a good solar heater!

Orient door and window openings to facilitate natural ventilation from prevailing summer breezes.

Outdoor breezes create air movement through the house interior by the "push-pull" effect of positive air pressure on the windward side and negative pressure (suction) on the leeward side. Good natural ventilation requires locating openings in opposing pressure zones. The greatest volume of air movement will occur when windows or screened doors are located in the portions of the facade that experience the greatest pressure differential between them.

The pressure field around a square building subject to perpendicular and oblique winds can be diagrammed as in *FIG. 47a*. The arrows represent the direction of pressure at the building surface; the length of the pressure diagrams indicates the relative magnitude of the force. Although the optimum location for inlet and outlet windows shifts with changes in the wind direction, good window locations are common to all and are illustrated in *FIG. 47b*.

The driving force of ventilation is increased (i.e., more of the wall is exposed to maximum pressure) if the building is elongated and sited with the long facade facing into the wind. With a rectangular plan, the rear wall will experience greater suction and will consequently be a more suitable location for outlet windows. Windows in the side walls will not be as significant in the rectangular plan as the square plan, except for the interior effects described below.

Effect of window location on internal air flow.

Placing windows in line with the direction of airflow in opposite walls of the structure will result in a narrow airstream of high velocity. If the windows are located in the center of the windward and leeward facades, airflow will pass straight through the middle of the interior space. If the windows are offset from center, the airflow will be diverted away from the long dimension of the facade, in this case, washing the interior side wall. In either case, although velocity of the internal air current will be high, overall ventilation of the space will be poor.

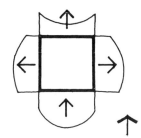
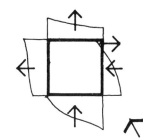
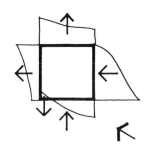

FIG. 47a. These diagrams indicate pressure areas around a square building. Note how the pressure distribution shifts as the wind direction changes, effect is most pronounced at positive pressure walls—suction walls are more uniform.

FIG. 47b. Shows good inlet locations. The relatively uniform negative pressure distribution suggests that outlets could be well located anywhere on walls C and D. The choice should be based on internal flow effects.

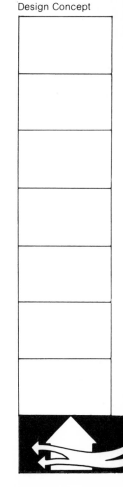

Design Concept

Orient door and window openings to facilitate natural ventilation from prevailing summer breezes.

FIG. 47c. Flow path of air thru the interior is determined by location of inlet opening with respect to exterior wall, and location of outlet with respect to direction of incoming air current. Diagrams may represent individual rooms or floor plan, provided wall opening relationships are maintained. Compare to relative driving pressure differences for these locations indicated in FIG. 48c.

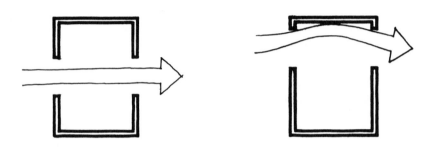

DIAGRAMS ABOVE INDICATE PLANS IN WHICH INTERNAL WIND SPEED WILL BE HIGH, BUT IN WHICH MUCH OF THE SPACE REMAINS UNAFFECTED. NOTE THAT OFFSETTING INLET PRODUCES DEFLECTION OF AIRSTREAM.

Design Concept

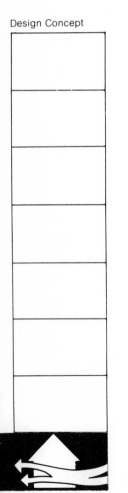

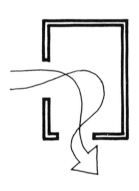
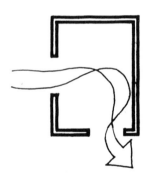

RELOCATING OUTLETS TO SIDE-WALL PRODUCES BETTER WASH OF THE INTERIOR

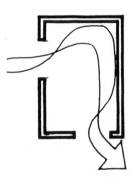
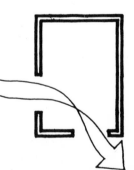

SHIFTING INTAKE WINDOW FROM CENTER OF FACADE RESULTS IN DEFLECTION OF INTERIOR AIR CURRENT. DEPENDING ON LOCATION OF OUTLET, THIS MAY IMPROVE OR SHORT-CIRCUIT THE CURRENT WITH RESPECT TO ITS WASH OF THE INTERIOR.

Orient door and window openings to facilitate natural ventilation from prevailing summer breezes.

Improved "air wash" of the interior space can be achieved by locating primary inlet and outlet windows in opposite pressure areas, but askew from the axis of prevailing winds. *FIG. 47c* shows the internal air paths for different inlet-outlet relationships. In general, it can be said that if the wind has to change direction within the room, a larger volume of the space is affected by the air flow, creating higher overall circulation velocities.

Effect of window size on ventilation.

The volume of airflow (number of air changes) that passes through a structure is governed by the size of the openings; the greatest number of air changes are obtained when both inlets and outlets are as large as possible. The velocity of air movement through the structure is maximized, creating the greatest cooling, when the area of outlets is greater than the area of inlet openings. Air movement within a structure is influenced by a great number of factors, such as location of furniture, exterior planting and geometrics, and shifting wind directions, making the prediction of airflow and ventilation rates practically impossible. Based on the foregoing

principles, however, some decisions can be made as to the best locations for window openings in a simply partitioned rectangular plane.

Effect of window type on cross ventilation.

Insofar as airflow is concerned, there are two distinguishable types of operable window sash: (1) pivoted and hinged windows (hopper, awning, casement), which exert a deflecting, turning effect on incoming air current, and (2) sliding and double-hung windows which operate in the plane of the wall and therefore do not steer incoming flows.

The deflecting ability of vane-type windows can be used to direct air streams which course along the ceiling down into the living zone. However, poor window selection can rob the room of ventilation benefit *(FIG. 47d).* Louvers and venetian blinds have similar influence on air flow patterns, and can be used in conjunction with sliding and double-hung windows to control the airstream path. One means of ensuring good and uniform ventilation is to utilize louvers (under a band of windows, for example) across the length of exterior walls.

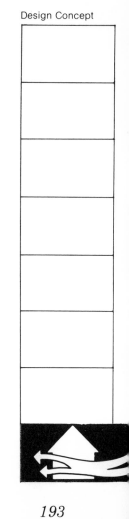

Promote Ventilation

Design Concept

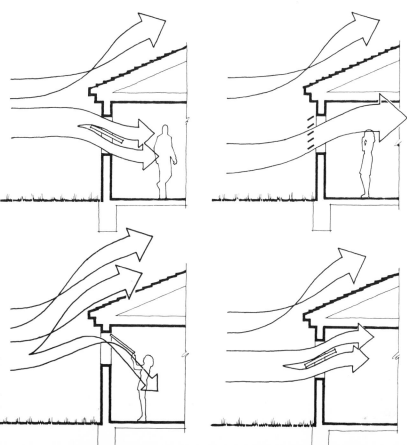

FIG. 47d. **The type of window can have considerable effect on the path of internal air flow and its cooling value. There is little mystery in the effect produced by these windows— they are just as expected.**

Use wingwalls, overhangs, and louvers to direct summer wind flow into the interior.

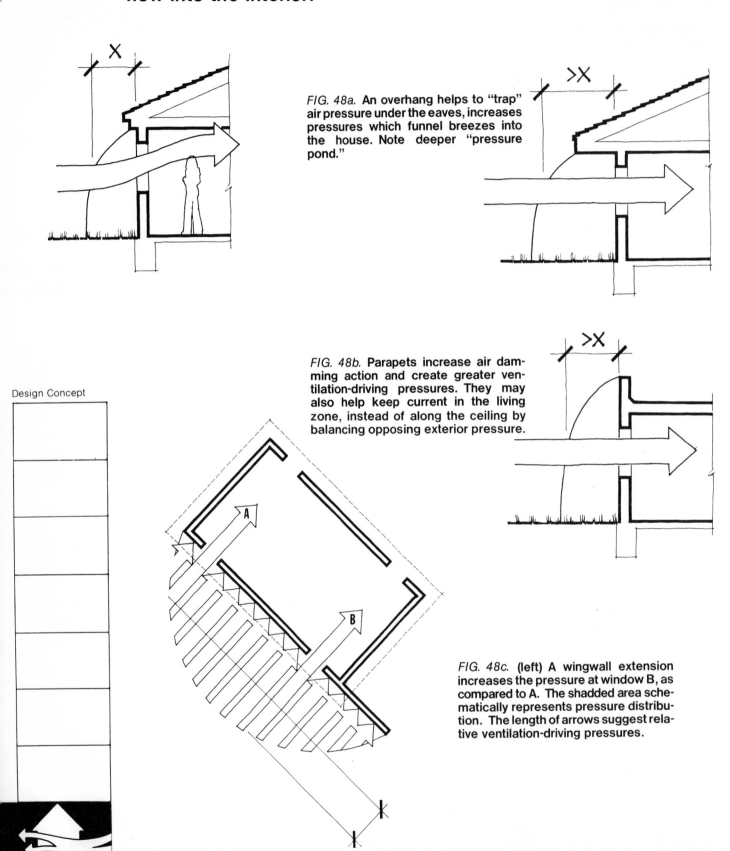

FIG. 48a. An overhang helps to "trap" air pressure under the eaves, increases pressures which funnel breezes into the house. Note deeper "pressure pond."

FIG. 48b. Parapets increase air damming action and create greater ventilation-driving pressures. They may also help keep current in the living zone, instead of along the ceiling by balancing opposing exterior pressure.

Design Concept

FIG. 48c. (left) A wingwall extension increases the pressure at window B, as compared to A. The shadded area schematically represents pressure distribution. The length of arrows suggest relative ventilation-driving pressures.

Use wingwalls, overhangs, and louvers to direct summer wind flow into the interior.

Projections from the surface of the building shell can be used to increase the volume and velocity of air flow into the structure and to alter the flow pattern of air as it enters and travels through the interior. Appendages to the building shell are not a substitute for proper site planning and orientation, but these devices can enhance the building's ventilating ability under both normal and less than ideal conditions.

Wind-directing devices may be either horizontal or vertical, and either fixed or operable. Both vertical and horizontal devices act through a damming effect that blocks and changes the direction of air flow, as shown in FIG 48a & 48b.

Projections from the building surface can be used in four ways to affect air movement. These are to (1) increase inflow of head-on breezes, (2) intercept and increase admittance of oblique breezes, (3) channel direction of incoming flow current, and (4) to create "artificial" cross ventilation.

1. Increase inflow of head-on breezes

The amount of airflow admitted from head-on breezes can be increased by enlarging the area of the windward facade, by trapping air in pressure pockets formed by overhangs, and by building shapes that funnel air into the interior (FIGS. 48b & 48c).

2. Intercept and increase admittance of oblique breezes

No significant ventilating breeze will enter an opening that is parallel to the wind flow, particularly if the opening is narrow. Placing a vertical baffle or fin on the downwind side of the window, however, will dam up the passing air, increasing pressure enough to admit some through the window. This effect may be produced through use of baffles or panels, wingwalls (in the case of corner windows), and even hinged shutters that can be fixed in a perpendicular open position (FIGS. 48d & 48e).

3. Rechannel direction of incoming flow current

In many cases the path of the internal air flow current will be as important, if not more so, than the velocity or volume of the current itself. If upon entering, the airflow is immediately driven against the ceiling, the breeze will be of little comfort to occupants of the interior. The location of external fins can deflect the air stream into useful directions. Proper detailing of these devices can also usually remedy any possible disadvantages.

A solid, planar sunshade attached to a windward wall, for example, will deflect the internal airstream up and out of the living zone—an undesirable effect of a projection used for non-ventilating reasons. This effect can be corrected by leaving a gap between the structure and the projection (FIG. 48f).

In plan, wall projections can be used to direct windflow into particular areas of the interior. This principle is illustrated in FIG. 48d. Practical applications of this effect are found in the use of exterior shutters in the open position, casement windows, doors propped open, and fixed fins such as blinder-type vertical sunshades. Like the overhead cantilevered sunshade, the deflection effect can be cancelled by leaving a gap between the wall and the fin.

4. Artificial cross-ventilation

Good ventilation always requires air inlets and outlets in zones of different air pressures. Very little air circulation will occur, therefore, in a room which has windows in a single outside wall. This is frequently the case in the design of garden apartments and other isolated back-to-back type units, such as hotels and motels.

Where prevailing breezes are parallel or oblique to the exterior windowed facade, zones of relatively high and low pressure can be "artificially" induced by projecting baffles on opposite sides of the windows.

Design Concept

Use wingwalls, overhangs, and louvers to direct summer wind flow into the interior.

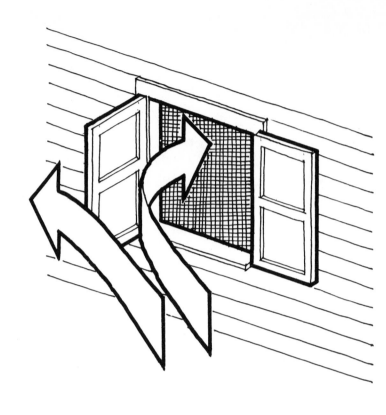

FIG. 48d. **Devices as simple as window shutters—if operable and knowledgeably used—can provide an aid to natural ventilation.**

Design Concept

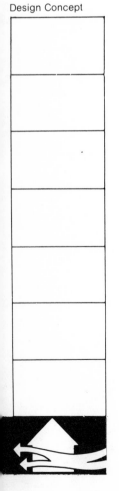

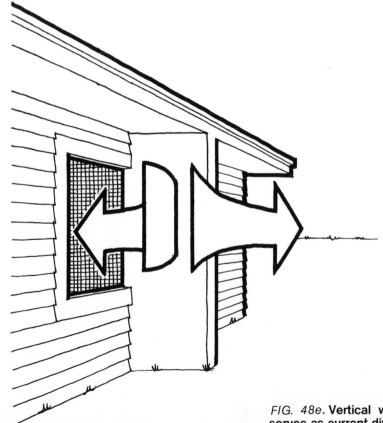

FIG. 48e. **Vertical wall panel serves as current diverter into window.**

Use wingwalls, overhangs, and louvers to direct summer wind flow into the interior.

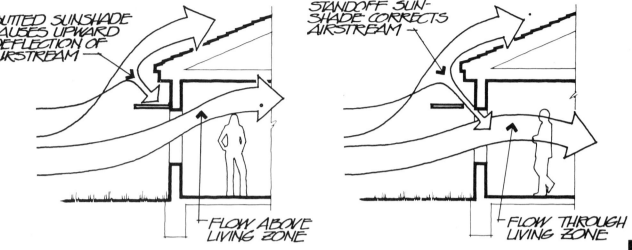

FIG. 48f. Devices installed for one climatic purpose (sunshade) may have other counterproductive consequences. An informed designer can recognize these and work with them. The example above shows a small detail with big implications for comfort—yet no difference in cost.

Promote Ventilation

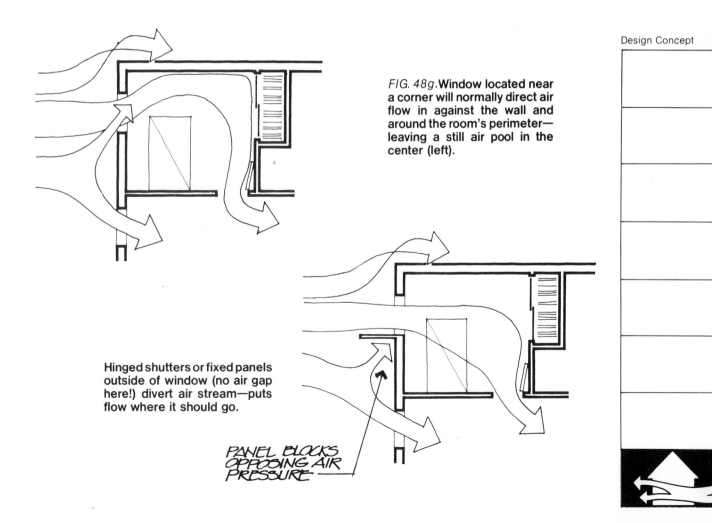

Design Concept

FIG. 48g. **Window located near a corner will normally direct air flow in against the wall and around the room's perimeter— leaving a still air pool in the center (left).**

Hinged shutters or fixed panels outside of window (no air gap here!) divert air stream—puts flow where it should go.

Use louvered wall for maximum ventilation control.

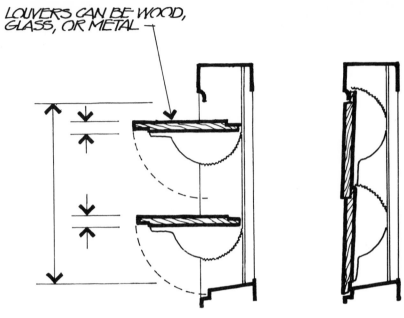

LOUVERS CAN BE WOOD, GLASS, OR METAL

FIG. 49b. **A major advantage of the jalousie window is the almost unrestricted free area in the open position. For glass louvers, a free area of up to 86% is obtainable in the open position, 46% at 30°, 22% at 15°**

Design Concept

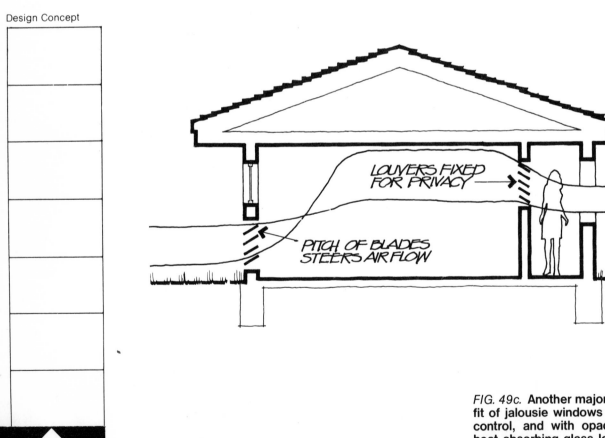

LOUVERS FIXED FOR PRIVACY

PITCH OF BLADES STEERS AIR FLOW

FIG. 49c. **Another major benefit of jalousie windows is rain control, and with opaque or heat absorbing glass louvers, sun screening is also achieved.**

Use louvered wall for maximum ventilation control.

At times when cross ventilation is capable of producing comfort conditions within a dwelling, the best wall will be no wall. This is commonplace in many tropical climates, where a house may consist of little more than a raised platform and roof to block the sun. Few parts of the United States are so mild or consistently overheated to make this strategy fully appropriate for even a short time of the year although examples can be found in the southern United States built before air conditioning became commonplace. Houses in these climates could still benefit from a device that facilitates nearly unrestricted air movement, but which can be closed at will to exclude the elements. The louvered wall is such a device.

The most familiar wall louver device is the jalousie window, units of which are prefabricated in a variety of sizes by several manufacturers. Glass jalousies and other adjustable louvers incur the penalty of excessive infiltration during the heating season, however, so their use is commonly restricted to very mild regions or to seasonal rooms (porches, sunrooms, breezeways) in more northerly zones or in combination with storm sash. Louvered walls—of wooden slats for example—can also help facilitate air flow when used as interior partitions. See *FIGS. 49a & 49c.*

Promote Ventilation

Design Concept

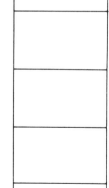

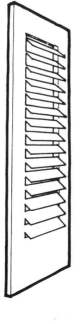

FIG. 49a. **A louvered door is ideal for porches, exterior rooms, and spaces where openess is desirable without sacrificing security. Louvered doors are also well suited for interior use (with opaque or translucent louvers) for ventilation and privacy.**

Use roof monitors for "stack effect" ventilation.

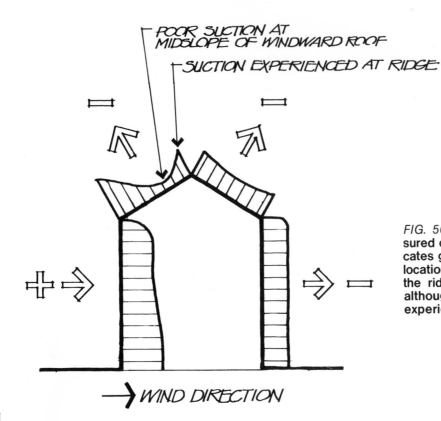

FIG. 50a. Suction field measured on a test structure indicates good negative pressure location for venting devices at the ridgeline (30° roof pitch, although entire area of roof experiences uplift).

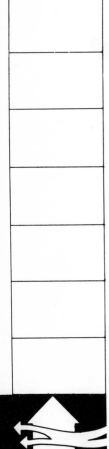

Design Concept

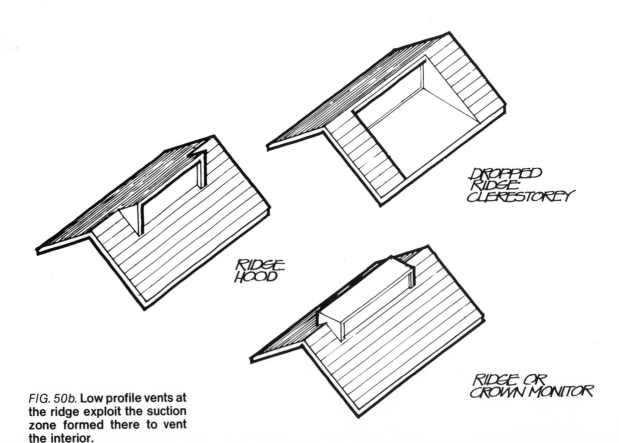

FIG. 50b. Low profile vents at the ridge exploit the suction zone formed there to vent the interior.

Use roof monitors for "stack effect" ventilation.

Roof outlets for stack ventilation may take a variety of forms. These range from non-descript devices such as ridge vents to vertically projecting monitors that are an integral feature of the design. The two general rules for internal stack outlet design are: (a) the outlet should be located as high as possible in the interior space; (b) the geometry of the outlet should offer minimum resistance to upward air flow.

Various forms of stack outlets may behave very differently in the presence of wind; some may augment uplift, while others may cancel out the driving flow with opposing pressures due to turbulent flow over the roof. Two basic options for the exterior design of the stack aperture are: (1) locate these in the negative pressure zone created by the building itself or, (2) use the form of the roof and the outlet to create secondary suction areas to increase uplift in the stack. ASHRAE *Fundamentals* [1981] recommends that a roof ventilator...

> should be located on that part of the roof where it receives the full wind without interference. If it is installed in the suction region created by the wind passing over the building...its performance will be seriously impaired and its ejector action, if any, may be lost.

Other ASHRAE recommendations are that the ventilator should be so located to be acted upon by winds from any direction and that the area of air inlets at lower levels should be greater than the throat area of the (combined) roof ventilator(s). *FIG. 50a* illustrates the cross sectional pressure field surrounding a simple house shape of low roof pitch. Both roof planes experience suction, although it is minimized near midpoint of the windward roof plane. For maximum stack height, and for independence of wind orientation, vent outlets should be located at the ridge. Some possible types of vents and monitors that fall within the suction envelope are illustrated in *FIG. 50b*.

Low pitched roofs, of course, limit the height of the interior stack, thereby limiting its effectiveness. Steeper roofs and projecting ventilating devices increase the effective chimney height, and in doing so, may project into the air stream. A steep roof which projects into the air stream receives a positive pressure over much of its windward side although it still experiences uplift at the ridge line *(FIG. 50c)*. A high-reaching projection above the roof can create its own secondary suction zone. As long as there are vent openings in the secondary suction zone created by roof or roof monitor, the projection can increase the draw of the stack. If the openings are on the windward side or in a turbulent (eddy) area, the stack action may be reversed, with the intended outlets serving as poorly located intakes. Three solutions to avoiding a counter-effect involve, (a) cross ventilating the exit openings, (b) designing openings for selective orientation, and (c) designing the opening to be independent of wind direction.

Traditional solutions to cross ventilating roof stack openings include the cupola, belvedere, and the ridge monitor (the latter usually being limited to industrial applications). Most other possibilities are variations on these basic forms *(FIG. 50d)*.

Obviously, any type of ventilating stack is well suited for power assisting by an electric fan. This aids in the exhausting of interior hot air as well as encouraging air circulation within the dwelling. On the other hand, winter conditions warrant nullifying the stack effect as much as possible. All vents should therefore be fixed with positive closures and insulating dampers wherever possible.

Design Concept

Use roof monitors for "stack effect" ventilation.

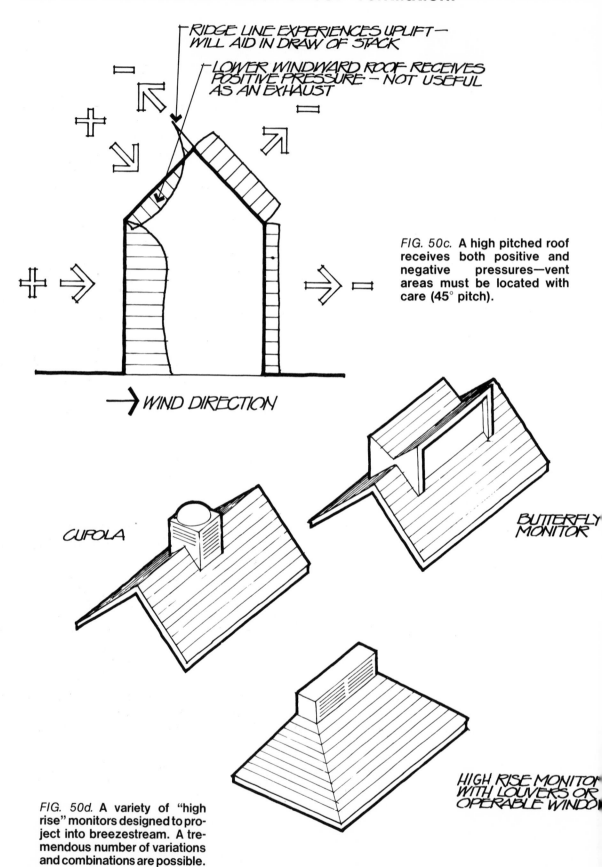

RIDGE LINE EXPERIENCES UPLIFT—
WILL AID IN DRAW OF STACK

LOWER WINDWARD ROOF RECEIVES
POSITIVE PRESSURE—NOT USEFUL
AS AN EXHAUST

FIG. 50c. A high pitched roof receives both positive and negative pressures—vent areas must be located with care (45° pitch).

→ WIND DIRECTION

CUPOLA

BUTTERFLY MONITOR

HIGH RISE MONITOR WITH LOUVERS OR OPERABLE WINDOW

Design Concept

FIG. 50d. A variety of "high rise" monitors designed to project into breezestream. A tremendous number of variations and combinations are possible.

Sunspace in built-for-sale house in Communico Development, Sante Fe, NM. Susan and Wayne Nichols, Architects and Builders.

Part III: Climatic Data

Climatic Data

In Part III, the basis for climatic analysis is explained, with reference to the Building Bioclimatic Chart and to climatic data, compiled in tables for 29 representative locations in the United States. The climatic data contained in the tables provide sufficient reference for climatic design in any region of the contiguous United States. These data can be supplemented by the narrative summaries of local climate that are published by the National Climatic Center (listed in Part IV) for most cities with weather stations throughout the U.S. The procedure for charting such climatic data on the Building Bioclimatic Chart and thereby determining the predominant climatic design strategies for each locale is explained and illustrated below.

The psychrometric chart has been partitioned into a series of subzones which, when added together in different combinations, closely describe the zones of the Building Bioclimatic Chart delineated by Givoni (see discussion on bioclimatic analysis in Part I). The combinations of subzones representing each of the climate control strategies are listed in *Table 1* corresponding to the boundaries depicted in *Figure 1*.

Long-term records of hourly weather observations from twenty-nine locations in the United States (*Table 9*) have been analyzed by computer to determine the average fraction of the year for which the local climate falls within the parameters described by each of the seventeen subzones of *Figure 1*. These are reported to the nearest 0.1 percent (8.7 hours) in *Tables 11-39*, in which subzones are also totalled to give a direct reading of the annual percentage fraction of effectiveness for each climatic design strategy. Daily average solar radiation for each month from Balcomb [1980a] and ASHRAE HVAC design temperatures [*1981 Fundamentals*] are included in the tables to provide a single, succinct climatic data summary for the designer. The specific contents of the data tables are explained in *Table 10*.

Wind speeds and directions have been computed for different subzones of the Building Bioclimatic Chart, and the most useful of these are presented in *Tables 5 – 8*. *Table 5* indicates wind directions for the hours of the year when air temperatures are less than 50F; these can be used as a guide for windbreak planning. *Table 7* gives wind directions during temperature-humidity conditions for which ventilation is a sufficient means of climate control. Wind direction distribution during comfort hours is given in *Table 6*; although ventilation is not necessary under these conditions for bodily comfort, structural ventilation is often desirable to prevent overheating of the interior. Wind directions for periods when relative humidity exceeds 80 percent are given in *Table 8*. Air movement is not considered capable of producing comfort during these conditions, but it is the only passive means of providing some relief from excess humidity.

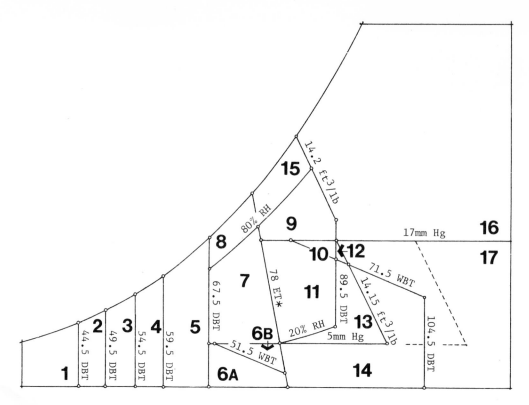

Figure 1. Building Bioclimatic Chart (after Givoni), showing psychrometric limits used for analysis reported in *Tables 11-39*.

Table 1   **Control Strategies**

Identification of climate control strategies on the Building Bioclimatic Chart (adapted after Givoni).

BIOCLIMATIC NEEDS ANALYSIS

Total heating (< 68F)	1-5
Total cooling (> 78ET*)	9-17
Total comfort (68F — 78ET*, 5mm Hg — 80% RH)	7
Dehumidification (> 17mm Hg or 80% RH)	8-9, 15-16
Humidification (< 5mm Hg)	6A, 6B (14)

STRATEGIES OF CLIMATE CONTROL

Restrict conduction	1-5; 9-11, 15-17
Restrict infiltration	1-5; 16-17
Promote solar gain	1-5
Restrict solar gain	6-17
Promote ventilation	9-11
Promote Evaporative cooling	11, 13-14 (6B)
Promote radiant cooling	10-13
Mechanical cooling	17
Mechanical cooling & dehumidification	15-16

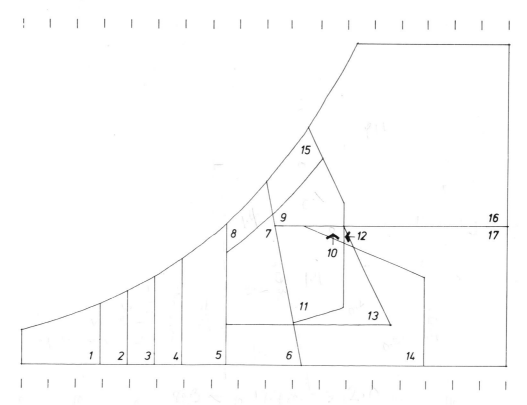

Figure 2. Blank Building Bioclimatic Chart for recording local weather data.

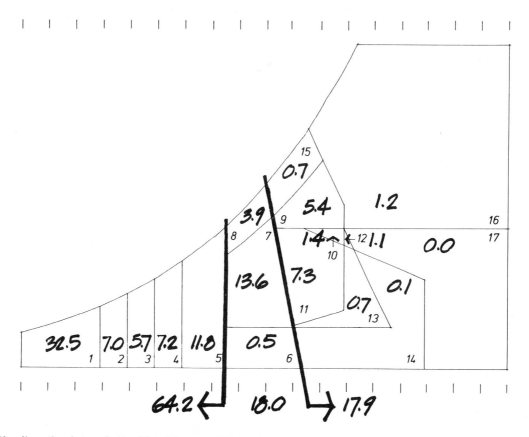

Figure 3. Bioclimatic data plotted for Kansas City, Missouri.

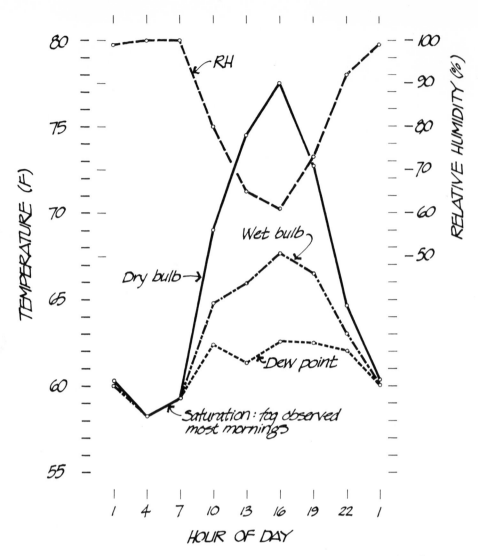

Figure 4 Coincident dry-bulb, wet-bulb, and dew point temperatures and corresponding relative humidity averaged over a nine day period in June 1979 for Asheville, North Carolina (see *Table 2*).

Table 2					Temperature-Humidity Correlations, June 22-30, Asheville, NC[a]				
Indicator				**Hour of day**					**Avg.**
	1[b]	4[b]	7[b]	10	13	16	19	22	
Dry-Bulb Temperature	60.3	58.3	59.3	69.1	74.5	77.5	72.7	64.7	67.0
Wet-Bulb Temperature	60.1	58.3	59.3	64.8	65.9	67.7	66.5	63.1	63.8
Dew Point Temperature	60.1	58.3	59.3	62.4	61.3	62.6	62.5	62.1	61.1
Relative Humidity	99	100	100	80	65	61	73	92	75

a Averaged values for the nine day period, 1979

b Fog was observed during seven of the nine days in early morning, and rain fell during two of these seven foggy mornings.

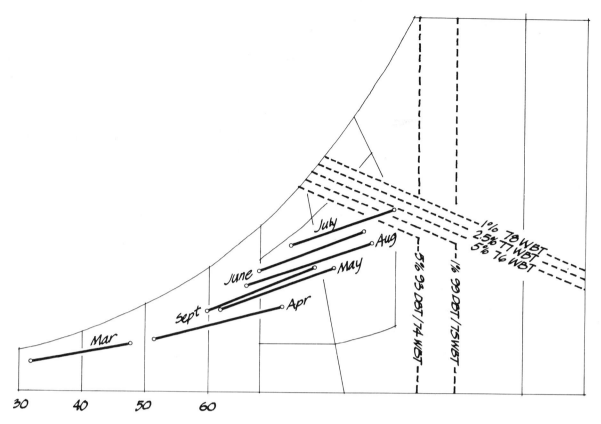

Figure 5. Plotting of fictitious average monthly conditions in Kansas City, Missouri for selected months of the year. Dashed lines plot ASHRAE summer design temperatures.

Graphic Representation of Data

It is often helpful to plot weather data directly on the psychrometric chart in order to obtain a better understanding of local climatic conditions. Data from *Tables 11 – 39* can be filled in on a blank psychrometric chart (*Figure 2*), as illustrated by example in *Figure 3* for selected months in Kansas City, MO.

If one has access to hourly or tri-hourly records of simultaneous dry-bulb and wet-bulb or dew point temperatures (or relative humidity), the daily progress of changing conditions can be traced on the chart. Average daily or monthly temperature and humidity data cannot accurately portray daily temperature cycles, because maximum, average, and minimum dry-bulb, wet-bulb, and dew point temperatures do not necessarily coincide with one another. Dew point temperature is really a measure of the amount of moisture contained in—or the absolute humidity of—the air, and this is governed only indirectly by air temperature. For an elaboration of cyclical temperature-humidity relationships, see Geiger [1965]. A brief explanation follows.

The moisture content of the air near the ground (usually measured at a height of 5 feet at U.S. weather stations) rises during the daylight morning hours, as the sun's heating of the soil drives off moisture contained under or upon its surface. Vapor accumulates in the air layer near the ground throughout the morning, reaching its maximum density—which produces a peak in dew point temperature—usually shortly before noon. This moisture is carried aloft by convection currents in the afternoon, resulting in a dip in dew point temperature during midday, by which time the ground surface region has given up most of its available moisture. As the ground surface cools in the evening, moisture from higher levels of the atmosphere returns to the region near the ground; this produces a second, usually lesser, peak in dew point temperature as the dry-bulb temperature falls. Air moisture content and dew point temperature fall during the evening as vapor returns to storage in—or condensate (dew or frost) upon—the ground. In temperate zones, the two peaks in dew point temperature are said to follow typically 3 to 4 hours each after sunrise and sunset [Pettersson 1969]. Evidence of this can be seen in the correlated weather data for the last nine days of June 1977 in Asheville, NC in *Table 2* and *Figure 4*. The data are especially interesting because they show the nighttime saturation of air, which was accompanied by fog for seven out of the nine mornings.

Appreciating the fact that uncorrelated data cannot portray daily cycles in temperature-humidity conditions, it is nevertheless graphically useful to plot the range of daily dry-bulb temperature against the range of absolute humidity for different times of the year. This is easily done by coupling the minimum dry-bulb and dew point temperatures on one hand and the maximum dry-bulb and dew point temperatures on the other. The former coupling is, in fact, a likely correlation, while the latter is likely to exaggerate the moisture content of the air. Coupling maximum wet-bulb temperature with maximum dry-bulb temperature is more realistic, but it fails to indicate the overall range in moisture content. Besides, the dew point temperature persists near its maximum value throughout much of the day, and this fact should be acknowledged in the graphic portrayal of the data. In either case, the line which may be drawn between these two extremes is fictitious, and does not necessarily represent conditions at any hour of the day. An example representing average monthly conditions for Kansas City, MO is depicted in *Figure 5*.

Bin-hour dry-bulb and wet-bulb temperature correlations produce a revealing graphic representation of monthly weather conditions. Data in the format of *Table 3* are available from a number of U.S. air bases in Air Force *Engineering Manual 88-8*. These tables are computed for each month and give the wet-bulb temperature most likely to be found within each 4F dry-bulb temperature range. What is especially useful is that the actual number of hours represented by each correlation is reported. These can be graphed as shown in *Figure 6*, plotting the mean persisting wet-bulb temperature against the midpoint of each dry-bulb temperature range.

Regional Characterization.
Regional trends in the effectiveness of various climate control strategies can be seen by mapping data from *Tables 11-39*. *Figures 7* and *8* illustrate indices of overheatedness in terms of cooling degree days to base 78ET*, and the percentage fraction of the year for which temperature-humidity conditions exceed the 78ET* comfort limit. *Figures 9* through *11* indicate the total annual fraction for which ventilation, thermal mass, and evaporative space cooling are capable of maintaining thermal comfort. *Figure 12* indicates the percentage of hours when dry-bulb temperatures fall within comfort limits, but during which relative humidity exceeds the desirable limit of 80 percent. In all maps, the exact station location falls under the decimal point.

Regional zones of climatic similarity have been suggested by many writers in the past as a means of generalizing on the suitability of different climate control practices. These range from very simplistic schemes which divide the United States into four or six zones (often not based on relevant bioclimatic parameters at all) to sophisticated analysis with dozens of zones and subzones. Notable among the latter are: Terjung [1966, 1967], who examined outdoor weather data on the basis of human comfort; Willmott and Vernon [1980], with an analysis of "solar climates,"; and Crow and Holladay [1976a, 1976b] who have partitioned the State of California into fifteen zones related to their requirements for heating, ventilating, and air conditioning. Papers describing other classification schemes are listed in the bibliography. The value of any classification can be judged quickly by the utility or relevance to building design of the parameters used in delineating zones. Even some very sophisticated mapping schemes fail by this criterion.

Table 3   July Bin-Hour Weather Data for Kansas City, MO

Temperature Range (F)	Observations/Hour Gp 02-09	Observations/Hour Gp 10-17	Observations/Hour Gp 18-01	Total Observations	Mean Coincident WBT
105-109	0	3	1	4	73
100-104	0	9	2	11	74
95-99	0	27	8	35	75
90-94	3	43	21	67	74
85-89	12	56	42	110	73
80-84	45	60	62	167	71
75-79	68	33	59	160	70
70-74	65	13	37	115	67
65-69	48	4	15	67	64
60-64	6	0	1	7	59
55-59	1	0	0	1	55

Source: Air Force Manual 88-8

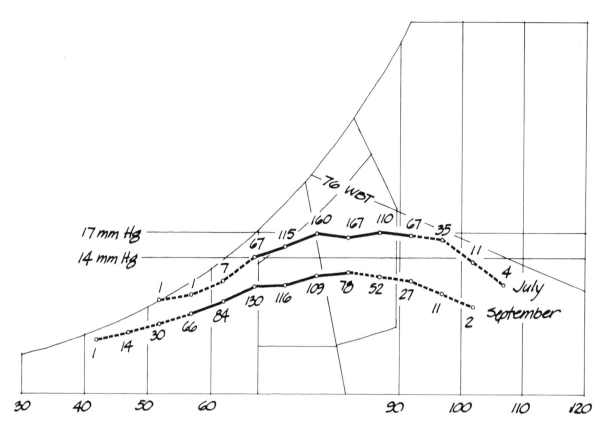

Figure 6. Mean coincident wet-bulb temperature associated with 5F dry-bulb temperature increments are plotted from bin-hour data for July and September in Kansas City. Solid segments represent 90% or more of all monthly observations; actual number of hours persisting in each 5F dry-bulb temperature range is given.

Table 4   Synopsis of Grogger's description of the interaction and importance of microclimatic elements with specific non-climatic elements.

Climatic Elements	Latitude	Elevation	Relief	Topography	Distance from water	Vegetation	Human Structures	Total
Cloud Cover	8	5	9	2	6	3	6	39
Sunshine	8	4	6	6	5	10	8	47
Solar Radiation	9	5	6	8	5	10	8	51
Airflow	5	5	8	7	7	7	7	46
Temperature	11	6	5	6	8	9	9	54
Rel Humidity	7	2	5	7	8	9	9	47
Evapotrans.	8	4	1	5	5	7	10	40
Albedo	2	1	2	5	4	6	7	27
Precipitation	7	4	8	1	9	5	4	38
Total	65	36	50	47	57	66	68	

Note: Individual categories are rated on a scale of 0 (not important) to 12 (very important).
Source: P.K. Grogger [1979].

Table 5   Heating Zones 1 + 2 : > 5 mph Wind Direction in Annual Hours

Location	Daylight Hours (7-18)				Nighttime Hours (19-6)			
	N	E	S	W	N	E	S	W
Albuquerque	281	131	131	184	622	351	167	167
Atlanta	167	184	44	342	254	237	79	596
Boston	482	281	210	631	675	219	412	973
Chicago	377	219	342	622	508	272	508	815
Dallas	281	96	88	96	394	175	149	114
Denver	421	219	412	202	386	263	1220	316
Houston	175	35	0	9	298	53	9	18
Indianapolis	272	263	289	570	394	386	421	736
Jackson	245	79	35	96	307	70	70	105
Kansas City	386	175	263	368	517	219	421	430
Little Rock	184	167	88	219	403	175	167	333
Los Angeles	18	18	0	0	105	61	9	9
Medford	377	35	79	140	587	105	263	131
Miami	18	0	0	0	44	0	0	9
Midland	298	131	149	114	368	228	281	210
Minneapolis	447	351	224	587	587	473	403	728
Nashville	263	61	131	298	351	61	228	368
New Orleans	158	79	0	35	272	79	26	88
New York City	456	149	281	535	666	175	386	815
Oklahoma City	508	105	316	140	675	193	552	193
Phoenix	35	61	0	9	88	202	17	26
Raleigh	315	79	105	237	456	175	237	272
Salt Lake City	333	167	430	254	386	552	430	254
San Antonio	237	35	17	35	386	35	26	88
San Francisco	44	44	44	289	70	140	88	245
Seattle	438	272	666	105	710	614	929	61
Tucson	17	96	70	44	17	342	272	79
Washington, DC	298	105	228	394	456	140	333	552
Windsor Locks *	508	61	272	543	640	53	482	789

* Windsor Locks, CT is listed in the climatic data as Hartford table 17.

Limitations.

The usefulness of any analysis technique is limited by the quality of weather data available. In the U.S., detailed hourly weather data are recorded mostly only at airports; the data, therefore, truly describe only the microclimate of the airport. This is problematic because many major metropolitan airports are located immediately adjacent to large bodies of water which significantly influence temperatures, humidity levels, and wind speeds and directions, compared to sites only a few miles inland. The analyzed data presented here and in handbooks such as ASHRAE *Fundamentals* all share this deficiency; they should not be taken as representative of a region unless the microclimate of the airport itself can be regarded as typical of the region. For any given building site, the influences of topography, difference in altitude, and proximity to bodies of water and other physiographic features must be considered.

Temperatures are measured in shade at height of 5 feet above the ground in open terrain over a grass-covered surface. Because air temperature is closely related to the temperature of the surface over which it is found (under still conditions), air temperatures in forests and in urban areas differ significantly from those measured at airports. Wind speed and direction also is measured over open terrain at airports, and these are observed at a height of 30 feet. Local winds are much influenced by physiographic features and wind speed varies with height, so the wind data presented in *Tables 5-8* should be regarded as a general guide. The influence of various microclimatic elements has been summarized by Grogger in *Table 4*. These relationships should be considered when interpreting any "regional" data for design purposes.

Table 6   **Comfort Zone (7) : > 5 mph Wind Direction In Annual Hours**

Location	Daylight Hours (7-18)				Nighttime Hours (19-6)			
	N	E	S	W	N	E	S	W
Albuquerque	61	79	131	70	70	210	114	26
Atlanta	53	61	44	96	105	158	131	167
Boston	88	175	184	193	35	0	114	105
Chicago	149	140	210	149	53	26	149	44
Dallas	123	88	263	44	70	149	219	9
Denver	131	96	140	61	26	26	149	26
Houston	114	131	79	44	61	70	26	9
Indianapolis	114	96	175	140	53	35	105	35
Jackson	96	70	202	70	44	35	123	9
Kansas City	149	123	245	70	79	96	219	18
Little Rock	105	131	184	79	61	35	131	18
Los Angeles	61	26	114	815	18	9	18	53
Medford	166	9	26	237	88	9	26	35
Miami	210	351	96	70	149	394	61	35
Midland	105	149	307	61	96	219	289	26
Minneapolis	123	114	123	158	35	53	96	26
Nashville	123	44	210	79	35	9	123	18
New Orleans	167	123	210	61	61	44	44	9
New York City	158	88	316	123	61	9	96	70
Oklahoma City	96	61	333	35	53	105	289	9
Phoenix	53	131	35	61	61	96	18	35
Raleigh	167	79	158	105	44	26	105	9
Salt Lake City	131	88	167	79	44	105	184	9
San Antonio	123	105	210	35	88	140	114	9
San Francisco	18	9	18	254	61	9	26	79
Seattle	202	9	79	140	35	0	0	0
Tucson	18	88	96	44	9	158	140	44
Washington, DC	149	105	210	131	61	35	149	61
Windsor Locks *	105	44	228	158	9	0	70	35

* Windsor Locks, CT is listed in the climatic data as Hartford table 17.

Table 7   **Ventilation Zones 9 + 11 : > 5 mph Wind Direction in Annual Hours**

Location	Daylight Hours (7-18)				Nighttime Hours (19-6)			
	N	E	S	W	N	E	S	W
Albuquerque	35	61	123	105	9	26	18	0
Atlanta	202	272	44	245	35	26	44	79
Boston	35	61	202	167	0	0	35	53
Chicago	88	105	324	193	9	9	96	18
Dallas	202	281	903	79	105	342	684	0
Denver	70	53	53	35	0	0	0	0
Houston	579	763	771	465	175	219	298	131
Indianapolis	79	70	377	219	9	9	70	18
Jackson	210	210	473	298	35	53	140	27
Kansas City	193	184	535	88	61	96	342	9
Little Rock	175	368	649	184	26	61	228	35
Los Angeles	0	0	9	105	0	0	0	0
Medford	88	0	9	202	18	0	0	9
Miami	561	2040	982	228	175	622	123	35
Midland	96	219	430	35	61	175	228	9
Minneapolis	61	105	272	105	9	18	61	0
Nashville	202	96	430	324	18	9	105	26
New Orleans	666	482	877	237	61	53	114	18
New York City	88	53	351	96	9	0	35	35
Oklahoma City	105	175	842	44	26	175	394	0
Phoenix	70	175	35	61	88	149	26	35
Raleigh	219	149	456	237	18	9	70	0
Salt Lake City	88	44	123	44	9	9	18	0
San Antonio	123	447	815	35	44	526	377	9
San Francisco	0	0	0	26	9	0	0	9
Seattle	61	0	0	26	0	0	0	0
Tucson	53	96	114	96	18	114	140	61
Washington, DC	123	123	473	184	9	18	149	35
Windsor Locks *	53	18	280	123	0	0	18	9

* Windsor Locks, CT is listed in the climatic data as Hartford table 17.

Table 8

Location	Daylight Hours (7-18)				Nighttime Hours (19-6)			
	N	E	S	W	N	E	S	W
Albuquerque	0	0	0	0	0	0	0	0
Atlanta	44	79	53	70	70	114	88	96
Boston	18	26	44	18	18	9	88	35
Chicago	18	9	44	26	18	18	61	18
Dallas	35	44	88	9	44	79	131	0
Denver	0	0	0	0	0	0	0	0
Houston	123	123	175	35	175	140	254	44
Indianapolis	18	18	79	26	26	26	96	35
Jackson	18	44	105	18	26	70	167	9
Kansas City	26	26	35	9	26	26	44	9
Little Rock	35	44	70	18	53	61	114	35
Los Angeles	0	0	0	18	9	9	9	18
Medford	0	0	0	0	0	0	0	0
Miami	79	53	53	26	202	175	149	70
Midland	26	35	35	0	26	35	26	0
Minneapolis	18	26	26	9	9	35	35	9
Nashville	18	18	70	26	26	26	105	26
New Orleans	61	70	114	26	140	140	254	44
New York City	35	44	114	18	53	35	158	35
Oklahoma City	35	44	149	9	35	96	184	9
Phoenix	0	0	0	0	0	9	0	0
Raleigh	70	44	96	18	96	79	193	18
Salt Lake City	0	0	0	0	0	0	0	0
San Antonio	61	88	114	9	61	158	140	9
San Francisco	0	0	0	0	0	0	0	0
Seattle	0	0	0	0	0	0	0	0
Tucson	0	9	9	0	0	18	18	0
Washington, DC	35	35	88	18	44	35	131	26
Windsor Locks *	18	9	79	9	9	9	105	18

* Windsor Locks, CT is listed in the climatic data as Hartford table 17.

Location and Site	Record	Latitude	Longitude	Elevation*
Albuquerque, NM (Municipal AP)	1949 - 1964	35°03'	106°37'	5310'
Atlanta, GA (Municipal AP)	1948 - 1964	33°39'	84°25'	975'
Boston, MA (Logan International AP)	1948 - 1964	42°22'	71°01'	15'
Chicago, IL (Midway International AP)	1948 - 1964	41°47'	87°45'	607'
Dallas, TX (Love Field)	1949 - 1964	32°51'	96°51'	481'
Denver, CO (Stapleton International AP)	1948 - 1964	39°46'	104°53'	5292'
Houston, TX (Houston International AP)	1957 - 1964	29°39'	95°17'	50'
Indianapolis, IN (Weir Cook Municipal AP)	1948 - 1964	39°44'	86°16'	792'
Jackson, MS (Municipal AP)	1949 - 1964	32°20'	90°13'	305'
Kansas City, MO (Municipal "Downtown" AP)	1948 - 1964	39°07'	94°35'	741'
Little Rock, AR (Adams Field)	1949 - 1964	34°44'	92°14'	257'
Los Angeles, CA (LA International AP)	1948 - 1964	33°56'	118°23'	97'
Medford, OR (Medford-Jackson Co AP)	1948 - 1964	42°22'	122°52'	1312'
Miami, FL (Miami International AP)	1948 - 1964	25°49'	80°17'	8'
Midland, TX (Midland-Odessa Reg'l. AP)	1949 - 1964	31°56'	102°12'	2854'
Minneapolis, MN (Minn. - St. Paul Int'l. AP)	1948 - 1964	44°53'	93°13'	830'
Nashville, TN (Barry Field)	1948 - 1964	36°07'	86°41'	577'
New Orleans, LA (Moisant Field)	1949 - 1964	29°59'	90°15'	4'
New York City, NY (La Guardia Field)	1949 - 1964	40°46'	73°52'	19'
Oklahoma City, OK (Will Rogers World AP)	1945 - 1964	35°24'	97°36'	1280'
Phoenix, AZ (Sky Harbor Int'l. AP)	1948 - 1964	33°26'	112°01'	1108'
Raleigh, NC (Raleigh - Durham AP)	1949 - 1964	35°52'	78°47'	438'
Salt Lake City, UT (SLC International AP)	1948 - 1964	40°46'	111°58'	4220'
San Antonio, TX (SA International AP)	1949 - 1964	29°32'	98°28'	792'
San Francisco, CA (SF International AP)	1948 - 1964	37°37'	122°23'	8'
Seattle, WA (Seattle - Tacoma AP)	1948 - 1964	47°27'	122°18'	379'
Tucson, AZ (Tucson International AP)	1949 - 1964	32°08'	110°57'	2558'
Washington, DC (National AP)	1948 - 1964	38°51'	77°03'	14'
Windsor Locks, CT (Bradley Int'l. AP)	1948 - 1957	41°56'	72°41'	169'

*Instruments have frequently been moved from place to place at an airport site; specific data given is for longest duration of siting during record years.

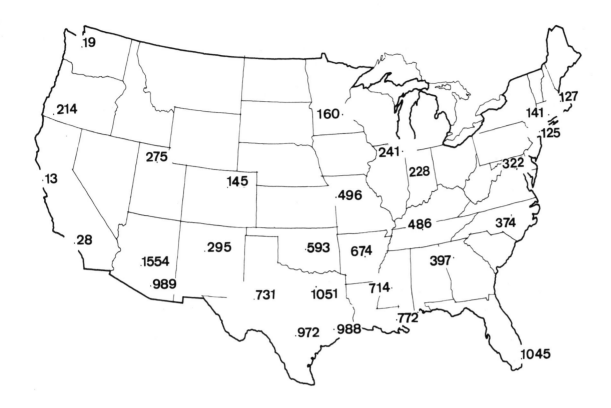

Figure 7. Annual cooling degree days, base 78ET*

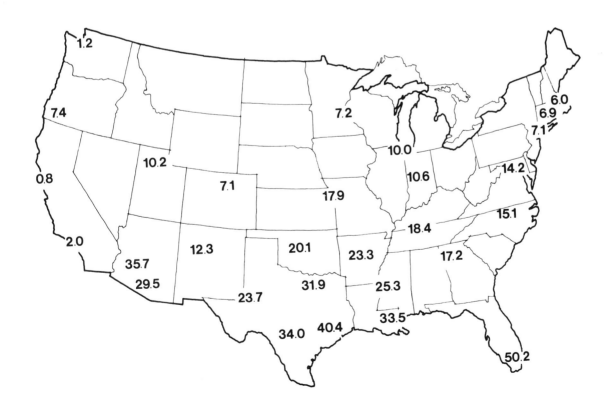

Figure 8. Overheated period as percentage of annual hours when conditions exceed 78ET*.

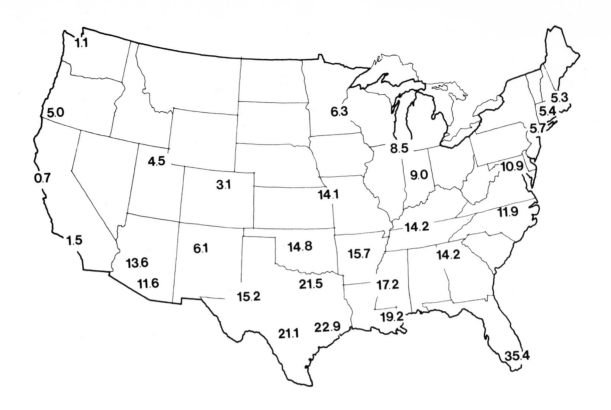

Figure 9. Percentage of annual hours during which ventilation is an effective climate control strategy.

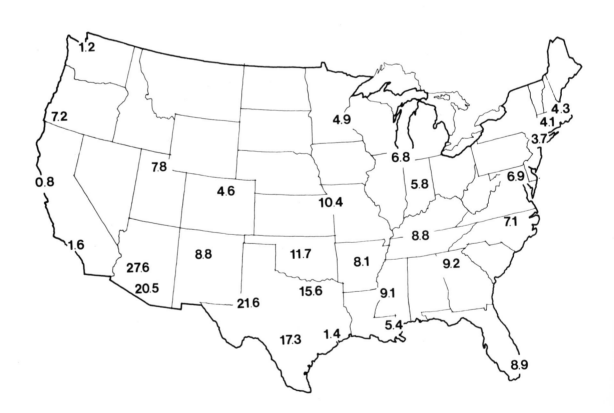

Figure 10. Percentage of annual hours during which massive building construction is an effective climate control strategy.

Figure 11. Percentage of annual hours during which evaporative space cooling is an effective climate control strategy.

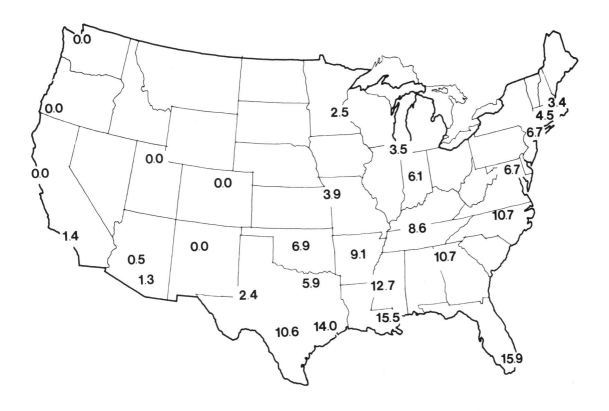

Figure 12. Percentage of annual hours during which dehumidification alone is sufficient to produce comfort.

Key To Bioclimatic Data Tables

1 Identification of region.

2
3 Mean daily maximum, average, and minimum dry-bulb temperature (DBT), dew point temperature
4 (DPT), and wet-bulb temperature (WBT), computed from long term records (usually 16 or 17 years). Each of these is determined independently of the others, so no necessary correlation exists between them. For instance, in contrast to dry-bulb temperature, dew point temperature peaks twice in a 24-hour period. The first usually occurs 3-4 hours after sunrise, and a second, lesser peak occurs 3-4 hours after sunset.

5 Solar radiation data given as the monthly average value for a single day, in Btu/sq ft(day). Data given here are reproduced from J.D. Balcomb, editor *Passive Solar Design Handbook*.

6 Annual heating and cooling degree days, computed for this study for base 65F. Also computed are cooling degree days for base 78ET*, which is the upper limit of the ASHRAE 55-74 comfort zone for light office clothing (.6 clo). This is used as an index of overheatedness.

7
8 Winter and summer design temperatures reproduced from the ASHRAE *1977 Fundamentals*
9 *Handbook*. These values are used principally for sizing mechanical systems, but are of more general interest as an indicator of the severity of normal winter and summer conditions; they represent extreme conditions which prevail for a significant number of seasonal hours. Winter design dry-bulb temperatures represent values which are equalled or exceeded 99% and 97.5% of total hours in the months of December, January, and February (total of 2160 hours). In a normal winter, there would be approximately 22 hours at or below the 99% value, and approximately 54 hours at or below the 97.5% design value. Summer design DBT are given with their corresponding coincident wet-bulb temperatures. The DBT represent values which are equalled or exceeded by 1%, 2.5%, and 5% of total hours during the summer months of June through September (total of 2928 hours). The coincident WBT listed with each design DBT is the mean of all WBT occurring at the specific design DBT. Summer design WBT represent values which are equalled or exceeded by 1%, 2.5%, and 5% of summer hours. These design WBT were selected independently of the design DBT values and should not be considered as coincident with the summer design DBT.

10 Building bioclimatic analysis. Data are presented as the % annual hours (1% annual hr = 87.66 hr) which fall within each designated zone or combination of zones delineated on the psychrometric chart. The outside column describes the fraction of a year that each climate control strategy is effective according to limits described by Givoni. Individual zones comprising each strategy are identified by Roman numerals. The inner column gives % annual hours that each zone of the chart is occupied. This allows the overlap between strategies to be identified.

Table 10

1 ALBUQUERQUE, NM   TEMPERATURE (F) ON 21ST DAY OF:

	J	F	M	A	M	J	J	A	S	O	N	D
2 DAILY MAX DBT	46.9	50.4	58.9	74.2	81.0	92.5	91.4	86.8	80.4	69.0	52.0	46.5
DAILY AVE DBT	34.7	38.6	47.0	60.0	66.5	78.2	77.8	73.7	67.9	55.9	41.2	34.7
DAILY MIN DBT	24.6	28.0	34.9	45.1	53.1	64.3	65.9	62.7	56.1	43.6	31.2	25.6
3 DAILY MAX DPT	22.2	23.6	23.7	30.3	38.3	44.1	55.4	55.9	45.1	33.6	27.8	22.9
DAILY AVE DPT	17.4	18.1	17.1	21.6	31.8	35.2	49.4	49.8	38.6	28.7	22.9	19.1
DAILY MIN DPT	11.8	12.7	10.1	14.1	23.7	24.8	42.3	43.3	32.0	22.9	18.5	14.2
4 DAILY MAX WBT	35.6	37.6	40.9	49.1	53.9	60.0	63.8	62.8	57.0	49.4	39.6	35.8
DAILY AVE WBT	28.5	31.1	35.4	43.1	49.0	55.4	60.5	59.4	52.3	43.5	33.7	29.1
DAILY MIN WBT	21.5	24.6	29.2	35.9	43.4	50.3	56.6	56.1	46.8	37.1	27.6	22.8

NORMAL DAILY SOLAR RADIATION (MONTHLY AVG) BTU/SQ FT(DAY)

	J	F	M	A	M	J	J	A	S	O	N	D
5 HORIZONTAL	1016.	1342.	1768.	2228.	2538.	2679.	2489.	2290.	1972.	1547.	1134.	928.
SO VERTICAL	1530.	1623.	1586.	1379.	1196.	1137.	1098.	1241.	1523.	1687.	1608.	1477.

6

HEATING DEGREE DAYS BASE 65F	4578.2
COOLING DEGREE DAYS BASE 65F	1575.8
COOLING DEGREE DAYS BASE 78F ET*	294.6

7

WINTER DESIGN DBT	99.0%	12.0
	97.5%	16.0

8

SUMMER DESIGN DBT / COINCIDENT WBT	1%	96.0 / 61.0
	2.5%	94.0 / 61.0
	5%	92.0 / 61.0

9

SUMMER DESIGN WBT	1%	66.0
	2.5%	65.0
	5%	64.0

10

% TOTAL HEATING HOURS (LESS THAN 68F) I-V		68.1
HEATING I	29.7	
HEATING II	8.8	
HEATING III	7.1	
HEATING IV	8.2	
HEATING V	14.3	
% HUMIDIFICATION HOURS VI.A + VI.B		7.0
% TOTAL COMFORT HOURS (SHADING REQUIRED) VII		12.5
% DEHUMIDIFICATION HOURS VIII		0.0
% TOTAL COOLING HOURS (GREATER THAN 78F ET*) IX - XVII		12.3
% PASSIVE COOLING HOURS IX - XIV		12.3
COOLING IX	0.0	
COOLING X	0.0	
COOLING XI	6.1	
COOLING XII	0.0	
COOLING XIII	2.7	
COOLING XIV	3.5	
% VENTILATION EFFECTIVENESS HOURS IX + X + XI		6.1
% MASS EFFECTIVENESS HOURS X + XI + XII + XIII		8.8
% EVAPORATIVE COOLING EFFECTIVENESS HOURS XI + XIII + XIV + VI.B		14.0
% HOURS BEYOND PASSIVE EFFECTIVENESS VIII + XV + XVI + XVII		0.0
DEHUMIDIFICATION VIII	0.0	
DEHUMIDIFICATION AND COOLING XV	0.0	
DEHUMIDIFICATION AND COOLING XVI	0.0	
COOLING XVII	0.0	

Table 11

221

```
ATLANTA, GA                    TEMPERATURE (F) ON 21ST DAY OF:

                J     F     M     A     M     J     J     A     S     O     N     D

DAILY MAX DBT  50.4  55.4  59.2  72.9  81.1  87.1  87.8  85.1  81.8  69.6  58.7  52.3
DAILY AVE DBT  42.8  45.7  50.8  63.0  71.8  77.0  78.4  75.5  72.2  59.3  49.0  42.3
DAILY MIN DBT  35.1  36.0  42.9  53.0  62.5  69.1  71.2  67.9  63.0  50.4  40.5  33.9

DAILY MAX DPT  41.4  40.1  45.3  54.0  62.4  70.4  72.1  68.9  64.8  50.2  42.7  40.0
DAILY AVE DPT  33.8  31.8  38.9  49.1  59.0  67.1  69.1  65.6  60.1  46.3  36.6  32.9
DAILY MIN DPT  27.6  24.6  31.6  44.2  55.4  63.5  66.1  61.9  54.6  41.1  30.2  25.9

DAILY MAX WBT  45.1  46.9  50.4  60.5  67.3  74.2  74.8  72.2  68.4  57.6  49.7  45.9
DAILY AVE WBT  39.2  40.1  45.5  55.6  63.8  70.4  71.9  68.8  64.5  52.8  43.6  38.7
DAILY MIN WBT  33.2  33.1  39.9  49.9  59.5  67.2  69.3  66.0  58.9  47.2  37.4  31.9
```

```
              NORMAL DAILY SOLAR RADIATION (MONTHLY AVG) BTU/SQ FT(DAY)

                J     F     M     A     M     J     J     A     S     O     N     D

HORIZONTAL    718.  969. 1304. 1686. 1854. 1914. 1812. 1708. 1422. 1200.  883.  674.
SO VERTICAL  1041. 1127. 1122. 1000.  843.  791.  775.  889. 1052. 1258. 1205. 1032.
```

```
HEATING DEGREE DAYS BASE 65F                      3126.9
COOLING DEGREE DAYS BASE 65F                      1785.7
COOLING DEGREE DAYS BASE 78F ET*                   397.2
```

```
WINTER DESIGN DBT                          99.0%  17.0
                                           97.5%  22.0
```

```
SUMMER DESIGN DBT / COINCIDENT WBT           1%   94.0 / 74.0
                                           2.5%   92.0 / 74.0
                                             5%   90.0 / 73.0
```

```
SUMMER DESIGN WBT                            1%   77.0
                                           2.5%   76.0
                                             5%   75.0
```

```
% TOTAL HEATING HOURS (LESS THAN 68F)  I-V                           59.3

     HEATING I                                     18.0
     HEATING II                                     8.2
     HEATING III                                    7.5
     HEATING IV                                     9.7
     HEATING V                                     15.9
```

```
% HUMIDIFICATION HOURS  VI.A + VI.B                                   0.4
```

```
% TOTAL COMFORT HOURS (SHADING REQUIRED)  VII                       12.5
```

```
% DEHUMIDIFICATION HOURS  VIII                                      10.7
```

```
% TOTAL COOLING HOURS (GREATER THAN 78F ET*) IX - XVII              17.2
```

```
% PASSIVE COOLING HOURS IX - XIV                                    15.2

     COOLING IX                                     6.0
     COOLING X                                      1.6
     COOLING XI                                     6.6
     COOLING XII                                    0.7
     COOLING XIII                                   0.3
     COOLING XIV                                    0.0
```

```
% VENTILATION EFFECTIVENESS HOURS  IX + X + XI                      14.2
```

```
% MASS EFFECTIVENESS HOURS  X + XI + XII + XIII                      9.2
```

```
% EVAPORATIVE COOLING EFFECTIVENESS HOURS  XI + XIII + XIV + VI.B    7.1
```

```
% HOURS BEYOND PASSIVE EFFECTIVENESS  VIII + XV + XVI + XVII        12.7

     DEHUMIDIFICATION  VIII                        10.7
     DEHUMIDIFICATION AND COOLING   XV              1.5
     DEHUMIDIFICATION AND COOLING   XVI             0.5
     COOLING XVII                                   0.0
```

Table 12

```
BOSTON, MA                    TEMPERATURE (F) ON 21ST DAY OF:

                J     F     M     A     M     J     J     A     S     O     N     D

DAILY MAX DBT  38.8  37.5  45.4  64.6  66.6  76.5  80.3  79.2  67.8  57.3  52.4  36.9
DAILY AVE DBT  30.6  30.7  38.6  53.3  58.0  68.7  72.7  70.9  60.6  49.8  44.7  30.9
DAILY MIN DBT  22.2  23.1  31.4  44.1  50.3  60.7  65.4  63.4  53.6  42.5  37.5  23.8

DAILY MAX DPT  31.8  24.9  34.4  41.6  50.2  61.4  65.4  64.2  54.2  42.5  40.9  26.9
DAILY AVE DPT  19.5  17.7  27.6  35.2  45.1  56.4  61.7  59.2  49.9  37.4  33.8  19.8
DAILY MIN DPT   6.8   9.5  20.8  28.9  39.6  50.8  57.0  54.3  45.4  31.8  26.6  11.9

DAILY MAX WBT  36.3  32.4  39.6  50.4  56.1  65.5  68.8  67.1  58.7  48.4  46.0  33.6
DAILY AVE WBT  28.0  27.3  34.6  45.1  51.4  61.3  65.6  63.7  54.8  44.2  40.3  28.1
DAILY MIN WBT  19.9  20.7  29.1  39.9  46.8  56.4  61.7  59.2  50.2  38.9  34.3  21.7
```

```
              NORMAL DAILY SOLAR RADIATION (MONTHLY AVG) BTU/SQ FT(DAY)

                J     F     M     A     M     J     J     A     S     O     N     D

HORIZONTAL    475.  710. 1016. 1326. 1620. 1817. 1749. 1486. 1260.  890.  503.  403.
SO VERTICAL   878. 1052. 1127. 1030.  944.  927.  940. 1009. 1212. 1192.  874.  788.
```

```
HEATING DEGREE DAYS BASE 65F                    5844.9
COOLING DEGREE DAYS BASE 65F                     789.8
COOLING DEGREE DAYS BASE 78F ET*                 126.5
```

```
WINTER DESIGN DBT                       99.0%    6.0
                                        97.5%    9.0

SUMMER DESIGN DBT / COINCIDENT WBT        1%    91.0 / 73.0
                                        2.5%    88.0 / 71.0
                                          5%    85.0 / 70.0

SUMMER DESIGN WBT                         1%    75.0
                                        2.5%    74.0
                                          5%    72.0
```

```
% TOTAL HEATING HOURS (LESS THAN 68F)  I-V                        79.6

      HEATING I                                         37.9
      HEATING II                                         9.5
      HEATING III                                        8.0
      HEATING IV                                         9.7
      HEATING V                                         14.6

% HUMIDIFICATION HOURS  VI.A + VI.B                               0.2

% TOTAL COMFORT HOURS (SHADING REQUIRED)  VII                    10.8

% DEHUMIDIFICATION HOURS  VIII                                    3.4

% TOTAL COOLING HOURS (GREATER THAN 78F ET*)  IX - XVII           6.0

% PASSIVE COOLING HOURS IX - XIV                                  5.6

      COOLING IX                                         1.3
      COOLING X                                          0.4
      COOLING XI                                         3.6
      COOLING XII                                        0.2
      COOLING XIII                                       0.1
      COOLING XIV                                        0.0

% VENTILATION EFFECTIVENESS HOURS   IX + X + XI                   5.3

% MASS EFFECTIVENESS HOURS   X + XI + XII + XIII                  4.3

% EVAPORATIVE COOLING EFFECTIVENESS HOURS  XI + XIII + XIV + VI.B  3.9

% HOURS BEYOND PASSIVE EFFECTIVENESS  VIII + XV + XVI + XVII      3.7

      DEHUMIDIFICATION  VIII                             3.4
      DEHUMIDIFICATION AND COOLING  XV                   0.2
      DEHUMIDIFICATION AND COOLING  XVI                  0.2
      COOLING  XVII                                      0.0
```

Table 13

CHICAGO, IL   TEMPERATURE (F) ON 21ST DAY OF:

	J	F	M	A	M	J	J	A	S	O	N	D
DAILY MAX DBT	30.4	33.8	46.6	66.9	72.7	81.1	85.9	82.8	73.7	63.3	43.5	33.1
DAILY AVE DBT	23.7	27.6	39.1	56.4	63.2	71.6	76.5	73.0	64.1	54.0	36.7	26.8
DAILY MIN DBT	15.9	21.3	31.4	46.4	51.7	62.1	68.4	62.8	55.6	44.8	29.5	21.2
DAILY MAX DPT	24.2	23.5	34.9	47.5	55.3	62.1	68.8	63.6	58.8	47.1	33.4	27.0
DAILY AVE DPT	16.8	18.5	28.8	39.3	47.0	57.1	64.9	58.6	52.3	42.3	27.0	20.8
DAILY MIN DPT	9.5	11.5	22.8	32.0	38.5	52.2	60.2	53.4	46.5	36.8	21.4	14.6
DAILY MAX WBT	27.9	29.8	40.6	54.1	60.8	67.0	72.5	67.6	63.2	52.9	37.8	30.4
DAILY AVE WBT	22.1	25.1	35.2	48.1	54.5	62.7	68.8	63.9	57.5	48.0	33.1	25.1
DAILY MIN WBT	14.9	19.5	29.1	41.7	46.6	58.2	64.9	58.9	51.5	41.7	27.2	20.1

NORMAL DAILY SOLAR RADIATION (MONTHLY AVG) BTU/SQ FT(DAY)

	J	F	M	A	M	J	J	A	S	O	N	D
HORIZONTAL	507.	759.	1107.	1459.	1789.	2007.	1944.	1719.	1354.	969.	566.	401.
SO VERTICAL	921.	1106.	1207.	1113.	1023.	1006.	1026.	1146.	1280.	1276.	967.	771.

HEATING DEGREE DAYS BASE 65F	6306.5
COOLING DEGREE DAYS BASE 65F	1143.7
COOLING DEGREE DAYS BASE 78F ET*	240.5

WINTER DESIGN DBT	99.0%	-5.0
	97.5%	0.0

SUMMER DESIGN DBT / COINCIDENT WBT	1%	94.0 / 74.0
	2.5%	91.0 / 73.0
	5%	88.0 / 72.0

SUMMER DESIGN WBT	1%	77.0
	2.5%	75.0
	5%	74.0

% TOTAL HEATING HOURS (LESS THAN 68F) I-V	73.7

HEATING I	41.0
HEATING II	6.8
HEATING III	5.9
HEATING IV	7.4
HEATING V	12.6

% HUMIDIFICATION HOURS VI.A + VI.B	0.3

% TOTAL COMFORT HOURS (SHADING REQUIRED) VII	12.5

% DEHUMIDIFICATION HOURS VIII	3.5

% TOTAL COOLING HOURS (GREATER THAN 78F ET*) IX - XVII	10.0

% PASSIVE COOLING HOURS IX - XIV	9.2

COOLING IX	2.5
COOLING X	0.8
COOLING XI	5.3
COOLING XII	0.4
COOLING XIII	0.3
COOLING XIV	0.0

% VENTILATION EFFECTIVENESS HOURS IX + X + XI	8.5

% MASS EFFECTIVENESS HOURS X + XI + XII + XIII	6.8

% EVAPORATIVE COOLING EFFECTIVENESS HOURS XI + XIII + XIV + VI.B	5.7

% HOURS BEYOND PASSIVE EFFECTIVENESS VIII + XV + XVI + XVII	4.3

DEHUMIDIFICATION VIII	3.5
DEHUMIDIFICATION AND COOLING XV	0.4
DEHUMIDIFICATION AND COOLING XVI	0.4
COOLING XVII	0.0

Table 14

DALLAS, TX   TEMPERATURE (F) ON 21ST DAY OF:

	J	F	M	A	M	J	J	A	S	O	N	D
DAILY MAX DBT	53.3	55.9	67.1	77.1	86.4	93.4	95.7	92.7	88.4	77.1	62.3	54.5
DAILY AVE DBT	42.8	47.1	55.5	68.0	76.9	83.5	86.5	82.5	78.1	66.8	52.5	44.9
DAILY MIN DBT	33.5	38.5	44.6	59.8	67.4	73.0	78.0	72.7	68.7	56.1	43.0	35.8
DAILY MAX DPT	40.8	39.4	48.6	61.8	65.9	70.6	72.1	69.8	68.6	57.4	45.2	40.5
DAILY AVE DPT	33.5	32.7	39.6	57.5	61.5	67.4	68.5	65.9	63.2	51.8	38.9	34.6
DAILY MIN DPT	27.2	26.4	32.1	52.2	57.3	63.4	64.4	61.4	57.8	45.2	31.6	28.5
DAILY MAX WBT	46.6	46.9	55.1	66.2	70.5	75.1	76.6	74.4	73.3	63.7	52.2	47.1
DAILY AVE WBT	39.1	41.0	47.8	61.7	67.1	72.6	74.0	71.3	68.5	58.4	46.3	40.6
DAILY MIN WBT	32.3	34.8	39.8	56.4	62.5	69.2	71.5	68.3	63.4	51.8	39.6	33.4

NORMAL DAILY SOLAR RADIATION (MONTHLY AVG) BTU/SQ FT(DAY)

	J	F	M	A	M	J	J	A	S	O	N	D
HORIZONTAL	821.	1071.	1422.	1627.	1888.	2135.	2122.	1950.	1587.	1276.	936.	780.
SO VERTICAL	1164.	1218.	1195.	942.	843.	870.	893.	992.	1146.	1308.	1250.	1168.

HEATING DEGREE DAYS BASE 65F	2567.0
COOLING DEGREE DAYS BASE 65F	2967.1
COOLING DEGREE DAYS BASE 78F ET*	1051.1

WINTER DESIGN DBT	99.0%	18.0
	97.5%	22.0

SUMMER DESIGN DBT / COINCIDENT WBT	1%	102.0 / 75.0
	2.5%	100.0 / 75.0
	5%	97.0 / 75.0

SUMMER DESIGN WBT	1%	78.0
	2.5%	78.0
	5%	77.0

% TOTAL HEATING HOURS (LESS THAN 68F) I-V		48.8
HEATING I	14.3	
HEATING II	7.0	
HEATING III	6.5	
HEATING IV	8.0	
HEATING V	13.0	
% HUMIDIFICATION HOURS VI.A + VI.B		0.7
% TOTAL COMFORT HOURS (SHADING REQUIRED) VII		12.7
% DEHUMIDIFICATION HOURS VIII		5.9
% TOTAL COOLING HOURS (GREATER THAN 78F ET*) IX - XVII		31.9
% PASSIVE COOLING HOURS IX - XIV		26.7
COOLING IX	11.1	
COOLING X	2.6	
COOLING XI	7.8	
COOLING XII	4.2	
COOLING XIII	1.0	
COOLING XIV	0.1	
% VENTILATION EFFECTIVENESS HOURS IX + X + XI		21.5
% MASS EFFECTIVENESS HOURS X + XI + XII + XIII		15.6
% EVAPORATIVE COOLING EFFECTIVENESS HOURS XI + XIII + XIV + VI.B		9.2
% HOURS BEYOND PASSIVE EFFECTIVENESS VIII + XV + XVI + XVII		11.1
DEHUMIDIFICATION VIII	5.9	
DEHUMIDIFICATION AND COOLING XV	2.2	
DEHUMIDIFICATION AND COOLING XVI	2.9	
COOLING XVII	0.1	

Table 15

DENVER, CO   TEMPERATURE (F) ON 21ST DAY OF:

	J	F	M	A	M	J	J	A	S	O	N	D
DAILY MAX DBT	40.7	43.1	52.1	65.0	71.2	81.3	85.5	82.2	69.2	65.5	48.9	45.6
DAILY AVE DBT	27.8	31.1	39.0	51.6	59.5	68.1	72.6	69.3	56.6	50.9	36.5	33.7
DAILY MIN DBT	15.2	20.8	26.2	38.8	47.6	54.8	59.2	57.3	44.9	37.8	25.6	23.4
DAILY MAX DPT	17.1	22.0	24.7	32.2	43.5	51.8	50.5	51.8	42.9	31.1	24.9	21.5
DAILY AVE DPT	8.9	15.3	18.1	25.1	36.4	44.7	44.5	44.9	37.8	24.9	18.5	14.2
DAILY MIN DPT	0.7	8.6	10.6	16.9	28.1	36.4	36.9	36.2	31.5	15.7	11.2	7.4
DAILY MAX WBT	30.6	33.2	37.3	45.9	53.1	59.3	60.2	59.4	52.2	45.8	36.8	33.9
DAILY AVE WBT	21.9	25.8	31.0	39.8	47.7	54.7	56.3	55.6	46.9	39.7	29.7	27.1
DAILY MIN WBT	12.6	18.5	23.2	33.5	41.5	49.4	51.2	51.4	40.8	33.0	22.5	19.8

NORMAL DAILY SOLAR RADIATION (MONTHLY AVG) BTU/SQ FT(DAY)

	J	F	M	A	M	J	J	A	S	O	N	D
HORIZONTAL	840.	1127.	1530.	1879.	2135.	2351.	2273.	2044.	1727.	1300.	883.	732.
SO VERTICAL	1440.	1551.	1572.	1344.	1147.	1114.	1130.	1277.	1535.	1616.	1424.	1327.

HEATING DEGREE DAYS BASE 65F	6416.6
COOLING DEGREE DAYS BASE 65F	967.1
COOLING DEGREE DAYS BASE 78F ET*	144.9

WINTER DESIGN DBT	99.0%	-5.0
	97.5%	1.0

SUMMER DESIGN DBT / COINCIDENT WBT	1%	93.0 / 59.0
	2.5%	91.0 / 59.0
	5%	89.0 / 59.0

SUMMER DESIGN WBT	1%	64.0
	2.5%	63.0
	5%	62.0

% TOTAL HEATING HOURS (LESS THAN 68F) I-V	79.1
HEATING I	40.2
HEATING II	8.7
HEATING III	7.1
HEATING IV	9.5
HEATING V	13.5
% HUMIDIFICATION HOURS VI.A + VI.B	4.5
% TOTAL COMFORT HOURS (SHADING REQUIRED) VII	9.3
% DEHUMIDIFICATION HOURS VIII	0.0
% TOTAL COOLING HOURS (GREATER THAN 78F ET*) IX - XVII	7.1
% PASSIVE COOLING HOURS IX - XIV	7.1
COOLING IX	0.0
COOLING X	0.0
COOLING XI	3.1
COOLING XII	0.0
COOLING XIII	1.5
COOLING XIV	2.5
% VENTILATION EFFECTIVENESS HOURS IX + X + XI	3.1
% MASS EFFECTIVENESS HOURS X + XI + XII + XIII	4.6
% EVAPORATIVE COOLING EFFECTIVENESS HOURS XI + XIII + XIV + VI.B	8.2
% HOURS BEYOND PASSIVE EFFECTIVENESS VIII + XV + XVI + XVII	0.0
DEHUMIDIFICATION VIII	0.0
DEHUMIDIFICATION AND COOLING XV	0.0
DEHUMIDIFICATION AND COOLING XVI	0.0
COOLING XVII	0.0

Table 16

WINDSOR LOCKS, CT   TEMPERATURE (F) ON 21ST DAY OF:

	J	F	M	A	M	J	J	A	S	O	N	D
DAILY MAX DBT	34.2	36.2	43.7	67.0	74.6	74.4	84.3	82.3	66.7	61.7	49.2	28.2
DAILY AVE DBT	24.4	26.7	34.5	53.7	62.0	66.1	73.0	71.8	58.2	49.0	41.0	22.1
DAILY MIN DBT	16.4	17.0	26.7	40.7	48.8	57.3	62.0	62.6	49.9	38.2	31.9	14.9
DAILY MAX DPT	25.9	22.1	30.4	42.9	54.4	61.7	68.0	67.9	54.3	43.4	38.0	22.0
DAILY AVE DPT	13.5	13.0	24.1	35.5	47.3	56.5	64.0	63.8	50.1	38.0	32.7	12.0
DAILY MIN DPT	7.1	5.2	18.1	28.3	39.6	52.1	59.1	59.1	45.8	31.9	26.4	4.3
DAILY MAX WBT	31.4	30.4	36.8	51.8	60.4	65.3	71.7	70.9	58.7	50.9	43.0	26.6
DAILY AVE WBT	22.5	23.2	31.0	45.4	54.1	60.4	67.1	66.5	54.0	43.9	37.6	20.1
DAILY MIN WBT	16.2	15.0	24.9	37.0	46.0	55.6	61.7	60.9	48.3	36.2	30.4	13.4

NORMAL DAILY SOLAR RADIATION (MONTHLY AVG) BTU/SQ FT(DAY)

	J	F	M	A	M	J	J	A	S	O	N	D
HORIZONTAL	477.	715.	978.	1315.	1568.	1686.	1649.	1422.	1154.	853.	497.	385.
SO VERTICAL	869.	1045.	1070.	1006.	900.	848.	873.	951.	1094.	1127.	852.	742.

HEATING DEGREE DAYS BASE 65F	6727.1
COOLING DEGREE DAYS BASE 65F	816.7
COOLING DEGREE DAYS BASE 78F ET*	141.4

WINTER DESIGN DBT	99.0%	0.0
	97.5%	4.0

SUMMER DESIGN DBT / COINCIDENT WBT	1%	91.0 / 74.0
	2.5%	88.0 / 72.0
	5%	85.0 / 71.0

SUMMER DESIGN WBT	1%	76.0
	2.5%	75.0
	5%	73.0

% TOTAL HEATING HOURS (LESS THAN 68F) I-V		79.5
HEATING I	43.3	
HEATING II	6.7	
HEATING III	7.6	
HEATING IV	8.2	
HEATING V	13.7	
% HUMIDIFICATION HOURS VI.A + VI.B		0.2
% TOTAL COMFORT HOURS (SHADING REQUIRED) VII		8.9
% DEHUMIDIFICATION HOURS VIII		4.5
% TOTAL COOLING HOURS (GREATER THAN 78F ET*) IX - XVII		6.9
% PASSIVE COOLING HOURS IX - XIV		5.7
COOLING IX	1.3	
COOLING X	0.3	
COOLING XI	3.8	
COOLING XII	0.0	
COOLING XIII	0.0	
COOLING XIV	0.2	
% VENTILATION EFFECTIVENESS HOURS IX + X + XI		5.4
% MASS EFFECTIVENESS HOURS X + XI + XII + XIII		4.1
% EVAPORATIVE COOLING EFFECTIVENESS HOURS XI + XIII + XIV + VI.B		4.2
% HOURS BEYOND PASSIVE EFFECTIVENESS VIII + XV + XVI + XVII		5.7
DEHUMIDIFICATION VIII	4.5	
DEHUMIDIFICATION AND COOLING XV	0.3	
DEHUMIDIFICATION AND COOLING XVI	0.6	
COOLING XVII	0.3	

Table 17

HOUSTON, TX   TEMPERATURE (F) ON 21ST DAY OF:

	J	F	M	A	M	J	J	A	S	O	N	D
DAILY MAX DBT	60.3	63.5	65.8	76.9	83.0	87.1	88.6	87.3	85.5	75.1	70.3	62.9
DAILY AVE DBT	53.5	56.6	59.9	71.3	78.1	82.0	83.5	82.6	81.2	70.5	65.5	55.0
DAILY MIN DBT	46.0	50.0	55.9	65.9	73.6	77.8	79.3	78.5	77.0	66.3	61.3	49.8
DAILY MAX DPT	53.0	56.9	58.8	70.5	74.5	75.5	77.3	78.0	76.8	66.3	63.6	56.0
DAILY AVE DPT	46.5	52.5	52.2	67.6	72.3	73.6	75.8	75.0	73.6	61.7	59.5	50.9
DAILY MIN DPT	39.4	47.3	47.4	64.6	69.6	71.0	74.0	71.3	69.6	58.0	55.4	47.4
DAILY MAX WBT	55.6	59.0	60.8	72.1	76.3	77.9	79.9	79.5	78.6	68.6	65.6	58.0
DAILY AVE WBT	50.2	54.3	55.7	68.9	74.0	75.9	77.9	77.0	75.8	65.0	61.9	52.8
DAILY MIN WBT	43.6	48.6	52.1	65.4	71.9	73.9	76.0	74.6	72.9	62.8	58.3	49.3

NORMAL DAILY SOLAR RADIATION (MONTHLY AVG) BTU/SQ FT(DAY)

	J	F	M	A	M	J	J	A	S	O	N	D
HORIZONTAL	772.	1034.	1297.	1522.	1775.	1898.	1828.	1686.	1471.	1276.	924.	730.
SO VERTICAL	1014.	1087.	1002.	810.	746.	744.	734.	795.	975.	1206.	1142.	1013.

HEATING DEGREE DAYS BASE 65F	1163.2
COOLING DEGREE DAYS BASE 65F	3028.9
COOLING DEGREE DAYS BASE 78F ET*	987.7

WINTER DESIGN DBT	99.0%	27.0
	97.5%	32.0

SUMMER DESIGN DBT / COINCIDENT WBT	1%	96.0 / 77.0
	2.5%	94.0 / 77.0
	5%	92.0 / 77.0

SUMMER DESIGN WBT	1%	80.0
	2.5%	79.0
	5%	79.0

% TOTAL HEATING HOURS (LESS THAN 68F) I-V	39.0

HEATING I	3.0
HEATING II	4.3
HEATING III	5.4
HEATING IV	9.5
HEATING V	16.7

% HUMIDIFICATION HOURS VI.A + VI.B	0.0

% TOTAL COMFORT HOURS (SHADING REQUIRED) VII	6.6

% DEHUMIDIFICATION HOURS VIII	14.0

% TOTAL COOLING HOURS (GREATER THAN 78F ET*) IX - XVII	40.4

% PASSIVE COOLING HOURS IX - XIV	22.9

COOLING IX	21.5
COOLING X	0.2
COOLING XI	1.2
COOLING XII	0.0
COOLING XIII	0.0
COOLING XIV	0.0

% VENTILATION EFFECTIVENESS HOURS IX + X + XI	22.9

% MASS EFFECTIVENESS HOURS X + XI + XII + XIII	1.4

% EVAPORATIVE COOLING EFFECTIVENESS HOURS XI + XIII + XIV + VI.B	1.2

% HOURS BEYOND PASSIVE EFFECTIVENESS VIII + XV + XVI + XVII	31.5

DEHUMIDIFICATION VIII	14.0
DEHUMIDIFICATION AND COOLING XV	17.2
DEHUMIDIFICATION AND COOLING XVI	0.3
COOLING XVII	0.0

Table 18

```
INDIANAPOLIS, IN            TEMPERATURE (F) ON 21ST DAY OF:

               J     F     M     A     M     J     J     A     S     O     N     D

DAILY MAX DBT  36.5  36.2  48.8  68.7  76.1  82.5  84.4  79.8  76.2  65.8  46.0  35.2
DAILY AVE DBT  28.4  29.8  41.7  58.2  65.1  72.5  75.2  70.6  65.4  54.5  37.8  28.5
DAILY MIN DBT  20.1  23.0  33.6  47.6  53.4  62.6  66.6  60.6  55.1  43.5  30.0  22.0

DAILY MAX DPT  30.5  28.5  38.4  50.3  60.2  66.3  70.5  64.4  61.4  47.5  35.5  28.8
DAILY AVE DPT  21.9  22.3  32.6  44.3  53.2  61.0  67.1  60.4  55.4  42.8  29.9  22.8
DAILY MIN DPT  13.9  15.9  27.0  38.1  46.2  55.4  62.8  55.2  48.8  38.7  24.4  16.4

DAILY MAX WBT  33.8  32.6  42.9  56.8  64.6  70.3  73.9  68.2  65.2  54.5  40.8  32.9
DAILY AVE WBT  26.6  27.6  37.9  51.1  58.2  65.2  69.6  64.2  59.6  48.6  34.7  27.0
DAILY MIN WBT  19.1  21.6  31.8  43.9  50.1  59.4  64.6  58.6  52.4  41.5  28.5  21.4
```

```
             NORMAL DAILY SOLAR RADIATION (MONTHLY AVG) BTU/SQ FT(DAY)

               J     F     M     A     M     J     J     A     S     O     N     D

HORIZONTAL    496.  747. 1037. 1398. 1688. 1868. 1806. 1643. 1324.  977.  579.  417.
SO VERTICAL   850. 1028. 1066. 1000.  907.  885.  898. 1027. 1177. 1215.  934.  756.
```

```
HEATING DEGREE DAYS BASE 65F                 5797.7
COOLING DEGREE DAYS BASE 65F                 1155.0
COOLING DEGREE DAYS BASE 78F ET*              228.1
```

```
WINTER DESIGN DBT                     99.0%  -2.0
                                      97.5%   2.0
```

```
SUMMER DESIGN DBT / COINCIDENT WBT      1%   92.0 / 74.0
                                      2.5%   90.0 / 74.0
                                        5%   87.0 / 73.0
```

```
SUMMER DESIGN WBT                       1%   78.0
                                      2.5%   76.0
                                        5%   75.0
```

```
% TOTAL HEATING HOURS (LESS THAN 68F)  I-V                72.4

    HEATING I                              37.6
    HEATING II                              7.0
    HEATING III                             6.3
    HEATING IV                              7.8
    HEATING V                              13.6
```

```
% HUMIDIFICATION HOURS  VI.A + VI.B                        0.1
```

```
% TOTAL COMFORT HOURS (SHADING REQUIRED)  VII            10.8
```

```
% DEHUMIDIFICATION HOURS  VIII                             6.1
```

```
% TOTAL COOLING HOURS (GREATER THAN 78F ET*) IX - XVII   10.6
```

```
% PASSIVE COOLING HOURS IX - XIV                          9.3

    COOLING IX                              3.6
    COOLING X                               0.7
    COOLING XI                              4.7
    COOLING XII                             0.2
    COOLING XIII                            0.2
    COOLING XIV                             0.0
```

```
% VENTILATION EFFECTIVENESS HOURS  IX + X + XI            9.0
```

```
% MASS EFFECTIVENESS HOURS  X + XI + XII + XIII           5.8
```

```
% EVAPORATIVE COOLING EFFECTIVENESS HOURS  XI + XIII + XIV + VI.B   4.9
```

```
% HOURS BEYOND PASSIVE EFFECTIVENESS  VIII + XV + XVI + XVII        7.4

    DEHUMIDIFICATION  VIII                  6.1
    DEHUMIDIFICATION AND COOLING  XV        1.0
    DEHUMIDIFICATION AND COOLING  XVI       0.3
    COOLING  XVII                           0.0
```

Table 19

JACKSON, MS   TEMPERATURE (F) ON 21ST DAY OF:

	J	F	M	A	M	J	J	A	S	O	N	D
DAILY MAX DBT	57.1	59.0	64.4	77.9	86.9	88.8	92.1	89.8	86.7	76.0	65.1	56.8
DAILY AVE DBT	46.4	48.3	55.1	67.6	75.7	79.2	81.5	77.9	75.6	63.3	52.3	45.8
DAILY MIN DBT	37.3	37.9	45.6	56.9	64.8	70.9	73.8	68.6	65.4	51.9	41.0	36.9
DAILY MAX DPT	45.7	43.3	49.6	60.9	66.7	72.3	74.4	71.2	68.9	55.1	48.2	41.9
DAILY AVE DPT	37.2	36.6	43.4	56.8	62.8	69.4	71.5	67.7	64.1	50.3	38.8	36.8
DAILY MIN DPT	29.3	28.7	35.6	52.0	58.8	66.4	68.6	63.8	58.1	45.2	32.3	32.1
DAILY MAX WBT	51.3	49.6	55.0	65.6	71.0	75.4	77.5	74.3	73.1	62.1	54.3	47.9
DAILY AVE WBT	42.7	43.3	49.5	61.2	67.2	72.4	74.4	71.0	68.2	56.3	46.0	42.0
DAILY MIN WBT	35.1	35.9	42.3	55.2	62.6	69.0	71.8	66.7	62.2	49.4	38.3	36.2

NORMAL DAILY SOLAR RADIATION (MONTHLY AVG) BTU/SQ FT(DAY)

	J	F	M	A	M	J	J	A	S	O	N	D
HORIZONTAL	753.	1026.	1369.	1708.	1941.	2024.	1909.	1780.	1509.	1271.	902.	709.
SO VERTICAL	1053.	1151.	1133.	974.	856.	818.	796.	893.	1073.	1284.	1188.	1048.

HEATING DEGREE DAYS BASE 65F	2545.7
COOLING DEGREE DAYS BASE 65F	2440.1
COOLING DEGREE DAYS BASE 78F ET*	714.1

WINTER DESIGN DBT	99.0%	21.0
	97.5%	25.0

SUMMER DESIGN DBT / COINCIDENT WBT	1%	97.0 / 76.0
	2.5%	95.0 / 76.0
	5%	93.0 / 76.0

SUMMER DESIGN WBT	1%	79.0
	2.5%	78.0
	5%	78.0

% TOTAL HEATING HOURS (LESS THAN 68F) I-V	50.7

HEATING I	13.9
HEATING II	7.2
HEATING III	6.5
HEATING IV	8.5
HEATING V	14.6

% HUMIDIFICATION HOURS VI.A + VI.B	0.3

% TOTAL COMFORT HOURS (SHADING REQUIRED) VII	10.9

% DEHUMIDIFICATION HOURS VIII	12.7

% TOTAL COOLING HOURS (GREATER THAN 78F ET*) IX - XVII	25.3

% PASSIVE COOLING HOURS IX - XIV	18.7

COOLING IX	9.6
COOLING X	1.5
COOLING XI	6.1
COOLING XII	1.1
COOLING XIII	0.4
COOLING XIV	0.0

% VENTILATION EFFECTIVENESS HOURS IX + X + XI	17.2

% MASS EFFECTIVENESS HOURS X + XI + XII + XIII	9.1

% EVAPORATIVE COOLING EFFECTIVENESS HOURS XI + XIII + XIV + VI.B	6.7

% HOURS BEYOND PASSIVE EFFECTIVENESS VIII + XV + XVI + XVII	19.3

DEHUMIDIFICATION VIII	12.7
DEHUMIDIFICATION AND COOLING XV	4.1
DEHUMIDIFICATION AND COOLING XVI	2.5
COOLING XVII	0.0

Table 20

```
KANSAS CITY, MO              TEMPERATURE (F) ON 21ST DAY OF:
```

	J	F	M	A	M	J	J	A	S	O	N	D
DAILY MAX DBT	35.4	40.6	47.5	71.5	79.6	84.2	89.2	85.5	76.5	69.6	51.5	37.6
DAILY AVE DBT	26.5	31.7	39.4	61.6	70.4	76.3	81.1	75.7	67.8	59.4	42.8	30.7
DAILY MIN DBT	17.4	23.6	31.8	51.1	61.5	67.7	72.8	65.8	59.6	49.2	34.4	23.4
DAILY MAX DPT	25.6	27.4	34.9	49.7	59.7	67.2	70.8	65.0	59.9	48.2	35.1	26.9
DAILY AVE DPT	16.8	20.3	28.7	43.4	54.4	63.2	67.6	60.7	53.8	42.7	29.1	21.4
DAILY MIN DPT	8.0	13.9	21.8	36.8	48.6	59.0	64.5	55.5	48.1	36.8	22.5	15.6
DAILY MAX WBT	31.3	34.5	40.7	57.8	64.9	70.9	74.6	69.6	64.1	56.4	42.8	33.2
DAILY AVE WBT	23.8	28.1	35.2	52.2	60.8	67.7	71.8	66.1	59.7	50.8	37.3	27.7
DAILY MIN WBT	15.8	21.9	29.7	45.5	56.5	64.1	68.8	62.0	55.6	45.0	31.2	21.7

```
              NORMAL DAILY SOLAR RADIATION (MONTHLY AVG) BTU/SQ FT(DAY)
```

	J	F	M	A	M	J	J	A	S	O	N	D
HORIZONTAL	648.	895.	1203.	1575.	1873.	2080.	2102.	1862.	1452.	1092.	737.	561.
SO VERTICAL	1098.	1218.	1222.	1113.	994.	975.	1034.	1149.	1276.	1343.	1176.	1005.

HEATING DEGREE DAYS BASE 65F		5034.8
COOLING DEGREE DAYS BASE 65F		1791.4
COOLING DEGREE DAYS BASE 78F ET*		496.2

WINTER DESIGN DBT	99.0%	2.0
	97.5%	6.0

SUMMER DESIGN DBT / COINCIDENT WBT	1%	99.0 / 75.0
	2.5%	96.0 / 74.0
	5%	93.0 / 74.0

SUMMER DESIGN WBT	1%	78.0
	2.5%	77.0
	5%	76.0

% TOTAL HEATING HOURS (LESS THAN 68F) I-V		64.2
HEATING I	32.5	
HEATING II	7.0	
HEATING III	5.7	
HEATING IV	7.2	
HEATING V	11.8	
% HUMIDIFICATION HOURS VI.A + VI.B		0.5
% TOTAL COMFORT HOURS (SHADING REQUIRED) VII		13.6
% DEHUMIDIFICATION HOURS VIII		3.9
% TOTAL COOLING HOURS (GREATER THAN 78F ET*) IX - XVII		17.9
% PASSIVE COOLING HOURS IX - XIV		15.9
COOLING IX	5.4	
COOLING X	1.4	
COOLING XI	7.3	
COOLING XII	1.1	
COOLING XIII	0.7	
COOLING XIV	0.1	
% VENTILATION EFFECTIVENESS HOURS IX + X + XI		14.1
% MASS EFFECTIVENESS HOURS X + XI + XII + XIII		10.4
% EVAPORATIVE COOLING EFFECTIVENESS HOURS XI + XIII + XIV + VI.B		8.3
% HOURS BEYOND PASSIVE EFFECTIVENESS VIII + XV + XVI + XVII		5.8
DEHUMIDIFICATION VIII	3.9	
DEHUMIDIFICATION AND COOLING XV	0.7	
DEHUMIDIFICATION AND COOLING XVI	1.2	
COOLING XVII	0.0	

Table 21

```
LITTLE ROCK, AR                    TEMPERATURE (F) ON 21ST DAY OF:
```

	J	F	M	A	M	J	J	A	S	O	N	D
DAILY MAX DBT	49.8	52.8	58.4	77.6	84.9	89.2	91.3	87.4	84.3	73.6	57.3	49.1
DAILY AVE DBT	39.4	43.7	50.7	66.9	74.2	80.3	81.5	77.3	73.7	61.5	47.4	40.0
DAILY MIN DBT	31.6	35.2	42.7	56.4	63.3	71.3	74.2	68.8	63.8	50.5	37.9	31.1
DAILY MAX DPT	38.9	37.3	45.2	59.5	64.7	71.5	75.3	69.0	66.5	56.1	41.1	35.1
DAILY AVE DPT	30.2	31.2	38.7	54.0	61.4	67.9	71.9	65.3	61.2	51.0	34.8	30.7
DAILY MIN DPT	23.8	25.7	31.1	48.6	57.2	63.8	69.1	62.0	56.7	45.6	29.2	25.6
DAILY MAX WBT	44.5	44.9	51.0	64.9	70.2	75.0	78.3	73.3	71.3	61.4	47.8	42.5
DAILY AVE WBT	36.1	38.8	45.5	59.6	66.0	71.9	74.7	69.4	66.1	55.8	42.0	36.5
DAILY MIN WBT	29.7	32.9	39.0	53.0	60.4	67.7	71.8	65.8	60.9	48.7	35.8	29.6

```
              NORMAL DAILY SOLAR RADIATION (MONTHLY AVG) BTU/SQ FT(DAY)
```

	J	F	M	A	M	J	J	A	S	O	N	D
HORIZONTAL	731.	1003.	1313.	1611.	1929.	2106.	2032.	1860.	1860.	1228.	847.	674.
SO VERTICAL	1092.	1203.	1167.	988.	902.	888.	890.	999.	1162.	1328.	1192.	1064.

HEATING DEGREE DAYS BASE 65F		3281.1
COOLING DEGREE DAYS BASE 65F		2207.2
COOLING DEGREE DAYS BASE 78F ET*		673.8

WINTER DESIGN DBT	99.0%	15.0
	97.5%	20.0

SUMMER DESIGN DBT / COINCIDENT WBT	1%	99.0 / 76.0
	2.5%	96.0 / 77.0
	5%	94.0 / 77.0

SUMMER DESIGN WBT	1%	80.0
	2.5%	79.0
	5%	78.0

% TOTAL HEATING HOURS (LESS THAN 68F) I-V	56.4

HEATING I	19.8
HEATING II	8.1
HEATING III	6.7
HEATING IV	8.3
HEATING V	13.5

% HUMIDIFICATION HOURS VI.A + VI.B	0.3

% TOTAL COMFORT HOURS (SHADING REQUIRED) VII	10.9

% DEHUMIDIFICATION HOURS VIII	9.1

% TOTAL COOLING HOURS (GREATER THAN 78F ET*) IX - XVII	23:3

% PASSIVE COOLING HOURS IX - XIV	17.2

COOLING IX	9.0
COOLING X	1.4
COOLING XI	5.4
COOLING XII	1.1
COOLING XIII	0.3
COOLING XIV	0.0

% VENTILATION EFFECTIVENESS HOURS IX + X + XI	15.7

% MASS EFFECTIVENESS HOURS X + XI + XII + XIII	8.1

% EVAPORATIVE COOLING EFFECTIVENESS HOURS XI + XIII + XIV + VI.B	5.9

% HOURS BEYOND PASSIVE EFFECTIVENESS VIII + XV + XVI + XVII	15.2

DEHUMIDIFICATION VIII	9.1
DEHUMIDIFICATION AND COOLING XV	3.6
DEHUMIDIFICATION AND COOLING XVI	2.5
COOLING XVII	0.0

Table 22

```
LOS ANGELES, CA              TEMPERATURE (F) ON 21ST DAY OF:
```

	J	F	M	A	M	J	J	A	S	O	N	D
DAILY MAX DBT	61.8	63.5	64.2	65.5	67.6	71.9	75.0	74.4	73.4	71.0	68.6	64.4
DAILY AVE DBT	53.5	54.8	56.2	59.5	61.4	65.3	68.4	67.7	66.6	63.2	58.7	55.7
DAILY MIN DBT	46.5	46.3	49.0	53.8	56.1	59.7	64.1	62.7	61.6	56.8	50.5	47.1
DAILY MAX DPT	47.2	49.1	51.4	53.2	54.9	58.7	61.2	61.2	60.3	57.3	49.4	48.6
DAILY AVE DPT	39.0	42.6	46.3	50.6	52.5	56.9	59.8	59.5	58.2	53.8	41.1	41.4
DAILY MIN DPT	30.9	35.1	39.6	47.8	49.6	54.6	57.8	57.8	55.9	49.4	32.4	34.3
DAILY MAX WBT	51.9	54.1	55.5	57.1	59.0	63.0	65.8	65.2	64.3	60.7	55.9	54.1
DAILY AVE WBT	46.9	49.0	51.3	54.4	56.3	60.1	63.1	62.6	61.5	57.7	50.3	49.2
DAILY MIN WBT	41.6	42.7	45.8	51.3	53.2	57.4	61.0	60.1	58.9	53.9	44.2	44.1

```
              NORMAL DAILY SOLAR RADIATION (MONTHLY AVG) BTU/SQ FT(DAY)
```

	J	F	M	A	M	J	J	A	S	O	N	D
HORIZONTAL	926.	1214.	1619.	1951.	2060.	2119.	2307.	2079.	1681.	1317.	1004.	848.
SO VERTICAL	1353.	1424.	1405.	1168.	944.	880.	993.	1091.	1255.	1392.	1382.	1309.

HEATING DEGREE DAYS BASE 65F	1985.4
COOLING DEGREE DAYS BASE 65F	581.2
COOLING DEGREE DAYS BASE 78F ET*	28.1

WINTER DESIGN DBT	99.0%	41.0
	97.5%	43.0
SUMMER DESIGN DBT / COINCIDENT WBT	1%	83.0 / 68.0
	2.5%	80.0 / 68.0
	5%	77.0 / 67.0
SUMMER DESIGN WBT	1%	70.0
	2.5%	69.0
	5%	68.0

% TOTAL HEATING HOURS (LESS THAN 68F) I-V	80.3
HEATING I	1.7
HEATING II	5.7
HEATING III	12.4
HEATING IV	23.8
HEATING V	36.7
% HUMIDIFICATION HOURS VI.A + VI.B	1.1
% TOTAL COMFORT HOURS (SHADING REQUIRED) VII	15.2
% DEHUMIDIFICATION HOURS VIII	1.4
% TOTAL COOLING HOURS (GREATER THAN 78F ET*) IX - XVII	2.0
% PASSIVE COOLING HOURS IX - XIV	2.0
COOLING IX	0.1
COOLING X	0.0
COOLING XI	1.4
COOLING XII	0.0
COOLING XIII	0.2
COOLING XIV	0.2
% VENTILATION EFFECTIVENESS HOURS IX + X + XI	1.5
% MASS EFFECTIVENESS HOURS X + XI + XII + XIII	1.6
% EVAPORATIVE COOLING EFFECTIVENESS HOURS XI + XIII + XIV + VI.B	2.4
% HOURS BEYOND PASSIVE EFFECTIVENESS VIII + XV + XVI + XVII	1.4
DEHUMIDIFICATION VIII	1.4
DEHUMIDIFICATION AND COOLING XV	0.0
DEHUMIDIFICATION AND COOLING XVI	0.0
COOLING XVII	0.0

Table 23

```
MEDFORD, OR                    TEMPERATURE (F) ON 21ST DAY OF:

               J     F     M     A     M     J     J     A     S     O     N     D

DAILY MAX DBT 45.2  53.1  56.7  66.2  72.1  82.5  91.2  86.8  79.9  63.9  48.8  43.2
DAILY AVE DBT 37.1  42.4  45.4  51.9  58.0  66.4  73.3  70.4  60.6  49.9  39.3  37.6
DAILY MIN DBT 30.9  34.2  35.8  39.1  44.9  50.1  55.4  54.6  44.2  39.7  31.7  32.2

DAILY MAX DPT 34.3  40.4  40.2  43.2  47.4  50.9  55.2  54.2  48.9  46.9  39.6  38.2
DAILY AVE DPT 30.8  35.8  35.2  39.2  43.1  46.1  50.0  49.8  43.3  42.3  35.3  34.9
DAILY MIN DPT 26.8  31.4  30.1  34.6  38.2  39.9  45.6  44.4  37.6  37.1  30.3  30.9

DAILY MAX WBT 39.6  45.5  46.5  52.4  56.0  61.6  66.4  64.7  59.6  53.1  43.2  40.6
DAILY AVE WBT 34.7  39.5  40.7  45.5  50.1  55.0  59.4  58.2  51.2  46.1  37.6  36.5
DAILY MIN WBT 30.1  33.5  34.6  37.9  43.3  47.3  51.6  50.9  42.5  38.9  31.6  31.9
```

```
            NORMAL DAILY SOLAR RADIATION (MONTHLY AVG) BTU/SQ FT(DAY)

               J     F     M     A     M     J     J     A     S     O     N     D

HORIZONTAL   407.  737. 1133. 1639. 2034. 2278. 2475. 2121. 1589.  982.  504.  337.
SO VERTICAL  752. 1092. 1257. 1273. 1185. 1162. 1330. 1440. 1528. 1315.  876.  659.
```

```
HEATING DEGREE DAYS BASE 65F                        5467.6
COOLING DEGREE DAYS BASE 65F                         926.6
COOLING DEGREE DAYS BASE 78F ET*                     214.3
```

```
WINTER DESIGN DBT                           99.0%  19.0
                                            97.5%  23.0
```

```
SUMMER DESIGN DBT / COINCIDENT WBT            1%  98.0 / 68.0
                                            2.5%  94.0 / 67.0
                                              5%  91.0 / 66.0
```

```
SUMMER DESIGN WBT                             1%  70.0
                                            2.5%  68.0
                                              5%  67.0
```

% TOTAL HEATING HOURS (LESS THAN 68F) I-V	81.9

HEATING I	35.3
HEATING II	13.4
HEATING III	10.5
HEATING IV	11.1
HEATING V	11.6

% HUMIDIFICATION HOURS VI.A + VI.B	0.5

% TOTAL COMFORT HOURS (SHADING REQUIRED) VII	10.1

% DEHUMIDIFICATION HOURS VIII	0.0

% TOTAL COOLING HOURS (GREATER THAN 78F ET*) IX - XVII	7.4

% PASSIVE COOLING HOURS IX - XIV	7.4

COOLING IX	0.0
COOLING X	0.0
COOLING XI	5.0
COOLING XII	0.1
COOLING XIII	2.1
COOLING XIV	0.2

% VENTILATION EFFECTIVENESS HOURS IX + X + XI	5.0

% MASS EFFECTIVENESS HOURS X + XI + XII + XIII	7.2

% EVAPORATIVE COOLING EFFECTIVENESS HOURS XI + XIII + XIV + VI.B	7.6

% HOURS BEYOND PASSIVE EFFECTIVENESS VIII + XV + XVI + XVII	0.1

DEHUMIDIFICATION VIII	0.0
DEHUMIDIFICATION AND COOLING XV	0.0
DEHUMIDIFICATION AND COOLING XVI	0.0
COOLING XVII	0.0

Table 24

```
MIAMI, FL                    TEMPERATURE (F) ON 21ST DAY OF:
```

	J	F	M	A	M	J	J	A	S	O	N	D
DAILY MAX DBT	74.8	75.9	79.4	80.0	85.9	87.8	87.9	88.8	85.6	82.4	78.6	75.6
DAILY AVE DBT	68.5	68.8	71.6	74.2	79.4	81.3	82.2	82.3	80.7	75.0	72.6	67.9
DAILY MIN DBT	62.6	60.6	63.7	68.1	72.4	75.4	76.8	76.4	75.9	68.6	66.4	60.7
DAILY MAX DPT	63.9	62.5	65.5	65.9	72.2	75.2	75.5	76.4	75.8	68.1	66.4	62.9
DAILY AVE DPT	60.3	58.1	60.0	62.4	69.7	72.8	73.3	73.9	73.4	64.6	63.4	58.7
DAILY MIN DPT	55.6	53.5	53.8	58.5	66.9	70.1	70.3	70.4	70.5	60.1	60.2	53.2
DAILY MAX WBT	66.6	66.1	69.0	69.2	74.9	77.7	77.9	78.4	77.6	71.1	69.2	66.2
DAILY AVE WBT	63.6	62.3	64.6	66.7	72.7	75.2	75.8	76.2	75.5	68.2	66.7	62.3
DAILY MIN WBT	59.6	57.6	59.6	63.8	69.8	72.6	73.5	73.8	72.7	65.1	63.6	58.0

```
              NORMAL DAILY SOLAR RADIATION (MONTHLY AVG) BTU/SQ FT(DAY)
```

	J	F	M	A	M	J	J	A	S	O	N	D
HORIZONTAL	1057.	1314.	1603.	1859.	1844.	1708.	1763.	1630.	1456.	1303.	1119.	1019.
SO VERTICAL	1236.	1224.	1088.	881.	725.	650.	677.	696.	848.	1088.	1231.	1260.

HEATING DEGREE DAYS BASE 65F		285.0
COOLING DEGREE DAYS BASE 65F		4022.3
COOLING DEGREE DAYS BASE 78F ET*		1045.3

WINTER DESIGN DBT	99.0%	44.0
	97.5%	47.0

SUMMER DESIGN DBT / COINCIDENT WBT	1%	91.0 / 77.0
	2.5%	90.0 / 77.0
	5%	89.0 / 77.0

SUMMER DESIGN WBT	1%	79.0
	2.5%	79.0
	5%	78.0

% TOTAL HEATING HOURS (LESS THAN 68F) I-V	15.7
HEATING I	0.3
HEATING II	0.8
HEATING III	1.5
HEATING IV	3.3
HEATING V	9.8
% HUMIDIFICATION HOURS VI.A + VI.B	0.1
% TOTAL COMFORT HOURS (SHADING REQUIRED) VII	18.1
% DEHUMIDIFICATION HOURS VIII	15.9
% TOTAL COOLING HOURS (GREATER THAN 78F ET*) IX - XVII	50.2
% PASSIVE COOLING HOURS IX - XIV	35.5
COOLING IX	26.6
COOLING X	1.4
COOLING XI	7.4
COOLING XII	0.1
COOLING XIII	0.0
COOLING XIV	0.0
% VENTILATION EFFECTIVENESS HOURS IX + X + XI	35.4
% MASS EFFECTIVENESS HOURS X + XI + XII + XIII	8.9
% EVAPORATIVE COOLING EFFECTIVENESS HOURS XI + XIII + XIV + VI.B	7.4
% HOURS BEYOND PASSIVE EFFECTIVENESS VIII + XV + XVI + XVII	30.6
DEHUMIDIFICATION VIII	15.9
DEHUMIDIFICATION AND COOLING XV	14.1
DEHUMIDIFICATION AND COOLING XVI	0.6
COOLING XVII	0.0

Table 25

MIDLAND, TEXAS   TEMPERATURE (F) ON 21ST DAY OF:

	J	F	M	A	M	J	J	A	S	O	N	D
DAILY MAX DBT	57.9	57.6	71.0	83.1	88.6	96.4	94.5	91.1	86.6	79.1	63.6	59.0
DAILY AVE DBT	43.2	45.5	56.0	68.4	75.7	83.7	82.6	80.0	75.7	65.0	50.5	44.5
DAILY MIN DBT	30.9	34.4	40.9	54.9	61.8	71.3	71.6	69.6	65.3	51.9	38.6	32.4
DAILY MAX DPT	31.4	35.5	35.8	49.3	59.1	64.0	65.4	64.2	60.3	52.3	39.9	34.6
DAILY AVE DPT	25.7	28.2	27.6	40.6	48.3	58.5	61.0	59.2	55.8	45.3	33.5	26.4
DAILY MIN DPT	19.1	20.9	21.1	28.9	38.5	52.5	56.8	54.1	50.1	38.3	27.1	20.3
DAILY MAX WBT	43.6	45.9	50.1	60.1	64.8	69.9	71.0	69.4	67.4	60.1	49.6	45.6
DAILY AVE WBT	36.1	38.6	43.2	54.3	59.8	66.9	68.0	66.4	63.4	54.1	42.9	37.1
DAILY MIN WBT	28.6	31.1	35.4	47.7	54.4	63.1	64.4	62.9	59.1	47.8	35.3	29.7

NORMAL DAILY SOLAR RADIATION (MONTHLY AVG) BTU/SQ FT(DAY)

	J	F	M	A	M	J	J	A	S	O	N	D
HORIZONTAL	1081.	1383.	1839.	2192.	2430.	2562.	2389.	2210.	1844.	1522.	1176.	1000.
SO VERTICAL	1496.	1534.	1504.	1235.	1062.	1030.	989.	1096.	1295.	1520.	1532.	1462.

HEATING DEGREE DAYS BASE 65F	2993.2
COOLING DEGREE DAYS BASE 65F	2578.4
COOLING DEGREE DAYS BASE 78F ET*	731.2

WINTER DESIGN DBT	99.0%	16.0
	97.5%	21.0

SUMMER DESIGN DBT / COINCIDENT WBT	1%	100.0 / 69.0
	2.5%	98.0 / 69.0
	5%	96.0 / 69.0

SUMMER DESIGN WBT	1%	73.0
	2.5%	72.0
	5%	71.0

% TOTAL HEATING HOURS (LESS THAN 68F) I-V	53.4

HEATING I	17.5
HEATING II	7.4
HEATING III	6.8
HEATING IV	8.4
HEATING V	13.3

% HUMIDIFICATION HOURS VI.A + VI.B	3.6

% TOTAL COMFORT HOURS (SHADING REQUIRED) VII	16.9

% DEHUMIDIFICATION HOURS VIII	2.4

% TOTAL COOLING HOURS (GREATER THAN 78F ET*) IX - XVII	23.7

% PASSIVE COOLING HOURS IX - XIV	23.7

COOLING IX	0.5
COOLING X	0.6
COOLING XI	14.1
COOLING XII	0.9
COOLING XIII	6.1
COOLING XIV	1.6

% VENTILATION EFFECTIVNESS HOURS IX + X + XI	15.2

% MASS EFFECTIVENESS HOURS X + XI + XII + XIII	21.6

% EVAPORATIVE COOLING EFFECTIVENESS HOURS XI + XII + XIV + VI.B	23.0

% HOURS BEYOND PASSIVE EFFECTIVENESS VIII + XV + XVI + XVII	2.5

DEHUMIDIFICATION VIII	2.4
DEHUMIDIFICATION AND COOLING XV	0.0
DEHUMIDIFICATION AND COOLING XVI	0.0
COOLING XVII	0.0

Table 26

```
MINNEAPOLIS, MN              TEMPERATURE (F) ON 21ST DAY OF:
```

	J	F	M	A	M	J	J	A	S	O	N	D
DAILY MAX DBT	14.8	24.1	37.4	60.6	69.7	76.7	83.6	80.4	65.3	59.8	36.6	24.4
DAILY AVE DBT	7.4	16.0	30.2	50.5	61.1	67.2	74.3	70.0	56.6	49.4	30.6	17.4
DAILY MIN DBT	-1.5	7.1	22.6	38.6	51.8	57.6	65.5	60.0	49.6	40.1	25.6	10.0
DAILY MAX DPT	7.0	15.4	27.5	41.7	52.4	60.1	67.6	60.2	52.2	44.0	27.5	17.9
DAILY AVE DPT	-0.4	7.6	21.9	36.2	46.1	55.2	62.9	56.6	47.5	37.8	22.7	11.2
DAILY MIN DPT	-8.6	-1.2	15.9	29.7	40.2	49.4	58.8	52.5	43.2	31.6	17.6	4.3
DAILY MAX WBT	13.5	21.4	33.5	49.8	57.6	64.5	71.0	65.2	56.4	50.6	32.9	22.8
DAILY AVE WBT	6.5	14.5	27.8	44.0	52.9	60.0	66.8	61.7	51.8	43.9	28.1	16.7
DAILY MIN WBT	-1.8	6.3	21.4	36.1	47.4	54.4	62.2	57.3	47.6	37.3	23.9	9.9

```
            NORMAL DAILY SOLAR RADIATION (MONTHLY AVG) BTU/SQ FT(DAY)
```

	J	F	M	A	M	J	J	A	S	O	N	D
HORIZONTAL	464.	764.	1103.	1442.	1737.	1927.	1970.	1687.	1255.	860.	480.	353.
SO VERTICAL	921.	1212.	1312.	1208.	1092.	1057.	1140.	1237.	1297.	1233.	895.	742.

HEATING DEGREE DAYS BASE 65F	8118.8
COOLING DEGREE DAYS BASE 65F	860.4
COOLING DEGREE DAYS BASE 78F ET*	160.3

WINTER DESIGN DBT	99.0%	-16.0
	97.5%	-12.0

SUMMER DESIGN DBT / COINCIDENT WBT	1%	92.0 / 75.0
	2.5%	89.0 / 73.0
	5%	86.0 / 71.0

SUMMER DESIGN WBT	1%	77.0
	2.5%	75.0
	5%	73.0

% TOTAL HEATING HOURS (LESS THAN 68F) I-V	79.4
HEATING I	47.3
HEATING II	6.0
HEATING III	5.8
HEATING IV	7.6
HEATING V	12.6
% HUMIDIFICATION HOURS VI.A + VI.B	0.3
% TOTAL COMFORT HOURS (SHADING REQUIRED) VII	10.7
% DEHUMIDIFICATION HOURS VIII	2.5
% TOTAL COOLING HOURS (GREATER THAN 78F ET*) IX - XVII	7.2
% PASSIVE COOLING HOURS IX - XIV	6.6
COOLING IX	1.7
COOLING X	0.5
COOLING XI	4.1
COOLING XII	0.2
COOLING XIII	0.1
COOLING XIV	0.0
% VENTILATION EFFECTIVENESS HOURS IX + X + XI	6.3
% MASS EFFECTIVENESS HOURS X + XI + XII + XIII	4.9
% EVAPORATIVE COOLING EFFECTIVENESS HOURS XI + XIII + XIV + VI.B	4.4
% HOURS BEYOND PASSIVE EFFECTIVENESS VIII + XV + XVI + XVII	3.1
DEHUMIDIFICATION VIII	2.5
DEHUMIDIFICATION AND COOLING XV	0.3
DEHUMIDIFICATION AND COOLING XVI	0.3
COOLING XVII	0.0

Table 27

NASHVILLE, TN   TEMPERATURE (F) ON 21ST DAY OF:

	J	F	M	A	M	J	J	A	S	O	N	D
DAILY MAX DBT	47.1	48.7	55.8	74.1	81.8	87.9	90.1	85.7	83.1	71.3	56.5	48.3
DAILY AVE DBT	38.5	39.3	48.3	63.8	71.2	77.4	80.0	75.8	71.0	58.9	45.9	39.0
DAILY MIN DBT	28.9	30.2	40.2	52.4	59.6	67.5	71.3	66.8	59.6	46.8	35.5	30.9
DAILY MAX DPT	39.0	37.1	45.1	55.4	63.5	71.1	72.2	68.0	64.4	52.5	40.2	38.5
DAILY AVE DPT	32.1	29.8	38.3	50.6	59.1	67.4	69.3	65.1	59.1	47.5	34.3	32.0
DAILY MIN DPT	23.7	23.9	31.4	44.9	54.8	63.0	66.6	61.7	54.2	42.2	29.1	26.3
DAILY MAX WBT	43.0	42.4	48.9	61.3	68.0	74.1	75.9	72.0	68.8	59.0	47.0	43.1
DAILY AVE WBT	36.1	35.7	43.8	56.3	63.6	70.5	72.5	68.6	63.5	52.8	40.8	36.2
DAILY MIN WBT	28.1	28.9	37.4	49.1	57.9	65.2	69.0	64.5	56.8	45.0	33.3	29.6

NORMAL DAILY SOLAR RADIATION (MONTHLY AVG) BTU/SQ FT(DAY)

	J	F	M	A	M	J	J	A	S	O	N	D
HORIZONTAL	580.	824.	1130.	1544.	1825.	1963.	1891.	1737.	1398.	1114.	711.	521.
SO VERTICAL	900.	1027.	1047.	989.	885.	853.	856.	973.	1116.	1253.	1039.	854.

HEATING DEGREE DAYS BASE 65F	3917.6
COOLING DEGREE DAYS BASE 65F	1824.0
COOLING DEGREE DAYS BASE 78F ET*	486.0

WINTER DESIGN DBT	99.0%	9.0
	97.5%	14.0

SUMMER DESIGN DBT / COINCIDENT WBT	1%	97.0 / 75.0
	2.5%	94.0 / 74.0
	5%	91.0 / 74.0

SUMMER DESIGN WBT	1%	78.0
	2.5%	77.0
	5%	76.0

% TOTAL HEATING HOURS (LESS THAN 68F) I-V	61.6

HEATING I	24.4
HEATING II	7.6
HEATING III	6.6
HEATING IV	8.8
HEATING V	14.1

% HUMIDIFICATION HOURS VI.A + VI.B	0.1

% TOTAL COMFORT HOURS (SHADING REQUIRED) VII	11.3

% DEHUMIDIFICATION HOURS VIII	8.6

% TOTAL COOLING HOURS (GREATER THAN 78F ET*) IX - XVII	18.4

% PASSIVE COOLING HOURS IX - XIV	15.4

COOLING IX	6.6
COOLING X	1.3
COOLING XI	6.3
COOLING XII	0.8
COOLING XIII	0.4
COOLING XIV	0.0

% VENTILATION EFFECTIVENESS HOURS IX + X + XI	14.2

% MASS EFFECTIVENESS HOURS X + XI + XII + XIII	8.8

% EVAPORATIVE COOLING EFFECTIVENESS HOURS XI + XIII + XIV + VI.B	6.7

% HOURS BEYOND PASSIVE EFFECTIVENESS VIII + XV + XVI + XVII	11.6

DEHUMIDIFICATION VIII	8.6
DEHUMIDIFICATION AND COOLING XV	1.9
DEHUMIDIFICATION AND COOLING XVI	1.1
COOLING XVII	0.0

Table 28

NEW ORLEANS, LA   TEMPERATURE (F) ON 21ST DAY OF:

	J	F	M	A	M	J	J	A	S	O	N	D
DAILY MAX DBT	64.0	64.6	70.3	79.3	85.4	89.4	89.5	87.7	86.6	76.6	68.0	61.9
DAILY AVE DBT	54.5	55.2	62.0	70.4	76.6	81.1	81.5	80.0	78.2	66.8	58.8	53.1
DAILY MIN DBT	44.9	47.4	53.1	61.3	67.9	73.5	74.3	73.9	71.0	57.3	49.6	45.4
DAILY MAX DPT	53.7	53.0	56.4	65.9	70.8	75.3	76.7	75.7	74.6	61.9	55.5	52.4
DAILY AVE DPT	46.6	46.2	50.8	61.5	67.3	71.8	74.1	72.4	70.6	57.1	48.0	45.6
DAILY MIN DPT	38.0	38.8	42.8	56.8	63.5	68.1	70.9	68.1	66.4	51.0	39.6	40.2
DAILY MAX WBT	57.3	56.4	60.4	68.6	73.5	77.4	78.9	77.7	76.7	65.4	59.4	55.3
DAILY AVE WBT	50.8	50.8	56.1	65.0	70.4	74.5	76.1	74.6	72.9	61.1	53.4	49.4
DAILY MIN WBT	42.3	44.4	49.3	59.6	65.9	71.1	72.9	71.4	68.6	54.9	46.2	43.9

NORMAL DAILY SOLAR RADIATION (MONTHLY AVG) BTU/SQ FT(DAY)

	J	F	M	A	M	J	J	A	S	O	N	D
HORIZONTAL	835.	1112.	1415.	1780.	1968.	2004.	1813.	1717.	1514.	1335.	973.	779.
SO VERTICAL	1097.	1169.	1093.	948.	827.	786.	728.	809.	1003.	1262.	1203.	1081.

HEATING DEGREE DAYS BASE 65F	1539.3
COOLING DEGREE DAYS BASE 65F	2763.2
COOLING DEGREE DAYS BASE 78F ET*	772.0

WINTER DESIGN DBT	99.0%	29.0
	97.5%	33.0
SUMMER DESIGN DBT / COINCIDENT WBT	1%	93.0 / 78.0
	2.5%	92.0 / 78.0
	5%	90.0 / 77.0
SUMMER DESIGN WBT	1%	81.0
	2.5%	80.0
	5%	79.0

% TOTAL HEATING HOURS (LESS THAN 68F) I-V	41.6
HEATING I	5.9
HEATING II	5.4
HEATING III	6.4
HEATING IV	8.7
HEATING V	15.2
% HUMIDIFICATION HOURS VI.A + VI.B	0.1
% TOTAL COMFORT HOURS (SHADING REQUIRED) VII	9.3
% DEHUMIDIFICATION HOURS VIII	15.5
% TOTAL COOLING HOURS (GREATER THAN 78F ET*) IX - XVII	33.5
% PASSIVE COOLING HOURS IX - XIV	19.5
COOLING IX	14.1
COOLING X	1.2
COOLING XI	4.0
COOLING XII	0.3
COOLING XIII	0.0
COOLING XIV	0.0
% VENTILATION EFFECTIVENESS HOURS IX + X + XI	19.2
% MASS EFFECTIVENESS HOURS X + XI + XII + XIII	5.4
% EVAPORATIVE COOLING EFFECTIVENESS HOURS XI + XIII + XIV + VI.B	4.1
% HOURS BEYOND PASSIVE EFFECTIVENESS VIII + XV + XVI + XVII	29.5
DEHUMIDIFICATION VIII	15.5
DEHUMIDIFICATION AND COOLING XV	12.5
DEHUMIDIFICATION AND COOLING XVI	1.6
COOLING XVII	0.0

Table 29

239

NEW YORK, NY   TEMPERATURE (F) ON 21ST DAY OF:

	J	F	M	A	M	J	J	A	S	O	N	D
DAILY MAX DBT	40.8	39.2	44.6	61.3	69.3	77.6	83.7	79.4	70.6	62.0	51.8	37.9
DAILY AVE DBT	33.4	33.4	39.4	52.2	60.5	70.7	75.7	72.1	63.8	54.3	46.0	32.8
DAILY MIN DBT	26.5	27.3	33.9	45.0	52.8	64.8	68.9	65.3	57.1	46.6	39.3	27.1
DAILY MAX DPT	32.7	29.4	35.3	44.6	54.9	65.3	69.1	67.6	58.6	48.0	42.3	30.4
DAILY AVE DPT	21.9	21.9	30.5	38.8	48.8	60.4	65.2	63.2	53.5	41.9	35.4	23.1
DAILY MIN DPT	12.2	14.2	23.9	32.4	42.0	54.8	61.0	58.8	49.0	35.2	28.3	15.6
DAILY MAX WBT	37.1	34.5	39.9	50.4	58.1	68.1	72.2	69.9	61.6	53.1	46.6	35.0
DAILY AVE WBT	30.0	30.1	36.1	46.0	54.4	64.2	68.8	66.5	58.1	48.3	41.7	30.1
DAILY MIN WBT	23.4	24.6	31.4	41.6	50.0	60.6	65.4	62.4	53.7	42.4	35.8	24.8

NORMAL DAILY SOLAR RADIATION (MONTHLY AVG) BTU/SQ FT(DAY)

	J	F	M	A	M	J	J	A	S	O	N	D
HORIZONTAL	500.	721.	1037.	1364.	1636.	1710.	1688.	1483.	1214.	895.	533.	404.
SO VERTICAL	884.	1023.	1100.	1009.	908.	834.	866.	959.	1115.	1147.	886.	755.

HEATING DEGREE DAYS BASE 65F	5162.0
COOLING DEGREE DAYS BASE 65F	927.4
COOLING DEGREE DAYS BASE 78F ET*	124.5

WINTER DESIGN DBT	99.0%	11.0
	97.5%	15.0

SUMMER DESIGN DBT / COINCIDENT WBT	1%	92.0 / 74.0
	2.5%	89.0 / 73.0
	5%	87.0 / 72.0

SUMMER DESIGN WBT	1%	76.0
	2.5%	75.0
	5%	74.0

% TOTAL HEATING HOURS (LESS THAN 68F) I-V		74.7
HEATING I	34.1	
HEATING II	9.8	
HEATING III	7.3	
HEATING IV	9.2	
HEATING V	14.2	
% HUMIDIFICATION HOURS VI.A + VI.B		0.1
% TOTAL COMFORT HOURS (SHADING REQUIRED) VII		11.4
% DEHUMIDIFICATION HOURS VIII		6.7
% TOTAL COOLING HOURS (GREATER THAN 78F ET*) IX - XVII		7.1
% PASSIVE COOLING HOURS IX - XIV		5.9
COOLING IX	2.2	
COOLING X	0.5	
COOLING XI	3.0	
COOLING XII	0.1	
COOLING XIII	0.1	
COOLING XIV	0.0	
% VENTILATION EFFECTIVENESS HOURS IX + X + XI		5.7
% MASS EFFECTIVENESS HOURS X + XI + XII + XIII		3.7
% EVAPORATIVE COOLING EFFECTIVENESS HOURS XI + XIII + XIV + VI.B		3.2
% HOURS BEYOND PASSIVE EFFECTIVENESS VIII + XV + XVI + XVII		7.9
DEHUMIDIFICATION VIII	6.7	
DEHUMIDIFICATION AND COOLING XV	1.1	
DEHUMIDIFICATION AND COOLING XVI	0.1	
COOLING XVII	0.0	

Table 30

```
OKLAHOMA CITY, OK          TEMPERATURE (F) ON 21ST DAY OF:

               J     F     M     A     M     J     J     A     S     O     N     D

DAILY MAX DBT  44.8  47.4  58.2  75.3  81.6  88.7  92.3  89.3  82.8  71.5  55.6  47.9
DAILY AVE DBT  34.2  38.3  47.2  64.8  70.7  78.3  81.6  78.0  72.4  60.9  45.4  36.9
DAILY MIN DBT  25.3  29.5  36.6  54.0  59.8  68.8  71.5  67.3  63.4  50.5  35.8  27.8

DAILY MAX DPT  31.6  34.6  41.3  59.1  62.3  69.1  70.0  67.6  64.1  54.7  39.4  33.2
DAILY AVE DPT  25.0  27.7  34.4  52.0  56.7  65.6  66.9  63.0  57.9  48.1  32.8  27.5
DAILY MIN DPT  17.8  20.9  27.4  45.4  50.8  61.9  63.3  58.5  52.4  41.4  25.5  21.1

DAILY MAX WBT  38.1  41.0  48.4  63.3  67.7  73.5  74.9  72.0  68.9  60.6  46.1  40.4
DAILY AVE WBT  31.0  34.4  41.5  57.5  62.1  69.7  71.5  68.1  63.6  54.0  40.1  33.4
DAILY MIN WBT  23.3  27.4  34.5  50.8  56.8  65.5  67.7  63.5  58.4  47.3  32.9  26.4
```

```
             NORMAL DAILY SOLAR RADIATION (MONTHLY AVG) BTU/SQ FT(DAY)

            J     F     M     A     M     J     J     A     S     O     N     D

HORIZONTAL  801. 1055. 1400. 1725. 1918. 2144. 2128. 1950. 1554. 1233.  901.  725.
SO VERTICAL 1220. 1290. 1270. 1081.  913.  917.  947. 1069. 1215. 1360. 1292. 1166.
```

```
HEATING DEGREE DAYS BASE 65F              4005.0
COOLING DEGREE DAYS BASE 65F              2015.0
COOLING DEGREE DAYS BASE 78F ET*          592.8
```

```
WINTER DESIGN DBT                   99.0%   9.0
                                    97.5%  13.0
```

```
SUMMER DESIGN DBT / COINCIDENT WBT     1% 100.0 / 74.0
                                     2.5%  97.0 / 74.0
                                       5%  95.0 / 73.0
```

```
SUMMER DESIGN WBT                      1%  78.0
                                     2.5%  77.0
                                       5%  76.0
```

% TOTAL HEATING HOURS (LESS THAN 68F) I-V	60.5
HEATING I	25.1
HEATING II	7.5
HEATING III	6.6
HEATING IV	8.2
HEATING V	13.1
% HUMIDIFICATION HOURS VI.A + VI.B	0.6
% TOTAL COMFORT HOURS (SHADING REQUIRED) VII	11.9
% DEHUMIDIFICATION HOURS VIII	6.9
% TOTAL COOLING HOURS (GREATER THAN 78F ET*) IX - XVII	20.1
% PASSIVE COOLING HOURS IX - XIV	17.9
COOLING IX	6.1
COOLING X	1.8
COOLING XI	6.9
COOLING XII	2.1
COOLING XIII	0.9
COOLING XIV	0.1
% VENTILATION EFFECTIVENESS HOURS IX + X + XI	14.8
% MASS EFFECTIVENESS HOURS X + XI + XII + XIII	11.7
% EVAPORATIVE COOLING EFFECTIVENESS HOURS XI + XIII + XIV + VI.B	8.1
% HOURS BEYOND PASSIVE EFFECTIVENESS VIII + XV + XVI + XVII	9.1
DEHUMIDIFICATION VIII	6.9
DEHUMIDIFICATION AND COOLING XV	1.1
DEHUMIDIFICATION AND COOLING XVI	1.1
COOLING XVII	0.0

Table 31

PHOENIX, AZ   TEMPERATURE (F) ON 21ST DAY OF:

	J	F	M	A	M	J	J	A	S	O	N	D
DAILY MAX DBT	62.2	69.4	74.5	83.4	92.8	104.2	102.2	99.4	95.1	85.0	70.3	65.2
DAILY AVE DBT	49.7	55.2	61.7	70.2	79.4	89.2	91.2	88.5	81.1	70.4	55.8	50.8
DAILY MIN DBT	38.0	42.3	47.9	55.5	62.9	71.8	80.3	77.2	66.8	56.5	42.8	39.0
DAILY MAX DPT	38.1	35.8	36.5	41.6	42.6	50.9	65.6	66.3	54.6	48.4	40.1	37.0
DAILY AVE DPT	32.8	31.2	31.3	36.2	36.7	43.6	60.7	61.9	49.5	42.2	34.3	32.0
DAILY MIN DPT	27.4	26.0	25.8	30.0	32.1	37.0	56.1	56.6	43.2	36.6	28.1	27.0
DAILY MAX WBT	48.6	51.3	53.4	58.6	61.5	68.0	74.2	74.1	67.0	60.5	52.9	49.6
DAILY AVE WBT	42.2	44.6	47.6	53.0	56.8	62.9	71.1	70.8	62.4	55.1	45.9	42.4
DAILY MIN WBT	35.1	37.9	40.4	46.2	50.4	55.8	67.9	67.6	56.5	48.6	38.3	35.4

NORMAL DAILY SOLAR RADIATION (MONTHLY AVG) BTU/SQ FT(DAY)

	J	F	M	A	M	J	J	A	S	O	N	D
HORIZONTAL	1021.	1374.	1814.	2355.	2676.	2739.	2486.	2293.	2015.	1576.	1150.	932.
SO VERTICAL	1472.	1589.	1552.	1388.	1211.	1128.	1059.	1186.	1482.	1643.	1561.	1419.

HEATING DEGREE DAYS BASE 65F	1864.1
COOLING DEGREE DAYS BASE 65F	3956.5
COOLING DEGREE DAYS BASE 78F ET*	1554.0

WINTER DESIGN DBT	99.0%	31.0
	97.5%	34.0

SUMMER DESIGN DBT / COINCIDENT WBT	1%	109.0 / 71.0
	2.5%	107.0 / 71.0
	5%	105.0 / 71.0

SUMMER DESIGN WBT	1%	76.0
	2.5%	75.0
	5%	75.0

% TOTAL HEATING HOURS (LESS THAN 68F) I-V	44.7

HEATING I	8.1
HEATING II	6.8
HEATING III	7.0
HEATING IV	9.3
HEATING V	13.4

% HUMIDIFICATION HOURS VI.A + VI.B	6.2
% TOTAL COMFORT HOURS (SHADING REQUIRED) VII	12.9
% DEHUMIDIFICATION HOURS VIII	0.5
% TOTAL COOLING HOURS (GREATER THAN 78F ET*) IX - XVII	35.7
% PASSIVE COOLING HOURS IX - XIV	33.4

COOLING IX	1.6
COOLING X	1.3
COOLING XI	10.7
COOLING XII	4.3
COOLING XIII	11.3
COOLING XIV	4.2

% VENTILATION EFFECTIVENESS HOURS IX + X + XI	13.6
% MASS EFFECTIVENESS HOURS X + XI + XII + XIII	27.6
% EVAPORATIVE COOLING EFFECTIVENESS HOURS XI + XIII + XIV + VI.B	29.6
% HOURS BEYOND PASSIVE EFFECTIVENESS VIII + XV + XVI + XVII	2.8

DEHUMIDIFICATION VIII	0.5
DEHUMIDIFICATION AND COOLING XV	0.2
DEHUMIDIFICATION AND COOLING XVI	0.3
COOLING XVII	1.8

Table 32

RALEIGH, NC   TEMPERATURE (F) ON 21ST DAY OF:

	J	F	M	A	M	J	J	A	S	O	N	D
DAILY MAX DBT	50.8	53.8	57.5	74.9	78.9	86.4	88.3	86.6	81.2	67.1	60.3	47.8
DAILY AVE DBT	41.7	44.0	48.0	61.9	68.5	76.2	78.3	75.5	69.1	56.1	48.5	38.6
DAILY MIN DBT	32.1	33.4	38.5	47.3	58.8	66.6	69.9	66.4	58.7	46.1	38.6	31.1
DAILY MAX DPT	41.1	38.2	42.2	51.5	62.9	70.5	73.0	70.1	64.4	51.7	43.5	35.0
DAILY AVE DPT	31.5	31.2	35.8	45.2	57.8	67.1	69.5	66.6	59.4	46.0	36.2	29.7
DAILY MIN DPT	22.3	22.1	29.4	39.6	52.6	63.3	65.9	62.8	53.9	40.5	27.9	23.4
DAILY MAX WBT	45.4	45.4	48.7	58.3	66.4	73.3	75.9	73.2	68.2	56.7	50.3	41.9
DAILY AVE WBT	38.0	39.1	42.9	53.2	61.8	70.1	72.2	69.4	63.1	50.8	43.3	35.5
DAILY MIN WBT	30.3	30.5	36.3	45.9	57.0	65.4	68.5	65.1	56.6	44.5	35.6	29.6

NORMAL DAILY SOLAR RADIATION (MONTHLY AVG) BTU/SQ FT(DAY)

	J	F	M	A	M	J	J	A	S	O	N	D
HORIZONTAL	694.	943.	1276.	1644.	1808.	1864.	1776.	1611.	1377.	1105.	812.	636.
SO VERTICAL	1071.	1169.	1175.	1046.	873.	806.	800.	897.	1093.	1236.	1180.	1037.

HEATING DEGREE DAYS BASE 65F	3712.9
COOLING DEGREE DAYS BASE 65F	1595.5
COOLING DEGREE DAYS BASE 78F ET*	374.2

WINTER DESIGN DBT	99.0%	16.0
	97.5%	20.0

SUMMER DESIGN DBT / COINCIDENT WBT	1%	94.0 / 75.0
	2.5%	92.0 / 75.0
	5%	90.0 / 75.0

SUMMER DESIGN WBT	1%	78.0
	2.5%	77.0
	5%	76.0

% TOTAL HEATING HOURS (LESS THAN 68F) I-V	63.8

HEATING I	22.6
HEATING II	8.4
HEATING III	7.1
HEATING IV	9.7
HEATING V	16.0

% HUMIDIFICATION HOURS VI.A + VI.B	0.3

% TOTAL COMFORT HOURS (SHADING REQUIRED) VII	10.1

% DEHUMIDIFICATION HOURS VIII	10.7

% TOTAL COOLING HOURS (GREATER THAN 78F ET*) IX - XVII	15.1

% PASSIVE COOLING HOURS IX - XIV	12.6

COOLING IX	5.4
COOLING X	1.1
COOLING XI	5.4
COOLING XII	0.5
COOLING XIII	0.2
COOLING XIV	0.0

% VENTILATION EFFECTIVENESS HOURS IX + X + XI	11.9

% MASS EFFECTIVENESS HOURS X + XI + XII + XIII	7.1

% EVAPORATIVE COOLING EFFECTIVENESS HOURS XI + XIII + XIV + VI.B	5.7

% HOURS BEYOND PASSIVE EFFECTIVENESS VIII + XV + XVI + XVII	13.2

DEHUMIDIFICATION VIII	10.7
DEHUMIDIFICATION AND COOLING XV	1.7
DEHUMIDIFICATION AND COOLING XVI	0.8
COOLING XVII	0.0

Table 33

SALT LAKE CITY, UT   TEMPERATURE (F) ON 21ST DAY OF:

	J	F	M	A	M	J	J	A	S	O	N	D
DAILY MAX DBT	35.5	43.6	51.3	63.2	71.8	80.8	91.5	87.5	72.5	63.2	44.6	38.5
DAILY AVE DBT	26.5	34.8	40.9	53.1	60.1	68.8	77.8	74.5	59.8	49.2	35.4	32.5
DAILY MIN DBT	16.7	26.5	30.6	42.3	47.9	54.6	62.5	61.7	46.2	37.2	27.5	26.3
DAILY MAX DPT	25.8	30.4	30.2	37.8	43.9	47.5	50.5	49.5	45.1	37.8	32.3	30.1
DAILY AVE DPT	19.3	24.8	25.3	31.9	38.1	40.7	44.8	44.3	38.5	32.0	27.5	24.9
DAILY MIN DPT	11.5	19.8	19.5	25.3	32.2	33.6	37.5	38.5	32.4	27.1	22.1	19.9
DAILY MAX WBT	31.2	36.2	39.4	48.1	52.8	57.6	62.8	61.3	53.9	47.5	38.2	33.9
DAILY AVE WBT	24.3	30.9	34.4	42.9	48.5	53.0	58.3	56.8	48.5	41.1	32.2	29.6
DAILY MIN WBT	15.9	25.1	28.6	37.0	42.5	47.4	51.9	51.8	41.4	34.4	26.1	24.7

NORMAL DAILY SOLAR RADIATION (MONTHLY AVG) BTU/SQ FT(DAY)

	J	F	M	A	M	J	J	A	S	O	N	D
HORIZONTAL	639.	989.	1454.	1894.	2362.	2561.	2590.	2254.	1843.	1293.	788.	570.
SO VERTICAL	1129.	1403.	1542.	1401.	1311.	1249.	1328.	1457.	1692.	1657.	1310.	1065.

HEATING DEGREE DAYS BASE 65F		6230.0
COOLING DEGREE DAYS BASE 65F		1273.0
COOLING DEGREE DAYS BASE 78F ET*		275.4

WINTER DESIGN DBT	99.0%	3.0
	97.5%	8.0

SUMMER DESIGN DBT / COINCIDENT WBT	1%	97.0 / 62.0
	2.5%	95.0 / 62.0
	5%	92.0 / 61.0

SUMMER DESIGN WBT	1%	66.0
	2.5%	65.0
	5%	64.0

% TOTAL HEATING HOURS (LESS THAN 68F) I-V	75.5

HEATING I	41.4
HEATING II	8.4
HEATING III	6.9
HEATING IV	7.8
HEATING V	11.0

% HUMIDIFICATION HOURS VI.A + VI.B	3.1

% TOTAL COMFORT HOURS (SHADING REQUIRED) VII	11.0

% DEHUMIDIFICATION HOURS VIII	0.0

% TOTAL COOLING HOURS (GREATER THAN 78F ET*) IX - XVII	10.2

% PASSIVE COOLING HOURS IX - XIV	10.2

COOLING IX	0.0
COOLING X	0.0
COOLING XI	4.5
COOLING XII	0.0
COOLING XIII	3.3
COOLING XIV	2.4

% VENTILATION EFFECTIVENESS HOURS IX + X + XI	4.5

% MASS EFFECTIVENESS HOURS X + XI + XII + XIII	7.8

% EVAPORATIVE COOLING EFFECTIVENESS HOURS XI + XIII + XIV + VI.B	11.6

% HOURS BEYOND PASSIVE EFFECTIVENESS VIII + XV + XVI + XVII	0.0

DEHUMIDIFICATION VIII	0.0
DEHUMIDIFICATION AND COOLING XV	0.0
DEHUMIDIFICATION AND COOLING XVI	0.0
COOLING XVII	0.0

Table 34

```
SAN ANTONIO, TX              TEMPERATURE (F) ON 21ST DAY OF:

             J     F     M     A     M     J     J     A     S     O     N     D

DAILY MAX DBT  59.5  59.9  76.4  80.7  87.1  93.6  94.9  93.8  91.1  80.5  68.0  62.1
DAILY AVE DBT  50.0  52.2  63.7  71.0  76.7  82.8  83.9  83.3  80.4  69.6  58.2  51.4
DAILY MIN DBT  40.6  44.3  50.4  63.3  66.8  73.6  74.8  74.3  71.4  59.3  49.0  40.9

DAILY MAX DPT  46.1  47.7  54.8  65.1  68.6  72.1  72.6  71.4  71.9  60.2  51.4  45.5
DAILY AVE DPT  37.9  40.6  44.1  60.4  64.0  67.8  68.7  66.9  66.8  54.7  43.9  38.8
DAILY MIN DPT  31.0  33.8  31.3  54.7  60.2  63.2  62.8  62.0  60.8  48.3  36.9  30.4

DAILY MAX WBT  51.3  52.3  60.3  68.6  71.6  75.1  75.2  74.7  74.5  65.1  57.5  52.6
DAILY AVE WBT  45.0  47.0  53.6  64.3  68.2  72.6  73.3  72.1  71.3  60.9  51.3  45.6
DAILY MIN WBT  37.8  40.6  43.6  59.4  64.3  70.0  71.1  69.3  67.5  55.2  45.2  37.8
```

```
              NORMAL DAILY SOLAR RADIATION (MONTHLY AVG) BTU/SQ FT(DAY)

             J     F     M     A     M     J     J     A     S     O     N     D

HORIZONTAL   895. 1154. 1450. 1612. 1894. 2069. 2121. 1947. 1638. 1350. 1009.  847.
SO VERTICAL 1160. 1196. 1103.  846.  788.  807.  845.  906. 1069. 1258. 1230. 1159.
```

```
HEATING DEGREE DAYS BASE 65F                 1765.0
COOLING DEGREE DAYS BASE 65F                 3111.3
COOLING DEGREE DAYS BASE 78F ET*              971.6
```

```
WINTER DESIGN DBT                      99.0%   25.0
                                       97.5%   30.0

SUMMER DESIGN DBT / COINCIDENT WBT        1%   99.0 / 72.0
                                        2.5%   97.0 / 73.0
                                          5%   96.0 / 73.0

SUMMER DESIGN WBT                         1%   77.0
                                        2.5%   76.0
                                          5%   76.0
```

```
% TOTAL HEATING HOURS (LESS THAN 68F)  I-V                42.1

      HEATING I                                  8.3
      HEATING II                                 5.6
      HEATING III                                6.0
      HEATING IV                                 8.2
      HEATING V                                 14.0

% HUMIDIFICATION HOURS  VI.A + VI.B                        1.4

% TOTAL COMFORT HOURS (SHADING REQUIRED)  VII            11.8

% DEHUMIDIFICATION HOURS  VIII                           10.6

% TOTAL COOLING HOURS (GREATER THAN 78F ET*) IX - XVII   34.0

% PASSIVE COOLING HOURS IX - XIV                         27.6

      COOLING IX                                10.0
      COOLING X                                  3.0
      COOLING XI                                 8.1
      COOLING XII                                4.4
      COOLING XIII                               1.9
      COOLING XIV                                0.2

% VENTILATION EFFECTIVENESS HOURS   IX + X + XI          21.1

% MASS EFFECTIVENESS HOURS  X + XI + XII + XIII          17.3

% EVAPORATIVE COOLING EFFECTIVENESS HOURS  XI + XIII + XIV + VI.B   10.9

% HOURS BEYOND PASSIVE EFFECTIVENESS  VIII + XV + XVI + XVII        17.1

      DEHUMIDIFICATION  VIII                    10.6
      DEHUMIDIFICATION AND COOLING  XV           5.4
      DEHUMIDIFICATION AND COOLING  XVI          1.1
      COOLING  XVII                              0.0
```

Table 35

```
SAN FRANCISCO, CA              TEMPERATURE (F) ON 21ST DAY OF:
```

	J	F	M	A	M	J	J	A	S	O	N	D
DAILY MAX DBT	53.9	58.4	58.1	62.0	65.5	73.8	70.3	70.6	74.0	68.4	61.1	55.0
DAILY AVE DBT	48.2	51.5	51.5	53.5	56.2	60.7	60.3	60.8	62.5	58.9	52.7	49.4
DAILY MIN DBT	42.6	44.6	46.1	46.9	49.5	52.1	53.7	54.8	54.4	51.8	45.2	43.6
DAILY MAX DPT	45.5	48.1	48.1	50.4	50.9	54.5	54.6	55.7	55.4	52.9	49.5	47.5
DAILY AVE DPT	41.6	44.3	44.1	46.6	47.7	50.5	52.3	53.1	51.3	48.1	43.7	43.2
DAILY MIN DPT	35.4	38.6	40.1	42.8	44.7	47.2	50.3	50.7	47.3	43.1	37.3	38.5
DAILY MAX WBT	48.6	51.6	51.6	54.9	56.4	61.0	60.2	60.6	61.1	57.4	53.2	50.2
DAILY AVE WBT	44.9	48.0	47.8	49.8	51.6	54.8	55.6	56.3	56.1	53.2	48.3	46.5
DAILY MIN WBT	39.1	43.2	44.1	45.3	47.5	49.9	52.1	52.8	51.4	49.2	42.7	42.2

```
              NORMAL DAILY SOLAR RADIATION (MONTHLY AVG) BTU/SQ FT(DAY)
```

	J	F	M	A	M	J	J	A	S	O	N	D
HORIZONTAL	708.	1009.	1455.	1920.	2226.	2377.	2392.	2116.	1742.	1226.	821.	642.
SO VERTICAL	1145.	1311.	1408.	1288.	1126.	1068.	1124.	1240.	1455.	1438.	1250.	1097.

HEATING DEGREE DAYS BASE 65F	3704.9
COOLING DEGREE DAYS BASE 65F	206.8
COOLING DEGREE DAYS BASE 78F ET*	12.9

WINTER DESIGN DBT	99.0%	35.0
	97.5%	38.0

SUMMER DESIGN DBT / COINCIDENT WBT	1%	82.0 / 64.0
	2.5%	77.0 / 63.0
	5%	73.0 / 62.0

SUMMER DESIGN WBT	1%	65.0
	2.5%	64.0
	5%	62.0

% TOTAL HEATING HOURS (LESS THAN 68F) I-V	93.1

HEATING I	7.1
HEATING II	13.6
HEATING III	26.9
HEATING IV	26.4
HEATING V	19.2

% HUMIDIFICATION HOURS VI.A + VI.B	0.2

% TOTAL COMFORT HOURS (SHADING REQUIRED) VII	5.9

% DEHUMIDIFICATION HOURS VIII	0.0

% TOTAL COOLING HOURS (GREATER THAN 78F ET*) IX - XVII	0.8

% PASSIVE COOLING HOURS IX - XIV	0.8

COOLING IX	0.0
COOLING X	0.0
COOLING XI	0.7
COOLING XII	0.0
COOLING XIII	0.1
COOLING XIV	0.0

% VENTILATION EFFECTIVENESS HOURS IX + X + XI	0.7

% MASS EFFECTIVENESS HOURS X + XI + XII + XIII	0.8

% EVAPORATIVE COOLING EFFECTIVENESS HOURS XI + XIII + XIV + VI.B	0.9

% HOURS BEYOND PASSIVE EFFECTIVENESS VIII + XV + XVI + XVII	0.0

DEHUMIDIFICATION VIII	0.0
DEHUMIDIFICATION AND COOLING XV	0.0
DEHUMIDIFICATION AND COOLING XVI	0.0
COOLING XVII	0.0

Table 36

SEATTLE, WA   TEMPERATURE (F) ON 21ST DAY OF:

	J	F	M	A	M	J	J	A	S	O	N	D
DAILY MAX DBT	41.2	47.8	49.6	55.9	63.6	68.7	75.3	73.4	67.5	55.6	47.6	44.5
DAILY AVE DBT	36.2	41.6	43.6	48.0	55.0	59.5	64.5	63.4	57.7	49.1	42.5	40.9
DAILY MIN DBT	31.9	35.6	37.4	40.2	47.1	51.0	54.3	53.6	48.8	43.9	37.6	37.0
DAILY MAX DPT	35.4	39.4	40.9	43.1	47.6	51.0	55.7	55.2	51.9	47.6	41.7	40.8
DAILY AVE DPT	30.8	34.9	37.4	39.5	44.6	48.0	52.9	52.3	48.5	44.6	38.2	37.6
DAILY MIN DPT	26.1	30.2	33.7	35.5	41.0	43.9	49.6	48.6	44.5	41.1	34.6	33.9
DAILY MAX WBT	38.1	42.9	44.6	48.4	53.8	57.0	61.8	61.4	57.2	50.7	44.1	42.5
DAILY AVE WBT	34.2	38.8	40.9	44.1	49.5	53.1	57.7	56.9	52.6	46.8	40.5	39.5
DAILY MIN WBT	30.4	34.4	36.4	39.2	45.2	48.6	52.8	52.1	46.9	42.9	36.4	36.3

NORMAL DAILY SOLAR RADIATION (MONTHLY AVG) BTU/SQ FT(DAY)

	J	F	M	A	M	J	J	A	S	O	N	D
HORIZONTAL	262.	495.	849.	1293.	1714.	1802.	2248.	1616.	1148.	656.	337.	211.
SO VERTICAL	559.	841.	1082.	1166.	1164.	1066.	1405.	1278.	1273.	1007.	674.	478.

HEATING DEGREE DAYS BASE 65F	5690.3
COOLING DEGREE DAYS BASE 65F	246.2
COOLING DEGREE DAYS BASE 78F ET*	19.0

WINTER DESIGN DBT	99.0%	21.0
	97.5%	26.0

SUMMER DESIGN DBT / COINCIDENT WBT	1%	84.0 / 65.0
	2.5%	80.0 / 64.0
	5%	76.0 / 62.0

SUMMER DESIGN WBT	1%	66.0
	2.5%	64.0
	5%	63.0

% TOTAL HEATING HOURS (LESS THAN 68F) I-V	92.7
HEATING I	33.4
HEATING II	18.1
HEATING III	14.8
HEATING IV	15.2
HEATING V	11.2
% HUMIDIFICATION HOURS VI.A + VI.B	0.1
% TOTAL COMFORT HOURS (SHADING REQUIRED) VII	6.0
% DEHUMIDIFICATION HOURS VIII	0.0
% TOTAL COOLING HOURS (GREATER THAN 78F ET*) IX - XVII	1.2
% PASSIVE COOLING HOURS IX - XIV	1.2
COOLING IX	0.0
COOLING X	0.0
COOLING XI	1.1
COOLING XII	0.0
COOLING XIII	0.1
COOLING XIV	0.0
% VENTILATION EFFECTIVENESS HOURS IX + X + XI	1.1
% MASS EFFECTIVENESS HOURS X + XI + XII + XIII	1.2
% EVAPORATIVE COOLING EFFECTIVENESS HOURS XI + XIII + XIV + VI.B	1.2
% HOURS BEYOND PASSIVE EFFECTIVENESS VIII + XV + XVI + XVII	0.0
DEHUMIDIFICATION VIII	0.0
DEHUMIDIFICATION AND COOLING XV	0.0
DEHUMIDIFICATION AND COOLING XVI	0.0
COOLING XVII	0.0

Table 37

TUCSON, AZ   TEMPERATURE (F) ON 21ST DAY OF:

	J	F	M	A	M	J	J	A	S	O	N	D
DAILY MAX DBT	61.3	66.6	72.6	82.5	89.4	100.8	96.9	93.9	93.1	83.0	69.6	64.8
DAILY AVE DBT	48.9	52.6	59.2	68.5	76.5	87.5	84.9	83.4	79.3	68.6	54.8	50.7
DAILY MIN DBT	38.3	38.9	45.0	53.1	60.3	72.6	74.4	73.7	66.2	55.2	41.8	39.3
DAILY MAX DPT	34.4	30.1	29.9	33.4	34.4	44.9	63.9	64.5	51.9	41.6	35.6	30.5
DAILY AVE DPT	28.1	23.9	22.3	27.3	27.7	35.8	59.1	59.7	45.1	36.7	29.6	25.6
DAILY MIN DPT	21.1	16.8	15.3	19.2	21.3	26.5	53.3	53.8	38.8	31.2	23.0	19.8
DAILY MAX WBT	46.1	46.6	49.6	54.8	57.5	64.4	70.3	69.8	64.3	57.6	50.0	46.8
DAILY AVE WBT	40.0	40.7	43.7	49.2	52.4	59.8	67.9	67.7	59.4	52.1	43.5	40.2
DAILY MIN WBT	33.8	33.6	36.8	42.1	45.2	53.8	65.3	64.9	53.6	46.0	36.6	34.3

NORMAL DAILY SOLAR RADIATION (MONTHLY AVG) BTU/SQ FT(DAY)

	J	F	M	A	M	J	J	A	S	O	N	D
HORIZONTAL	1099.	1432.	1864.	2363.	2671.	2730.	2341.	2183.	1979.	1602.	1208.	996.
SO VERTICAL	1529.	1597.	1534.	1339.	1173.	1100.	973.	1089.	1399.	1609.	1583.	1464.

HEATING DEGREE DAYS BASE 65F	2085.9
COOLING DEGREE DAYS BASE 65F	3265.2
COOLING DEGREE DAYS BASE 78F ET*	988.6

WINTER DESIGN DBT	99.0%	28.0
	97.5%	32.0

SUMMER DESIGN DBT / COINCIDENT WBT	1%	104.0 / 66.0
	2.5%	102.0 / 66.0
	5%	100.0 / 66.0

SUMMER DESIGN WBT	1%	72.0
	2.5%	71.0
	5%	71.0

% TOTAL HEATING HOURS (LESS THAN 68F) I-V	47.3

HEATING I	9.6
HEATING II	7.5
HEATING III	7.3
HEATING IV	9.3
HEATING V	13.6

% HUMIDIFICATION HOURS VI.A + VI.B	11.2

% TOTAL COMFORT HOURS (SHADING REQUIRED) VII	10.7

% DEHUMIDIFICATION HOURS VIII	1.3

% TOTAL COOLING HOURS (GREATER THAN 78F ET*) IX - XVII	29.5

% PASSIVE COOLING HOURS IX - XIV	29.2

COOLING IX	0.4
COOLING X	0.3
COOLING XI	11.0
COOLING XII	0.3
COOLING XIII	8.9
COOLING XIV	8.3

% VENTILATION EFFECTIVENESS HOURS IX + X + XI	11.6

% MASS EFFECTIVENESS HOURS X + XI + XII + XIII	20.5

% EVAPORATIVE COOLING EFFECTIVENESS HOURS XI + XIII + XIV + VI.B	32.0

% HOURS BEYOND PASSIVE EFFECTIVENESS VIII + XV + XVI + XVII	1.6

DEHUMIDIFICATION VIII	1.3
DEHUMIDIFICATION AND COOLING XV	0.0
DEHUMIDIFICATION AND COOLING XVI	0.0
COOLING XVII	0.3

Table 38

WASHINGTON, DC   TEMPERATURE (F) ON 21ST DAY OF:

	J	F	M	A	M	J	J	A	S	O	N	D
DAILY MAX DBT	45.1	45.9	54.5	70.2	75.7	82.2	88.1	84.2	75.7	64.6	54.9	42.5
DAILY AVE DBT	37.0	37.9	46.0	59.4	66.4	74.4	79.5	75.5	66.8	55.5	47.1	36.0
DAILY MIN DBT	30.0	30.5	37.7	48.6	57.9	66.5	71.5	67.8	58.7	47.1	39.5	30.1
DAILY MAX DPT	34.7	33.2	38.5	48.8	58.2	67.8	70.9	68.2	61.4	48.0	41.5	30.4
DAILY AVE DPT	25.7	24.1	32.7	42.2	52.0	62.7	66.9	64.4	56.5	43.1	33.9	24.7
DAILY MIN DPT	17.5	15.0	26.8	35.7	46.3	58.4	63.3	59.8	50.8	38.5	27.2	17.8
DAILY MAX WBT	40.3	39.2	45.2	55.6	62.5	70.9	74.0	71.4	65.4	54.1	47.2	36.9
DAILY AVE WBT	33.2	33.4	40.3	50.7	58.2	66.8	71.0	68.2	60.8	49.3	41.8	32.4
DAILY MIN WBT	26.9	26.9	34.6	44.3	53.4	63.3	67.8	65.1	55.5	43.9	35.7	27.4

NORMAL DAILY SOLAR RADIATION (MONTHLY AVG) BTU/SQ FT(DAY)

	J	F	M	A	M	J	J	A	S	O	N	D
HORIZONTAL	572.	815.	1125.	1459.	1718.	1901.	1817.	1617.	1340.	1004.	651.	481.
SO VERTICAL	959.	1097.	1130.	1019.	902.	882.	884.	986.	1163.	1221.	1027.	852.

HEATING DEGREE DAYS BASE 65F	4407.7
COOLING DEGREE DAYS BASE 65F	1474.2
COOLING DEGREE DAYS BASE 78F ET*	322.0

WINTER DESIGN DBT	99.0%	14.0
	97.5%	17.0

SUMMER DESIGN DBT / COINCIDENT WBT	1%	93.0 / 75.0
	2.5%	91.0 / 74.0
	5%	89.0 / 74.0

SUMMER DESIGN WBT	1%	78.0
	2.5%	77.0
	5%	76.0

% TOTAL HEATING HOURS (LESS THAN 68F) I-V	66.4

HEATING I	29.0
HEATING II	7.9
HEATING III	7.7
HEATING IV	7.7
HEATING V	14.0

% HUMIDIFICATION HOURS VI.A + VI.B	0.3

% TOTAL COMFORT HOURS (SHADING REQUIRED) VII	12.4

% DEHUMIDIFICATION HOURS VIII	6.7

% TOTAL COOLING HOURS (GREATER THAN 78F ET*) IX - XVII	14.2

% PASSIVE COOLING HOURS IX - XIV	11.4

COOLING IX	4.3
COOLING X	0.8
COOLING XI	5.8
COOLING XII	0.1
COOLING XIII	0.2
COOLING XIV	0.2

% VENTILATION EFFECTIVENESS HOURS IX + X + XI	10.9

% MASS EFFECTIVENESS HOURS X + XI + XII + XIII	6.9

% EVAPORATIVE COOLING EFFECTIVENESS HOURS XI + XIII + XIV + VI.B	6.3

% HOURS BEYOND PASSIVE EFFECTIVENESS VIII + XV + XVI + XVII	9.5

DEHUMIDIFICATION VIII	6.7
DEHUMIDIFICATION AND COOLING XV	1.3
DEHUMIDIFICATION AND COOLING XVI	1.0
COOLING XVII	0.5

Table 39

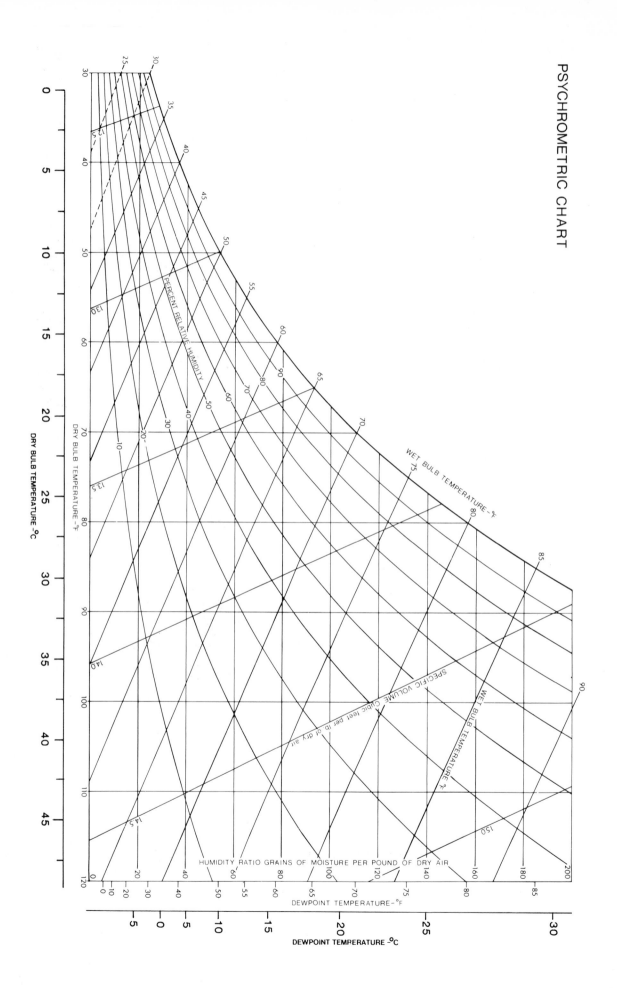

PSYCHROMETRIC CHART

Virginia residence, designed by William Turnbull of MLTW/Turnbull Associates, features shading trellis and interior light-court.

Part IV: Bibliography

The sections of the bibliography parallel those of the text:

Principles
Tools and Techniques
Texts and Overview
Site Planning
Building Massing
Building Plan
Building Envelope
Building Openings
Information Sources

The first three headings, *Principles, Tools and Techniques,* and *Texts and Overview,* correspond in content to the subjects discussed in Part I. References cited in Part I almost always are found in the *Principles* listing of the bibliography, although sometimes these appear under the appropriate heading of *Practices.*

The reader who is interested in more detail about any design technique or practice described in Part II will find technical references listed under the corresponding headings *Site Planning, Building Massing, Building Plan, Building Envelope,* and *Building Openings.* A listing of information sources abbreviated throughout the bibliography concludes it.

Individual titles appear only once in the bibliography. However, because some of the most valuable discussions of specific subjects appear in textbooks, many chapters of both single-author and edited texts are listed under the appropriate topical headings. The texts so divided are listed in full under the heading *Texts and Overview.*

Principles

Abrams, D.W. and C.C. Benton, J.M. Akridge, 1980. "Simulated and Measured Performance of Earth Cooling Tubes", *5th NPSC, AS/ISES,* pp. 737-741.

Air Force Manual (AFM) 88-8, "Engineering Weather Data", Chapter 6, June 1976.

Akins, R.E. and J.A. Peterka, J.E. Cermak, 1980. "Averaged Pressure Coefficients for Rectangular Buildings," J.E. Cermak, ed., *Wind Engineering,* Proceedings of the 5th International Conference, Pergamon Press, New York, vol. I, pp. 369-380.

Akridge, J.M., 1981. "A Decremented Average Ground Temperature Method for Estimating the Thermal Performance of Underground Houses," *Passive Cooling '81,* AS/ISES, pp. 141-145.

Arens, E.A. and L.E. Flynn, D.N. Nall, K. Ruberg, 1980. *Geographical Extrapolation of Typical Hourly Weather Data for Energy Calculation in Buildings,* NBS Building Science Series 126, 121 pp. (U.S. GPO LC No. 80-600059).

Arens, E. and R. Gonzalez, L. Berglund, P. McNall, L. Zeren, 1980. "A New Bioclimatic Chart for Passive Solar Design," in J. Hayes and R. Snyder, eds., *5th NPSC,* AS/ISES, Univ. of Delaware, Newark, pp. 1202-1206.

Arens, E.A. and P.B. Williams, 1977. "The Effect of Wind on Energy Consumption in Buildings," *Energy and Buildings* (Elsevier Sequoia), vol. 1, no. 1, May 1977, pp. 77-84.

Arnold, F. and R. Barlow, K. Collier, 1981. *Dehumidification in Passively Cooled Buildings,* SERI/TR-631-995, May 1981, 49 pp. (draft).

Atwater, M.A. and J.T. Ball, 1978. "Computation of IR Sky Temperature and Comparison with Surface Temperature," *Solar Energy* (Pergamon), vol. 21, no. 3, pp. 211-216.

Aynsley, R.M., 1980a. "Tropical Housing Comfort by Natural Airflow," *Building Research and Practice* (U.K.), July/August 1980, pp. 242-252.

Aynsley, R.M., 1980b. "Wind-Generated Natural Ventilation of Housing for Thermal Comfort in Hot Humid Climates," J.E. Cermak, ed., *Wind Engineering,* Proceedings of the 5th International Conference, Pergamon Press, New York, pp. 243-254.

Aynsley, R. and W. Melbourne, B.J. Vickery, 1977. *Architectural Aerodynamics,* Applied Science Publishers, Ltd., London.

Beckman, W.A. *et al.,* 1978. "Units and Symbols in Solar Energy," *Solar Energy* (Pergamon), vol. 21, no. 1, pp. 65-68.

Beranek, W.J., 1980. "General Rules for Determination of Wind Environment," J.E. Cermak, ed., *Wind Engineering,* Proceedings of the 5th International Conference, Pergamon Press, New York, vol. I, pp. 225-234.

Berdahl, P. and M. Martin, 1981. "Thermal Radiance of Skies with Low Clouds," *Passive Cooling '81,* AS/ISES, pp. 266-269.

Berglund, L.G. and A.P. Gagge, 1979. "Thermal Comfort and Radiant Heat," *3rd NPSC,* AS/ISES, pp. 260-265.

Bliss, R.W., 1961. "Atmospheric Radiation Near the Surface of the Ground: A Summary for Engineers," *Solar Energy,* vol. 5, no. 3, pp. 103-120.

Bowen, A., 1981. "Classification of Air Motion Systems and Patterns," *Passive Cooling '81,* AS/ISES, pp. 743-763.

Boyd, D.W., 1975. *Climatic Information for Building Design in Canada,* Supplement No. 1 to the National Building Code of Canada, NRCC No. 13986, DBR/NRCC, Ottawa, 47 pp.

Brown, G.Z. and B.J. Novitski, 1979. "Architectural Response to Climatic Patterns," *4th NPSC,* AS/ISES, pp. 259-262.

Building Climatology: Proceedings of the Symposium on Urban Climates and Building Climatology (Volume II), 1969. Technical Note No. 109, World Meteorological Organization, Geneva (UNIPUB).

Burns, A.P., 1980. "Wind-Induced Heat Loss from Buildings," J.E. Cermak, ed., *Wind Engineering,* Proceedings of the 5th International Conference, Pergamon Press, New York, vol. I, pp. 255-265.

Bush, A.W. and R.D. Gibson, 1979. "A Theoretical Investigation of Thermal Contact Conductance," *Appl. Energy,* vol. 5:11.

Businger, J.A., 1963. "The Glasshouse Climate," Chapter 9, W.R. van Wijk, ed., *Physics of Plant Environment,* Wiley-Interscience/John Wiley and Sons, Inc., New York, pp. 277-318. Note: discusses climate and climate control of commercial greenhouses.

Campbell, G.S., 1977. *An Introduction to Environmental Biophysics,* Springer-Verlag, New York, 159 pp.

Chandler, T.J., 1969. *The Air Around Us,* The Natural History Press/Doubleday and Co., Garden City, NY, 156 pp.

Chandler, T.J., 1976. *Urban Climatology and its Relevance to Urban Design,* Technical Note No. 149, World Meteorological Organization, Geneva, 61 pp. (UNIPUB).

Chang, Jen-hu, 1958. *Ground Temperature,* vol. I, Blue Hill Meteorological Observatory, Harvard Univ., Milton, MA.

Chen, Z. and J.W. Mitchell, 1981. "The Cooling Effect of Night Ventilation," *6th NPSC,* AS/ISES, pp. 832-836.

Clark, G., 1981. "Passive/Hybrid Comfort Cooling by Thermal Radiation," *Passive Cooling '81,* AS/ISES, pp. 682-714.

Clark, G. and F. Loxsom, B. Shutt, J. Faultersack, 1981. "Simple Estimation of Temperature and Nocturnal Heat Loss for a Radiantly Cooled Roof Mass," *Passive Cooling '81,* AS/ISES, pp. 244-248.

"Climate and Architecture," Theme issue, *Research and Design,* Quarterly of the AIA Research Corp., vol. 2, no. 2, Spring 1979, 16 pp. (AIA/RC, 1735 New York Ave. NW, Washington, DC 20006).

The Climate of Canada. Meteorological Branch, Air Services, Dept. of Transport, Toronto, Ontario, 1962, 74 pp. (distribution by Information Canada, Ottawa KIA 059).

Climates of the States, NOAA Narrative Summaries, Tables and Maps for each State, 2nd ed. (1980), Gale Research Co., Book Tower, Detroit, MI 48226, 2 vol., 1175 pp.

Climatic Atlas of the United States. Environmental Data Service, Environmental Science Services Administration (reprinted in 1977 by NOAA), National Climatic Center, Federal Building, Asheville, NC 28801, 1968, 80 pp.

Collins, J.O., 1981. "Air Leakage Measurement and Reduction Techniques on Electrically Heated Homes," *Thermal Performance of the Exterior Envelope of Buildings,* Special Publication No. 28, ASHRAE, pp. 195-209.

Cook, J., 1979. "Evaporative Cooling Systems," *Cool Houses for Desert Suburbs,* Arizona Solar Energy Commission, 1700 West Washington, Phoenix, AZ 85007, pp. 28-38.

Crawford, C.B., 1952. "Soil Temperatures, A Review of Published Records," *Frost Action in Soils,* Highway Research Board Special Report No. 2, NAS/NRC Pub. No. 213, National Research Council, Washington, D.C.

Crow, L.W., 1976. "Weather Data Related to Evaporative Cooling," ASHRAE Paper No. 2223, January 23, 1972.

Crow, L.W., 1976. "The Wind Blows This-a-way and That-a-way," ASHRAE *Transactions,* vol. 82, part I, p. 1039.

Crow, L.W. and W.L. Holladay, 1976a. *Report on the Approach and Methodology used to Prepare the State Climate Zone Maps,* Report P400-76-017, California Energy Commission, 1111 Howe Avenue, Sacramento, CA 95825, July 1976, 12 pp.

Crow, L.W. and W.L. Holladay, 1976b. *California Climatic Thermal Zones Related to Energy Requirements for Heating, Ventilating and Air Conditioning,* Report P400-76-018, California Energy Commission, 1111 Howe Ave., Sacramento, CA 95825, November 1976, 59 pp.

Dalgliesh, W.A. and D.W. Boyd, 1962. *Wind on Buildings,* Canadian Building Digest CBD 28, April 1962, 4 pp. (DBR/NRCC).

Danby, M., 1973. "The Design of Buildings in Hot-dry Climates and the Internal Environment," *Build International* (Applied Science Publishers, Ltd., London) vol. 6, no. 1, January/February 1973, pp. 55-76.

Dick, J.B., 1950. "The Fundamentals of Natural Ventilation in Houses," *Jrnl. of the Institution of Heating and Ventilating Engineers,* vol. 18, no. 179, June 1950, pp. 123-134.

Evans, M., 1980. "Climatic Types in Relation to Building Needs," Chapter 5, *Housing, Climate and Comfort,* Halsted Press/John Wiley and Sons, Inc., New York, pp. 44-53.

Evans, M., 1980. "The Basic Data for Climatic Design," Chapter 2, *Housing, Climate and Comfort,* Halsted Press/John Wiley and Sons, Inc., New York, pp. 4-16.

Evans, M., 1980. "Thermal Comfort," Chapter 3, *Housing, Climate and Comfort,* Halsted Press/John Wiley and Sons, Inc., New York, pp. 17-32.

Fanger, P.O., 1972. *Thermal Comfort: Analysis and Applications in Environmental Engineering,* McGraw Hill Book Co., New York, 244 pp.

"Fenestration," Chapter 27, *1981 Fundamentals Handbook,* ASHRAE, pp. 27.1-27.48.

Fourt, L. and R.S. Hollies, 1970. *Clothing: Comfort and Function,* Marcel Dekker, Inc., New York, 254 pp.

Gagge, A.P. and Y. Nishi, 1976. "Physical Indices of the Thermal Environment," ASHRAE *Journal,* vol. 18, no. 1, January 1976, pp. 47-51.

Geiger, R., 1965. *The Climate Near the Ground,* Harvard Univ. Press, Cambridge, 611 pp.

Giesecke, F.E., 1950. "Night-Air Cooling," Paper 1379, *Transactions* of the American Society of Heating and Ventilating Engineers, vol. 56, pp. 45-50.

Givoni, B., 1976. "Design Factors Affecting Ventilation," Chapter 15, *Man, Climate and Architecture,* 2nd ed., Applied Science Publishers, Ltd., London (1981 by Van Nostrand Reinhold, New York), pp. 289-306.

Givoni, B., 1976. "Heating and Cooling of Buildings by Natural Energies—An Overview," Chapter 18, *Man Climate and Architecture,* 2nd ed., Applied Science Publishers, Ltd., London (1981 by Van Nostrand Reinhold, New York), pp. 373-413.

Givoni, B., 1976. "Principles of Design and Choice of Materials to Adapt Building to Climate," Chapters 16 and 17, *Man, Climate and Architecture,* 2nd ed., Applied Science Publishers, Ltd., London (1981 by Van Nostrand Reinhold, New York), pp. 307-372.

Goldman, R.F., 1978. "The Role of Clothing in Achieving Acceptability of Environmental Temperatures between 65F and 85F," J.A.J. Stolwijk, ed., *Energy Conservation Strategies in Buildings: Comfort, Acceptability, and Health,* J.B. Pierce Fndn. of CT, Inc., Yale Univ., New Haven, pp. 38-52.

Gonzalez, R.R., 1977. "Experimental Analysis of Thermal Acceptability," B.W. Mangum and J.E. Hill, eds., *Thermal Analysis—Human Comfort—Indoor Environments,* NBS Special Publication (SP) 491, pp. 131-151, (U.S. GPO SN 003-003-01849-0).

Gordon, J.M. and Y. Zarmi, 1981. "Transient Behavior in Radiative Cooling," *Solar Energy* (Pergamon), vol. 27, no. 1, pp. 77-79.

Grogger, P.K., 1979. "The Utilization of Microclimate to Conserve Energy in Single and Multi-Family Dwellings," *Technology for Energy Conservation* conference proceedings, Tucson (Information Transfer).

Hall, E.T., 1979. "Let's Heat People Instead of Houses," *Human Nature,* January 1979, pp. 45-47.

Handa, K., 1979. *Wind Induced Natural Ventilation,* D10: 1979, Svensk Byggtjanst, Stockholm.

Hans, G.E., 1981. "Proposed Streamlined Residential Heating Energy Budget Analysis by a Variable Temperature Design Method," *Thermal Performance of the Exterior Envelopes of Buildings,* Special Publication No. 28, ASHRAE, pp. 16-32.

Harrje, D.T. and G.S. Dutt, J. Beyea, 1979. "Locating and Eliminating Obscure but Major Energy Losses in Residential Housing," ASHRAE *Transactions,* vol. 85, pt. II.

Haves, P. and D. Bentley, G. Clark, 1980. *Heat Transfer in Passively Cooled Buildings,* Publication 80-HT-65, American Society of Mechanical Engineers, 345 E. 47th Street, New York, NY 10017, 7 pp.

"Heat Transfer," Chapter 2, *1981 Fundamentals Handbook,* ASHRAE pp. 2.1 - 2.34.

Hess, A.J., 1950. "Plant Research in the Phytotron," Paper No. 1397, *Transactions,* Am. Society of Heating and Ventilating Engineers, vol. 56, pp. 341-354.

Ingersoll, L.R. and O.J. Zobel, A.C. Ingersoll, 1954. "Theory of Earth Heat Exchangers," Chapter 13, *Heat Conduction, with Engineering, Geological and Other Applications,* Revised Edition, Univ. of Wisconsin Press, Madison.

Jennings, B.H. and J.A. Armstrong, 1971. "Ventilation Theory and Practice," ASHRAE *Transactions,* part I.

Kaushik, N.D. and A. Srivastava, 1980. "Temperature Distribution in Ground: Response Function Technique," *International Journal of Heat and Mass Transfer* (Pergamon), vol. 23, no. 6, pp. 903-906.

Keast, D.N. *Acoustic Isolation of Infiltration Openings in Buildings,* Report No. BNL 50952, prepared by Bolt, Beranek and Newman, Inc., Brookhaven National Laboratories, (Attn: Mr. M.L. Woodworth), Upton, NY 11978.

Keast, D.N. and H.S. Pei, 1981. "The Use of Sound to Locate Infiltration Openings in Buildings," *Thermal Performance of the Exterior Envelope of Buildings,* Special Publication No. 28, ASHRAE, pp. 85-93.

Knowles, R., 1967. *The Sun: Heat and Light,* School of Architecture and Fine Arts, Univ. of Southern California, Los Angeles. CA 90007.

Kusuda, T., 1968. *Least Squares Analysis of Annual Earth Temperature Cycles for Selected Stations in the Contiguous United States,* NBS Report 9493, 150 pp.

Kusuda, T. and K. Ishii, 1977. *Hourly Solar Radiation Data for Vertical and Horizontal Surfaces on Average Days in the U.S. and Canada,* NBS Building Science Series 96, April 1977, 412 pp. (U.S. GPO SD No. C13.29/2:96).

Labs, K., 1979. "Underground Building Climate," *Solar Age,* vol. 4, no. 10, October 1979, pp. 44-50.

Labs, K., 1981a. *Regional Analysis of Ground and Above-Ground Climate,* U.S. DOE/Oak Ridge National Laboratory, 410 pp. (NTIS No. ORNL/Sub-81/40451/1); republished in *Underground Space* (Pergamon), vol. 6, no. 6, May/June 1982.

Labs, K. and K. Harrington, 1982. "A Comparison of Ground and Above Ground Climates for Identifying Appropriate Cooling Strategies," *Passive Solar Journal* (AS/ISES), vol. 1, no. 1, January 1982, pp. 4-11.

Labs, K. and D. Watson, 1981. "Regional Suitability of Earth Tempering," L. L. Boyer, ed., *Earth Shelter Performance and Evaluation* conference proceedings, Oklahoma State Univ. Arch. Extension, Stillwater, OK 74078, October 1981, pp. 39-51.

Lamm, L.O., 1981. "A New Analytic Expression for the Equation of Time," *Solar Energy* (Pergamon), vol. 26, no. 5, p. 465.

Landro, B. and P.G. McCormick, 1980. "Effect of Surface Characteristics and Atmospheric Conditions on Radiative Loss to a Clear Sky," *International Journal of Heat and Mass Transfer* (Pergamon), vol. 23, no. 5, pp. 613-620.

Landsberg, H.E., 1947. "Microclimatology," *Architectural Forum,* vol. 86, no. 3, March 1947, pp. 114-119.

Landsberg, H.E., 1972. *The Assessment of Human Bioclimate: A Limited Review of Physical Parameters,* Technical Note No. 123, World Meteorological Organization, Geneva, 36 pp. (UNIPUB).

Lee, D.H.K., 1963. "Basic Climatology of Tropical and Subtropical Regions," Chapter 2, *Physiological Objectives in Hot Weather Housing,* HUD Office of International Affairs, Room 2118, Washington, D.C. 20410, pp. 11-26.

Lee, M., 1978. "Evaporative Cooling: Energy-Saving Alternative with a Record of Consumer Acceptance," *Technology for Energy Conservation* conference proceedings, Albuquerque, pp. 556-564 (Information Transfer); Note: includes map of effectiveness index for the U.S.

Lee, R., 1973. "The 'Greenhouse' Effect," *Journal of Applied Meteorology* (Am. Met. Soc.), vol. 12, no. 3, April 1973, pp. 556-557.

Lee, R., 1974. Reply to Comment by E.X. Berry, *Journal of Applied Meteorology* (Am. Met. Soc.), vol. 13, no. 5, August 1974, pp. 603-606.

Leisy, J., 1981. *Calories In/Calories Out: The Energy Budget Way to Weight Control,* Stephen Greene Press, Brattleboro, VT 05301, 116 pp.

List, R.J., 1958. *Smithsonian Meteorological Tables,* 6th ed., Publication 4014, Smithsonian Institution, Washington, DC.

Loftness, V., 1977. *Identifying Climatic Design Regions and Assessing Climatic Impact on Residential Building Design,* Technical Paper No. 1, AIA Research Corp., Washington, DC., 25 pp.

Loxsom, F.M. and B. Kelly, 1980. "Evaporative Cooling: Roofs as Dissipators," *Passive Cooling Handbook,* prepared for the Passive Cooling Workshop held in conjunction with the 5th NPSC, Pub-375, Lawrence Berkeley Laboratory, Berkeley, CA, pp. 81-110.

Mangum, B.W. and J.E. Hill, eds., 1977. *Thermal Analysis—Human Comfort—Indoor Environments,* NBS Special Publication (SP) 491, 191 pp. (U.S. GPO SN 003-003-01849-0).

Markus, T.A. and E.N. Morris, 1980. "People's Response to the Thermal Environment," Chapter 3, *Buildings, Climate and Energy,* Pitman Publishers, Marshfield, MA, pp. 33-139.

Markus, T.A. and E.N. Morris, 1980. "Climate," Chapter 4, *Buildings, Climate and Energy,* Pitman Publishers, Marshfield, MA, pp. 140-267.

Mather, J.R., 1974. *Climatology: Fundamentals and Applications,* McGraw-Hill Book Co., New York, 412 pp.

McIntyre, D.A., 1980. *Indoor Climate,* Applied Science Publishers, Ltd., London, 443 pp. (International Ideas, Inc., Philadelphia).

Menicucci, D.F., 1978. "A Method for Classifying Climate for Use in the Design of Thermally Efficient Structures," *Technology for Energy Conservation* conference proceedings, Albuquerque, pp. 528-535 (Information Transfer).

Mills, C.A., 1950a. "My House was in the Lab for 20 Years," *House Beautiful,* vol. 92, no. 10, October 1950, pp. 180-188+.

Mills, C.A., 1950b. "Experimental Cooling-Heating System," *Architectural Forum,* vol. 93, no. 5, November 1950, pp. 127-131.

Milne, M., 1976. "Sun Motion and Control of Incident Solar Radiation," Chapter 10, B. Givoni, *Man, Climate and Architecture,* 2nd ed., Applied Science Publishers, Ltd., London (1981 by Van Nostrand Reinhold, New York), pp. 181-212.

Milne, M. and G. Givoni, 1979. "Architectural Design Based on Climate," D. Watson, ed., *Energy Conservation Through Building Design,* Architectural Record Books/McGraw-Hill Book Co., New York, pp. 96-113.

Monteith, J.L., 1973. *Principles of Environmental Physics,* American Elsevier Publishing Co., New York, 241 pp.

Mosley, J. and G. Clark, 1981. "Quantitative Solar-Climatology with Space Heating Applications," *1981 Annual Meeting,* AS/ISES, pp. 1516-1520.

Nylund, P.O., 1980. *Infiltration and Ventilation,* D22:1980, Svensk Byggtjanst, Stockholm.

Oke, T.R., 1978. *Boundary Layer Climates,* Chapman and Hall (England), 372 pp.

Oke, T.R., 1979. *Review of Urban Climatology,* Technical Note No. 169, World Meteorological Organization, Geneva, 100 pp. (UNIPUB).

Olgyay, V. and A. Olgyay, 1953. *Application of Climatic Data to House Design,* U.S. Housing and Home Finance Agency, Washington, D.C., 2 vol.

Olgyay, V., 1963. "The Bioclimatic Approach," Chapter 2, *Design With Climate,* Princeton Univ. Press, Princeton, NJ, pp. 14-23.

Oliver, J.E., 1973. *Climate and Man's Environment: An Introduction to Applied Climatology,* J. Wiley and Sons, Inc., New York, 517 pp.

Page, J.K., 1976. *Application of Building Climatology to the Problems of Housing and Building for Human Settlements,* Technical Note No. 150, World Meteorological Organization No. 441, Geneva, 64 pp. (UNIPUB).

Page, J.K., 1979. "Systematic Classification of Climate for Solar House Design," A.A.M. Sayigh, ed., *Solar Energy Applications in Buildings,* Academic Press, New York, pp. 41-79.

Parmelee, G.V. and W.W. Aubele, 1952. "Radiant Energy Emission at the Atmosphere and Ground," Paper 1442, ASHVE *Transactions,* vol. 58, pp. 85-106 (much outdated, but nevertheless a useful discussion of the issues).

Passmore, R. and J.V.G.A. Durnin, 1967. *Energy, Work and Leisure,* Heinemann Educational Books, Ltd., London.

Petterssen, S., 1969. *Introduction to Meteorology,* 3rd ed., McGraw-Hill Book Co.

"Physiological Principles, Comfort, and Health," Chapter 8, *1981 Fundamentals Handbook,* ASHRAE, pp. 8.1 - 8.34.

Poulos, J., 1982. "Thermal Performance of Underground Structures: The Development of the Decremented Average Ground Temperature Method for Estimating the Thermal Performance of Underground Walls," Master of Architecture Thesis, Georgia Inst. of Technology, 113 pp.

Pound, R.V., 1980. "Radiant Heat for Energy Conservation," *Science,* vol. 208, no. 4443, May 2, 1980, pp. 494-495.

Principles of Natural Ventilation, BRE Digest No. 210, Building Research Establishment, Garston, Watford, WD27JR, February 1978, 7 pp.

Roseme, G.D. *et. al.,* 1981. "Residential Ventilation with Heat Recovery: Improving Indoor Air Quality and Saving Energy," *Thermal Performance of the Exterior Envelope of Buildings,* Special Publication No. 28, ASHRAE, pp. 609-623.

Rosenberg, N.J., 1974. *Microclimate: The Biological Environment,* John Wiley and Sons, Inc., New York.

Sachs, P., 1978. "Buildings," Chapter 7, *Wind Forces in Engineering,* 2nd ed., Pergamon Press, New York, pp. 212-243.

Sayigh, A.A.M., 1981 "Greenhouse and Arbor Cooling," *Passive Cooling '81,* AS/ISES, pp. 782-792.

Scott, N.R. and R.A. Parsons, T.A. Koehler, 1965. "Analysis and Performance of an Earth-Air Heat Exchanger," Paper No. 65-840, *American Society of Agricultural Engineers 1965 Winter Meeting,* ASAE, Box 229, St. Joseph, MI, 46 pp.

Seemann, J., ed. *Climate Under Glass,* Technical Note No. 131, World Meteorological Organization, Geneva, 1974, 40 pp. (UNIPUB).

Selective Guide to Climatic Data Sources Key to Meteorological Records Documentation No. 4.11). National Climatic Center, Federal Building, Asheville, NC, 28801, December 1979, 142 pp.

Sellers, W.D., 1965. *Physical Climatology,* Univ. of Chicago Press, Chicago, 272 pp.

Shah, M.M., 1981. "Estimation of Evaporation from Horizontal Surfaces," Paper No. 2612, ASHRAE *Transactions,* vol. 87. part I, pp. 35-51.

Shelton, J.W., 1976. "The Energy Cost of Humidification," ASHRAE *Journal*, vol. 18, no. 1, January 1976, pp. 52-55.

Shurcliff, W.A., 1981. *Air-to-Air Heat Exchangers for Houses,* published by the author, 19 Appleton St., Cambridge, MA 02138, 196 pp.

Silverstein, S.D., 1976. "Effect of Infrared Transparency on the Heat Transfer through Greenhouses: A Clarification of the Greenhouse Effect," *Science* (AAAS), vol. 193, no. 4249, July 16, 1976, pp. 229-231.

Siple, P., 1950. "Climatic Criteria for Building Construction," *Weather and the Building Industry,* BRAB Conference Report No. 1, National Academy of Sciences/National Research Council, pp. 5-22.

Smith, D.L., 1979. "Mean Radiant Temperature and its Effect on Energy Conservation," *4th NPSC,* AS/ISES, pp. 431-435.

Smith, W.T., ed., 1955. "Evaporative Cooling: A Symposium," *Heating, Piping and Air Conditioning* (ASHRAE Journal Section), vol. 27, no. 8, August 1955, pp. 141-147 (evaporative space cooling; contains maps of effectiveness in the U.S.).

Sodha, M.S. and A.K. Khatry, M.A.S. Malik, 1978. "Reduction of Heat Flux through a Roof by Water Film," *Solar Energy* (Pergamon), vol. 20, no. 2, pp. 189-191.

Stein, R.G., 1977. *Architecture and Energy,* Anchor Press, Garden City, NY, 322 pp.

Stolwijk, J.A.J., ed., 1978. *Energy Conservation Strategies in Buildings: Comfort, Acceptability, and Health,* John B. Pierce Fndn. of CT, Inc., Yale Univ., New Haven, 216 pp.

Sutton, G.E., 1950. "Roof Spray for Reduction in Transmitted Solar Radiation," *Heating, Piping and Air Conditioning* (ASHVE Journal Section), September, 1950, pp. 131-137.

Symposium on Weather Data for Building Design, ASHRAE (MF)PH-79-8, Philadelphia Meeting, January 1979 (microfiche edition).

Symposium on Weather Data Sources for Solar Design, ASHRAE (MF) DE-79-2, Detroit Meeting, June 1979 (microfiche edition).

Teaching the Teachers on Building Climatology, CIB Proceedings No. 25, National Swedish Institute for Building Research Document D20:1973, Svensk Byggtjanst, 306 pp.

Terjung, W.H., 1966. "Physiologic Climates of the Coterminous United States: A Bioclimatic Classification Based on Man," *Annals of the Association of American Geographers,* vol. 56, March 1966, pp. 141-179.

Terjung, W.H., 1967. "Annual Physioclimatic Stresses and Regimes in the United States," *The Geographic Review,* vol. 57, pp. 225-240.

Urban Climates: Proceedings of the Symposium on Urban Climates and Building Climatology (Volume I), Technical Note No. 108, World Meteorological Organization, Geneva, 1969 (UNIPUB).

Van Straaten, J.F., 1967. "Natural Ventilation," Chapter 14, *Thermal Performance of Buildings,* American Elsevier Publishing Co., Inc., New York, pp. 228-283.

van Wijk, W.R., ed., 1963. *Physics of Plant Environment,* Wiley-Interscience/John Wiley and Sons, Inc., New York, 382 pp. (note: a classic modern text on soil-radiation-atmospheric physics).

"Ventilation and Infiltration," Chapter 22, *1981 Fundamentals Handbook,* ASHRAE, pp. 22.1 - 22.20.

Visher, S.S., 1924. *Climatic Laws,* John Wiley & Sons, New York, 96 pp.

Visher, S.S., 1954. *Climatic Atlas of the United States,* Harvard Univ. Press, Cambridge, 403 pp. (1031 maps)

Watson, D., 1980. "The Energy Design Process," H.W. Richardson and D. Bremer, eds., *Energy Conscious Design,* AIA Research Corp., Washington, D.C., March 1980, pp. 6.1 - 6.30.

Watt, J.R., 1963. *Evaporative Air Conditioning,* The Industrial Press, New York.

Weather and the Building Industry, BRAB Conference Report No. 1, National Academy of Sciences/National Research Council, 1950, 158 pp.

"Weather Data and Design Conditions," Chapter 24, *1981 Fundamentals Handbook,* ASHRAE, pp. 24.1 - 24.28.

Weather Data Handbook for HVAC and Cooling Equipment Design, Ecodyne Corporation, McGraw-Hill Book Co., New York, 1980. (13 chapters)

Weichman, F.L., 1981. "Channelled Plastic for Greenhouse and Skylight Use," *Solar Energy* (Pergamon), vol. 27, no. 6, pp. 571-575.

Wilkinson, B.J., 1981. "An Improved FORTRAN Program for Calculating the Solar Position," *Solar Energy* (Pergamon), vol. 27, no. 1, pp. 67-68.

Willmott, C.J. and M.T. Vernon, 1980. "Solar Climates of the Coterminous U.S.: A Preliminary Investigation," *Solar Energy* (Pergamon), vol. 24, no. 3, pp. 295-303.

Wood, R.W., 1909. "Note on the Theory of the Greenhouse," The London, Edinburgh and Dublin *Philosophical Magazine* and Journal of Science, vol. XVII - 6th Series, no. 98, February 1909, pp. 319-320.

Yeang, K., 1974. "Energetics of the Built Environment," *Architectural Design* (London), vol. 44, no. 7, July 1974, pp. 446-451.

Yellott, J.I., 1966. "Roof Cooling with Intermittent Water Sprays," *ASHRAE Transactions,* vol. 72, part II, pp. III 1.2 - 1.10.

Yellott, J.I., 1978b. "Solar Energy Utilization for Heating and Cooling," Chapter 58, *1978 Applications Handbook,* ASHRAE.

Yellott, J.I., 1981a. "Passive and Hybrid Cooling Methods," *Passive and Hybrid Cooling Workshop Handbook,* prepared for the P & H Clg. Wkshop held in conjunction with the 1981 AS/ISES Annual Meeting, Philadelphia, May 26, 1981, pp. I.9 - I.28.

Yellott, J.I., 1981b. "Evaporative Cooling," *Passive Cooling '81,* AS/ISES, pp. 764-772.

Yoshino, M.M., 1975. *Climate in a Small Area,* Univ. of Tokyo Press, 549 pp.

Tools and Techniques

Abrams, D.W., 1981. "A Passive Solar Performance Fraction Relative to the Actual Building," *6th NPSC,* AS/ISES, pp. 343-347.

Alereza, T. and R.I. Hossli, 1979. "A Simplified Method of Calculating Heat Loss and Solar Gain through Residential Windows during the Heating Season," PH-79-6, no. 4, ASHRAE *Transactions,* vol. 85, part I, pp. 582-606.

Balcomb, J.D., ed., 1980a. *Passive Solar Design Handbook, vol. 2: Passive Solar Design Analysis,* Los Alamos Scientific Laboratory/U.S. Dept. of Energy, Los Alamos, 270 pp. + App. (NTIS No. DOE/CS-0127/2)

Balcomb, J.D., 1980b. "Rules of Thumb for Passive Solar Heating," *1980 Annual Meeting,* AS/ISES, pp. 889-893.

Balcomb, J.D., 1980c. "Conservation and Solar: Working Together," *5th NPSC,* AS/ISES, pp. 44-50.

Balcomb, J.D., 1981. "How to Balance Solar and Conservation in Passive Homes," *Solar Age,* vol. 6, no. 9, September 1981, pp. 38-45.

Barnett, J.P., *Energy Analysis of a Prototype Single-Family Detached Residence,* NBSIR 80-2184, 45 pp., NTIS.

Bennett, R., 1978. *Sun Angles for Design,* available from the author, 6 Snowden Road, Bala Cynwyd, PA 19004.

Brown, G.E. and B.J. Novitski, 1981. "A Design Methodology Based on Climate Characteristics," *6th NPSC,* AS/ISES, pp. 372-376.

Clark, G. *et al.,* 1981. *An Assessment of Passive Cooling Rates and Applications in the United States,* Final Report, U.S. DOE Contract No. DE-AC03-79CS30201 1600, Trinity University, San Antonio.

Cuplinskas, E.L., 1981. "A Rational Manual Method for Determination of Space Temperature Swing due to Solar Gains," *Thermal Performance of the Exterior Envelope of Buildings,* Special Publication No. 28, ASHRAE pp. 58-72.

Dean, E.T., 1979. "Graphic Methods for Determining the Solar Access Design Envelope for Sites with Irregular Topography," *4th NPSC,* AS/ISES, pp. 287-291.

Duffie, J.A. and W.A. Beckman, 1980. *Solar Engineering of Thermal Processes,* John Wiley and Sons, Inc., New York.

"Energy Estimating Methods," Chapter 28, *1981 Fundamentals Handbook,* ASHRAE, pp. 28.1 - 28.32.

Erbs, D.G. and W.A. Beckmann, S.A. Klein, 1981. "Degree Days for Variable Base Temperatures," *6th NPSC*, pp. 387-391.

Fisk, D.J., 1981. *Thermal Control fo Buildings,* Applied Science Publishers, Inc., Englewood, NJ, 245 pp.

Froehlich, D.P. and L.D. Albright, N.R. Scott, 1976. "Steady-Periodic Analysis of the Greenhouse Thermal Environment," Paper 76-4005, American Society of Agricultural Engineers 1976, *Annual Meeting* (Univ. of Nebraska), June 1976, 43 pp.

Gordon, J.M. and Y. Zarmi, 1981a. "Analytic Model for Passively Heated Solar Houses—I: Theory," *Solar Energy* (Pergamon), vol. 27, no. 4, pp. 331-342.

Gordon, J.M. and Y. Zarmi, 1981b. "Analytic Model for Passively Heated Solar Houses—II: Users Guide," *Solar Energy* (Pergamon), vol. 27, no. 4, pp. 343-347.

Gordon, J.M. and Y. Zarmi, 1981c. "Massive Storage Walls as Passive Solar Heating Elements: An Analytic Model," *Solar Energy,* (Pergamon), vol. 27, no. 4, pp. 349-355.

Guglielmini, G. and U. Magrini, 1971. "Periodic Heat Flow through Lightweight Walls: Influence of the Heat Capacity of Solid Bodies Upon Room Temperature," *Heat Transfer,* proceedings of a 1969 meeting in Liege, Belgium, of the International Institute of Refrigeration, Pergamon Press, Elmsford, New York, pp. 153-162.

Hannifan, M. and C. Christensen, R. Perkins, 1981. "Comparison of Residential Window Distributions and Effects of Mass and Insulation," *6th NPSC,* AS/ISES, pp. 213-217.

Harrison, P.L., 1974. "A Device for Finding True North," *Solar Energy* (Pergamon), vol. 14, no. 4, pp. 303-308.

Heat Gain Calculator (A slide rule for estimating window heat gains), Libbey Owens Ford Co., 811 Madison Ave., Toledo, OH 43695, 1975.

Hirshmann, J.R., 1974. "The Cosine Function as a Mathematical Expression for the Processes of Solar Energy," *Solar Energy* (Pergamon), vol. 16, no. 2, pp. 117-124.

Hoffman, M.E., 1976. "Prediction of Indoor Temperature," Chapter 19, B. Givoni, *Man, Climate and Architecture,* 2nd ed., Applied Science Publishers, Ltd., London (1981 by Van Nostrand Reinhold, New York), pp. 414-450.

Incropera, F.P. and D.P. DeWitt, 1981. *Fundamentals of Heat Transfer,* John Wiley and Sons, Inc., New York, 819 pp.

Jones, R.W. and R.D. McFarland, 1981. "Attached Sunspace Design Analysis," *6th NPSC,* AS/ISES, pp. 279-283.

Kluck, M., 1980. "Shadow Angle Charts for a North-South Profile," *5th NPSC,* AS/ISES, pp. 1188-1191.

Kreider, J.F. and F. Kreith, 1977. *Solar Heating and Cooling,* Rev. 1st ed., Hemisphere Pub. Co./McGraw-Hill Book Co., 342 pp.

Langdon, W.K., 1982. "Regional Guidelines for the Optimization of Passive South Wall Assemblies," *1982 Annual Meeting,* ASES, pp. 103-108.

Lebens, R.M., 1980. *Passive Solar Heating Design,* Halsted Press/John Wiley and Sons, Inc., New York, 234 pp.

Lord, D., 1980. "The Matchbox Sundial, A Design Aid for Architects," *5th NPSC,* AS/ISES, pp. 220-223.

Los, S., 1981. "Solar Perspective: A Graphic Technique to Draw Buildings as Seen by the Sun," *Passive Cooling '81,* AS/ISES, pp. 422-425.

Lunde, P.J., 1982. "Analytical Solutions for the Daily Performance of Direct Gain Passive Heating Systems," *1982 Annual Meeting,* ASES, pp. 763-766.

Markus, T.A. and E.N. Morris, 1980. "Experimental Methods," Chapter 10, *Buildings, Climate and Energy,* Pitman Publishers, Marshfield, MA, pp. 397-414.

Markus, T.A. and E.N. Morris, 1980. "Heat Gains due to Solar Radiation—Steady-State and Cyclic Conditions," Chapter 6, *Buildings, Climate and Energy,* Pitman Publishers, Marshfield, MA, pp. 310-343.

Mawson, K.H., 1981. "FIne Tuning Solar Access: Calculators and Computers as Design Tools," *6th NPSC,* AS/ISES, pp. 392-396.

McGeorge, W.D., 1980. "Energy Accounting—Methods of Analysis," Chapter 12, T.A. Markus and E.N. Morris, *Buildings, Climate and Energy,* Pitman Publishers, Marshfield, MA, pp. 479-506.

Monsen, W.A. and S.A. Klein, W.A. Beckman, 1981. "Prediction of Direct Gain Solar Heating System Performance," *Solar Energy* (Pergamon), vol. 27, no. 2, pp. 143-147.

Nall, D.H. and E.A. Arens, 1979. "The Influence of Degree Day Base Temperature on Residential Building Energy Prediction," PH-79-8, No. 2, ASHRAE *Transactions*, vol. 85, part I, pp. 722-734.

Novell, B., 1981a. "Low Cost Model Stand and Sunpath Stimulator," *Passive Cooling '81*, AS/ISES, pp. 387-391.

Novell, B., 1981b. "A Simple Design Method for Shading Devices and Passive Cooling Strategies based on Monthly Average Temperatures," *Passive Cooling '81*, AS/ISES, pp. 392-396.

Penrod, E.B., 1964. "Solar Load Analysis by Use of Orthographic Projections and Spherical Trigonometry," *Solar Energy*, vol. 18, no. 4, pp. 127-133.

Place, W. *et al.*, 1980. "Human Comfort and Auxiliary Control Considerations in Passive Solar Structures," *1980 Annual Meeting*, AS/ISES, pp. 821-825.

Powell, G.L. and J.I. Yellott, 1979. "A Procedure for Determining Solar Heat Gain Factors for South-facing Vertical Surfaces on Average Days," K.W. Boer and B.H. Glenn, eds., *Sun II*, ISES Silver Jubilee Congress, Pergamon Press, New York, vol. 2, pp. 1615-1619.

Reeves, G.A., 1981. "Degree Day Correction Factors—Basis for Values," CH-81-3, No. 4, ASHRAE *Transactions*, vol. 87, part I, pp. 507-513.

Robinson, D.A., 1981. "Life-Cycle Cost Economic Optimization of Insulation, Infiltration, and Solar Aperture in Passive Solar Homes," *6th NPSC*, AS/ISES, pp. 218-222.

Rosen, J. and J.A. Moore, 1981. "A Graphic Method for Predicting Temperature Swings," *6th NPSC*, AS/ISES, pp. 308-312.

Rudoy, W., Project Director, 1979. *Load Calculation Manual*, ASHRAE Publication No. SPLCM, 218 pp.

Saleh, A.M. "The Shadow Template: A New Method of Design of Sunshading Devices," *Solar Energy* (Pergamon), vol. 28, no. 3, 1982, pp. 239-256.

Scofield, S. and F. Moore, 1981. "Climatological Sundial," *6th NPSC* AS/ISES, pp. 382-386.

Siminovitch, M.J. and N.T. McFarland, 1980. "A Template System for Architects to Plot Building Shadows and Solar Penetration on Plan View Drawings," *5th NPSC*, AS/ISES, pp. 215-219.

Sinden, F.W., 1978. "Conservation of Domestic Space Heat and its Relation to Passive Solar Heating," *2nd NPSC*, AS/ISES, vol. 2, pp. 435-439.

Smith, E.G., 1951. *The Feasibility of Using Models for Predetermining Natural Ventilation*, Research Report No. 26, Texas Engineering Experiment Station, Texas A & M College, College Station.

Sonderegger, R.C., 1977. "Harmonic Analysis of Building Thermal Response Applied to the Optimum Location of Insulation within the Walls," *Energy and Buildings* (Elsevier Sequoia), vol. 1, no. 2, October 1977, pp. 131-140.

Sun Angle Calculator (a protractor or finding sun angles), Libbey Owens Ford Co., 811 Madison Ave., Toledo, OH 43695, 1974.

Watson, D., 1979. "Insolutions," *Progressive Architecture*, vol. 60, no. 11, November, 1979, pp. 102-107.

Watson, D. and K. Harrington, 1979. "Research on Climate Design for Home Builders: Brief Summary of Quantification Results," *4th NPSC*, AS/ISES, 1979, pp. 636-639.

Waugh, A.E., 1973. *Sundials: Their Theory and Construction*, Dover Publications, New York, 228 pp.

White, R.W., 1981a. "Pseudoshadows and Window Design," *Passive Cooling '81*, AS/ISES, pp. 412-416.

Winn, C.B., 1981. "A Simple Method for Determining the Average Temperature, the Temperature Variation, and the Solar Fraction for Passive Solar Buildings," *6th NPSC*, AS/ISES, pp 303-307.

Wray, W.O., 1981. "Performance of Low-Mass, Sun-Tempered Buildings," *1981 Annual Meeting*, AS/ISES, pp. 881-885.

Texts and Overview

AIA Research Corp., 1978. *Regional Guidelines for Building Passive Energy Conserving Homes*, U.S. Dept. of HUD and DOE, U.S. GPO, November 1978, 312 pp.

All-Weather Home Building Manual, Guideline 4, prepared by NAHB Research Foundation for U.S. Dept. HUD and Dept. of Labor, June 1976, 143 pp. (U.S. GPO SN 023-000-00339-2). Note: deals with weather and the construction process, not design.

Argue, R., 1980. *The Well-tempered House: Energy Efficient Building for Cold Climates,* Renewable Energy in Canada, 415 Parkside Drive, Toronto, Canada M6R 2Z7, 218 pp.

Aronin, J.E., 1953. *Climate and Architecture,* Reinhold Publishing Corp., New York, 304 pp.

ASHRAE Handbook, 1981 Fundamentals ASHRAE, 39 Chapters.

Atkinson, G.A., 1960. "Principles of Tropical Design," *Architectural Review,* vol. 128, no. 761, July 1960, pp. 81-83.

Bahadori, M.N., 1978. "Passive Cooling Systems in Iranian Architecture," *Scientific American,* vol. 238, no. 2, February 1978, pp. 144-154.

Bahadori, M.N., 1979. "Natural Cooling in Hot Arid Regions," A.A.M. Sayigh, ed., *Solar Energy Applications in Buildings,* Academic Press, New York, pp. 195-225.

Bentley, D. and G. Chabannes, 1982. "The Effectiveness of Traditional Climatic Responses in the Central Texas Region in Maintaining Thermal Comfort," *1982 Annual Meeting,* ASES, pp. 745-749.

Billington, N., 1967. *Building Physics: Heat,* Pergamon Press, Oxford.

Builders' Guide to Energy Efficiency in New Housing, Housing and Urban Development Association of Canada and Ontario Ministry of Energy, HUDAC National Office, 15 Toronto, Ontario M5C 2E3, 1980, 138 pp. (metric units).

Canadian Building Digests, a series of over 200 4-page pamphlets discussing issues and recommended practices in building construction, many concerning energy efficient design and detailing. Publications Section, Div. of Building Research, National Research Council of Canada, Ottawa, K1A 0R6 (individual copies free, bound sets at nominal cost).

Conklin, G., 1958. *The Weather Conditioned House,* Reinhold Publishing Co., New York, 238 pp.

The Conservation of Energy in Housing, Publication NHA 5149 81/02, Canada Mortgage and Housing Corp. (CMHC), Ottawa, 1981, 153 pp. (metric units).

Cook, J., 1979. *Cool Houses for Desert Suburbs,* Arizona Solar Energy Commission, 1700 West Washington, Phoenix, AZ 85007, 126 pp.

Energy Conservation Design Manual for New Residential Buildings (for California) California Energy Commission, 1111 Howe Ave., Sacramento, CA 95825, 1978.

Energy Research Development Group (Univ. of Saskatchewan), 1980. *Energy Efficient Housing: A Prairie Approach,* Office of Energy Conservation, Saskatchewan Mineral Resources, 1914 Hamilton Street, Regina, Saskatchewan, S4P 4V4, 31 pp. (metric units).

Evans, M., 1980. *Housing, Climate and Comfort,* Halsted Press/John Wiley and Sons, Inc., New York, 186 pp.

Fifth National Passive Solar Conference (5th NPSC), held in Amherst, MA, October 1980; Proceedings edited by J. Hayes and R. Snyder, published by AS/ISES, 2 vol., 1409 pp.

Fitch, J.M. and D.P. Branch, 1960. "Primitive Architecture and Climate," *Scientific American, vol. 203, no. 6, December 1960, pp. 134-144.*

Fitch, J.M., 1968. "The Control of the Luminous Environment," *Scientific American,* vol. 219, no. 3, September 1968, pp. 190-202.

Fitch, J.M., 1972. *American Building: The Environmental Forces That Shape It,* 2nd ed., Houghton Mifflin Co., Boston 349 pp.

Fourth National Passive Solar Conference (4th NPSC), held in Kansas City, MO, October 1979; Proceedings edited by G. Franta, published by AS/ISES, 772 pp.

Fry, M. and J. Drew, 1964. *Tropical Architecture in the Dry and Humid Zones,* Robt. E. Krieger Pub. Co., Inc., 645 New York Ave., Huntington, NY 11743, 264 pp.

Fullerton, R.L., 1977. *Building Construction in Warm Climates,* Oxford Tropical Handbooks vol. 3, Oxford Univ. Press.

Givoni, B., 1976. *Man, Climate and Architecture,* 2nd ed., Applied Science Publishers, Ltd., London (1981 by Van Nostrand Reinhold, New York), 496 pp.

Givoni, B., 1979a. "Passive Cooling of Buildings by Natural Energies," *Energy and Buildings* (Elsevier Sequoia), vol. 2, no. 4, December 1979, pp. 279-285.

Givoni, B., 1981a. "Cooling of Buildings by Passive Systems," *Passive Cooling '81,* AS/ISES, pp. 588-596.

Golany, G.S., ed., 1980. *Housing in Arid Lands: Design and Planning,* Halsted Press/John Wiley and Sons, Inc., New York, 257 pp.

Haysom, J.C. and J.W. Sawers, 1981. *Energy Conservation in New Small Residential Buildings,* Publication NHA 5417 81/06, Canada Mortgage and Housing Corp. (CMHC), Ottawa, 73 pp. (metric units).

House Beautiful Climate Control Project: Regional Climate Analysis and Design Data; originally published in the *Bulletin* of the American Institute of Architects, 1949-1952; reproduced on demand by University Microfilms International, 300 North Zeeb Road, Ann Arbor, MI 48106.

Housing and Building in Hot-Humid and Hot-Dry Climates, BRAB Research Conference Report No. 5, National Academy of Sciences/National Research Council, 1953, 177 pp.

Koenigsberger, O.H. and T.G. Ingersoll, A. Mayhew, S.V. Szokolay, 1974. *Manual of Tropical Housing and Building,* Longman Inc., New York, 320 pp.

Lee, D.H.K., 1963. *Physiological Objectives in Hot Weather Housing; An Introduction to Hot Weather Housing Design,* HUD Office of International Affairs, Room 2118, Washington, DC 20410, 79 pp.

Markus, T.A. and E.N. Morris, 1980. *Buildings, Climate and Energy,* Pitman Publishers, Marshfield, MA, 540 pp.

Mazria, E., 1979. *The Passive Solar Energy Book,* Rodale Press, Emmaus, PA, 687 pp.

Niles, P.W.B. and K.L. Haggard, 1980. *Passive Solar Handbook* (for California), P500-80-032, California Energy commission, 1111 Howe Avenue, Sacramento, CA 95825, 330 pp.

Olgyay, V., 1963. *Design With Climate: Bioclimatic Approach to Architectural Regionalism,* Princeton Univ. Press, Princeton, NJ, 190 pp.

Passive Solar Journal, quarterly publication of the Passive Systems Division of AS/ISES, beginning vol. 1, no. 1, January 1982.

Passive Cooling '81, International Technical Conference held in Miami Beach, FL, November 1981; Proceedings edited by A. Bowen, E. Clark and K. Labs, published by AS/ISES, 1052 pp.

Saini, B.S., 1980. *Building in Hot, Dry Climates,* Wiley-Interscience/John Wiley and Sons, Inc., New York, 176 pp.

Sayigh, A.A.M., ed., 1979. *Solar Energy Applications in Buildings,* Academic Press, New York (21 international contributors).

Second National Passive Solar Conference (2nd NPSC), held in Philadelphia, PA, March 1978; Proceedings edited by D. Prowler, published by AS/ISES, 3 vol., 690 pp.

Sixth National Passive Solar Conference (6th NPSC), held in Portland, OR, September 1981; Proceedings edited by J. Hayes and W. Kolar, published by AS/ISES, 893 pp.

Spielvogel, L.G., 1979. "How and Why Buildings Use Energy," D. Watson, ed., *Energy Conservation through Building Design,* Architectural Record Books/McGraw-Hill Book Co., New York, pp. 52-75.

Third National Passive Solar Conference (3rd NPSC), held in San Jose, CA, January 1979; Proceedings edited by H. Miller, M. Riordan, D. Richards, published by AS/ISES, 943 pp.

Van Straaten, J.F., 1967. *Thermal Performance of Buildings,* Applied Science Publishers, Ltd., London, 311 pp. (International Ideas, Inc., Philadelphia).

Watson, D., 1977. *Designing and Building a Solar House,* Garden Way Publishing, Charlotte, VT, 281 pp.

Watson, D., ed., 1979. *Energy Conservation Through Building Design,* Architectural Record Books/McGraw-Hill Book Co., New York, 308 pp.

Wilson, T., ed., 1981. *Home Remedies: A Guidebook for Residential Retrofit,* Mid-Atlantic Solar Energy Association, 2233 Gray's Ferry Ave., Philadelphia, PA 19146, 253 pp.

Winter Associates, Inc, 1981. *Passive Solar Construction Handbook,* Southern Solar Energy Center, 61 Perimeter Park, Atlanta, GA 30341.

Yellott, J.I., 1978a. "Passive Solar Heating and Cooling Systems," ASHRAE *Journal,* vol. 20, no. 1, January 1978, pp. 60-67.

Site Planning

Buckley, C.E., and D.T. Harrje, M.P. Knowlton, G.M. Heisler, 1978. *The Optimum Use of Coniferous Trees in Reducing Home Energy Consumption,* CES Report No. 71, Princeton Univ., 76 pp.

Center for Landscape Architectural Education and Research, 1978. *Options for Passive Energy Conservation in Site Design,* U.S. DOE, Washington, DC, 300+ pp. (NTIS NO. HCP/M5037-01/UC-95d).

Cook, J., 1980a. "Landscaping for Microclimatic Advantage in Arid-Zone Housing," G.S. Golany, ed., *Housing in Arid Lands: Design and Planning,* Halsted Press/John Wiley and Sons, Inc., New York, pp. 225-234.

Cook, J., 1980b. "Microclimates in Desert Housing," Clark and Paylore, eds., *Desert Housing,* Univ. of Arizona, Tucson.

Erley, D. and M. Jaffe, 1979. *Site Planning for Solar Access: A Guidebook for Residential Developers and Site Planners,* U.S. Dept. of HUD, 149 pp. (U.S. GPO 1979: 630-916/2704).

Evans, B.H., 1957. *Natural Air Flow Around Buildings,* Research Report No. 59, Texas Engineering Experiment Station, Texas A & M College, College Station.

Evans, M., 1980. "Site Selection," Chapter 6, *Housing, Climate and Comfort,* Halsted Press/John Wiley and Sons, Inc., New York, pp. 54-58.

Grist, C.R., 1978. "Assessing the Energy-Saving Benefits of Windbreaks for Residential Buildings," *Technology for Energy Conservation* conference proceedings, Albuquerque, pp. 522-527 (Information Transfer).

Heisler, G.M., 1982. "Reductions of Solar Radiation by Tree Crowns," *1982 Annual Meeting,* ASES, pp. 133-138.

Holzberlein, T.M., 1979. "Don't Let the Trees Make a Monkey of You," *4th NPSC,* AS/ISES, pp. 416-419.

Hunn, B.D. and D.O. Calafell, 1977. "Determination of Average Ground Reflectivity for Solar Collectors," *Solar Energy* (Pergamon), vol. 19, no. 1, pp. 87-89.

Mattingly, G.E. and E.F. Peters, 1975. *Wind and Trees: Air Infiltration Effects on Energy in Housing,* CES Report No. 20, Princeton Univ., 101 pp.

McPherson, E.G., 1981. "The Effects of Orientation and Shading from Trees on the Inside and Outside Temperatures of Model Homes," *Passive Cooling '81,* AS/ISES, pp. 369-373.

Microclimate, Architecture and Landscaping Relationships in an Arid Region: Phoenix, AZ, Research Paper No. 4, Center for Environmental Studies, Arizona State Univ., Tempe, 1977.

Moffat, A.S. and M. Schiler, 1981. *Landscape Design that Saves Energy,* Wm. Morrow and Co., Inc., New York.

Montgomery, D.A., 1981. "Landscape for Passive Solar Cooling," *Passive Cooling '81,* AS/ISES, pp. 360-364.

O'Hare, M. and R.E. Kronauer, 1969. "Fence Designs to Keep Wind from being a Nuisance," *Architectural Record,* vol. 146, no. 1, July 1969, pp. 151-156.

Olgyay, V., 1963, "Site Selection," Chapter 5, *Design with Climate,* Princeton Univ. Press, Princeton, NJ, pp. 44-52.

Olgyay, V., 1963. "Wind Effects and Air Flow Patterns," Chapter 9, *Design with Climate,* Princeton Univ. Press, Princeton, NJ., pp. 94-112.

Parker, J.H., 1979. "Precision Landscaping for Energy Conservation," *Technology for Energy Conservation* conference proceedings, Tucson (Information Transfer).

Parker, J.H., 1981a. *Uses of Landscaping for Energy Conservation* Final Report, STAR Project 78-012, Florida International University, Miami, 112 pp.

Parker, J.H., 1981b. "A Comparative Analysis of the Role of Various Landscape Elements in Passive Cooling in Warm, Humid Environments," *Passive Cooling '81,* pp. 365-368.

Read, R.A., 1964 *Tree Windbreaks for the Central Great Plains,* Agriculture Handbook No. 250, USDA Forest Service Washington, DC., 68 pp. (U.S. GPO).

Robinette, G.O., 1972. *Plants, People, and Environmental Quality,* U.S. Dept. of the Interior/American Society of Landscape Architects Foundation, 140 pp. (U.S. GPO No. 2405-0479).

Robinette, G.O. and C. McClenon, eds., 1977. *Landscape Planning for Energy Conservation,* Environmental Design Press, P.O. Box 2187, Reston, VA 22090, 224 pp.

Schramm, D. and C. R. Grist, 1978. "Building Site Selection to Conserve Energy by Optimizing Topoclimatic Benefits," *Technology for Energy Conservation* conference proceedings, Albuquerque, pp. 536-543 (Information Transfer).

Spirn, A.W., 1981. "Using Vegetation to Cool Small Structures," *1981 Annual Meeting,* AS/ISES, pp. 906-910.

Stoeckeler, J.H. and R.A. Williams, 1949. "Windbreaks and Shelterbelts," *Trees,* The Yearbook of Agriculture, USDA, pp. 191-199.

Subdivisions and Sun: 3 Design Studies. Ontario Ministry of Energy, 1979, 75 pp + illus. (Ontario Govt. Pub. Service, 880 Bay St., 5th Floor, Toronto, Ontario M5S 1Z8).

White, R.F., 1954. *Effects of Landscape Development on Natural Ventilation,* Research Report No. 45, Texas Engineering Experiment Station, Texas A & M College, College Station.

White, R.W., 1981b. "Pseudoshadows for Solar Design: Site Planning and Building Mass Studies," *Passive Cooling '81,* AS/ISES, pp. 407-411.

Windbreaks and Shelterbelts, Technical Note No. 59, World Meteorological Organization Geneva, 1964 (UNIPUB).

Woodruff, N.P., 1954. *Shelterbelt and Surface Barrier Effects on Wind Velocities, Evaporation, House Heating, and Snow Drifting,* Kansas Agricultural Experiment Station Technical Bulletin No. 77, 27 pp.

Building Massing

Berkoz, E.B., 1977. "Optimum Building Shapes for Energy Conservation," *Journal of Architectural Education,* D.Watson, ed., vol. 30. no. 3, February 1977, pp. 25-29, (available through Xerox University Microfilms, 300 North Zeeb Road, Ann Arbor, MI 48106).

Bircher, T.L., 1981. "Ground Coupling Techniques for Cooling in Desert Regions," *1981 Annual Meeting,* AS/ISES, pp. 981-985.

Bligh, T.P., 1976. "Energy Conservation by Building Underground," *Underground Space* (Pergamon), vol. 1, no. 1, May/June 1976, pp. 19-33, (AUA reprint 2).

Boileau, G.G. and J.K. Latta, 1968. *Calculation of Basement Heat Losses,* Technical Paper 292, NRCC 10477, DBR/NRCC, Ottawa, 19 pp.

Brown, G.Z. and B.J. Novitski, 1981. "Climate Responsive Earth Sheltered Buildings," *Underground Space* (Pergamon), vol. 5, no. 5, March/April 1981, pp. 299-305, (AUA reprint 31).

Claesson, J. and B. Eftring, 1980. *Optimal Distribution of Thermal Insulation and Ground Heat Losses,* D33:1980, Svensk Byggtjanst, Stockholm, 103 pp.

Cramer, R.D. and L.W. Neubauer, 1966. "Thermal Effectiveness of Shape," *Solar Energy,* vol. 10, no. 3, pp. 141-149.

Derickson, R.G. and K.S. Sadlon, 1981. "Flexibilities in Passive Design: Examining Some Limiting Solar Myths," *6th NPSC,* AS/ISES, pp. 333-337.

Evans, M. 1980. "The Form of Dwellings," Chapter 7, *Housing, Climate and Comfort,* Halsted Press/John Wiley and Sons, Inc., New York, pp. 59-75.

Givoni, B., 1976. "Orientation and its Effect on Indoor Climate," Chapter 11, *Man, Climate and Architecture,* 2nd ed., Applied Science Publishers, Ltd., London (1981 by Van Nostrand Reinhold Co.), pp. 213-231.

Givoni, G., 1979b. "Modifying the Ambient Temperature of Underground Buildings," F. Moreland, F. Higgs, J. Shih, eds., *Earth Covered Buildings: Technical Notes,* proceedings of the Earth Covered Settlements Conference, Fort Worth, Univ. Of Texas/U.S. DOE, pp. 123-138 (NTIS NO. CONF-7805138-P1).

Konzo, S. and W.L. Shick, W.H. Scheick, D.B. Lindsay, 1977. *Solar Orientation,* SHC/BRC Council Notes, vol. 2, Fall 1977, 8 pp.

Labs, K., 1980a. "Terratypes: Underground Housing for Arid Zones," G.S. Golany, ed., *Housing in Arid Lands: Design and Planning,* Halsted Press/John Wiley and Sons, Inc., New York, pp. 123-140.

Labs, K., 1980b. "Building Underground: A Tempered Climate, Earth as Insulation, and the Surface-Undersurface Interface," W.F. Wagner, ed., *Energy Efficient Buildings,* Architectural Record Books/McGraw-Hill Book Co., New York, pp. 82-90.

Labs, K., 1981b. "Direct-Coupled Ground Cooling: Issues and Opportunities," *Passive Cooling '81,* AS/ISES, pp. 131-135.

Labs, K., 1982. "Living Up to Underground Design," *Solar Age,* vol. 7, no. 8, August 1982, pp. 34-38.

Lee, D.H.K., 1963. "Housing as Climatic Protection in Hot Dry Environments," Chapter 3, *Physiological Objectives in Hot Weather Housing,* HUD Office of International Affairs, Room 2118, Washington, DC 20410, pp. 27-46.

Markus, T.A. and E.N. Morris, 1980. "Shape of Building," Chapter 9, *Buildings, Climate and Energy,* Pitman Publishers, Marshfield, MA, pp. 373-396.

Meixel, G.D. and P.H. Shipp, T.P. Bligh, 1981. "The Impact of Insulation Placement on the Seasonal Heat Loss through Basement and Earth-Sheltered Walls," *Thermal Performance of the Exterior Envelope of Buildings,* Special Publication No. 28, ASHRAE, pp. 987-1001 (AUA reprint 22).

Olgyay, V. and A. Olgyay, 1954a. "The Theory of Sol-Air Orientation," *Architectural Forum,* March 1954, pp. 133-137.

Olgyay, V. and A. Olgyay, 1954b. "Environment and Building Shape," *Architectural Forum,* August 1954, pp.104-108.

Olgyay, V., 1963. "Environment and Building Forms," Chapter 8, *Design with Climate,* Princeton Univ. Press, Princeton, NJ, pp. 84-93.

Olgyay, V., 1967. "Bioclimatic Orientation Method for Buildings," *International Journal of Biometeorology,* vol. 11, no. 2, July 1967, pp. 163-174.

Page, J.K., 1974. "Optimization of Building Shape to Conserve Energy," *Journal of Architectural Research,* vol. 3, no. 3, pp. 20-28.

Robinsky, E.I. and K.E. Bespflug, 1973. "Design of Insulated Foundations," Proceedings Paper 10009, *Jrnl. of the Soil Mechanics and Foundations Div.,* Am. Soc. of Civil Engineers, vol. 99, no. SM9, September 1973, pp. 649-667.

Sharma, M.R., 1969, *Orientation of Buildings,* Building Digest No. 74, Roorke U.P. (India) Central Building Research Institute.

Shick, W.L., 1977. "Effects of Building Orientation on Energy Savings," *Energy Efficiency in Wood Building Construction,* Proceedings No. P-77-18, Forest Products Research Society, 2801 Marshall Court, Madison, WI 53705, pp. 34-36.

Sterling, R.L. and G.D. Meixel, 1981. "Review of Underground Heat Transfer Research," L.L. Boyer, ed., *Earth Shelter Performance and Evaluation* conference proceedings, Oklahoma State Univ. Arch. Extension, Stillwater, OK 74078, October 1981, pp. 69-79

Szydlowski, R.F., 1980. "Analysis of Transient Heat Loss in Earth Sheltered Structures," Master of Science Thesis, Iowa State Univ., Ames, 196 pp.

Szydlowski, R.F. and T.H. Kuehn, 1980. "Analysis of Transient Heat Loss in Earth Sheltered Structures," L.L. Boyer, ed., *Earth Sheltered Buildings Design Innovations* conference proceedings, Oklahoma State Univ. Arch. Extension, Stillwater, OK 74078, April 1980, pp. III.27-III.37 (AUA reprint 28).

Underground Space Center, (Univ. of Minnesota), 1979. *Earth Sheltered Housing Design,* Van Nostrand Reinhold Co., New York, 318 pp.

Underground Space Center, 1981a. *Earth Sheltered Community Design,* Van Nostrand Reinhold Co., New York, 270 pp.

Underground Space Center, 1981b. *Insulation Materials and Placement,* Earth Sheltered Structures Fact Sheet 06, Univ. of Minnesota for U.S. DOE, May 1981, 6 pp. (AUA publication 103).

Underground Space Center, 1981c. *Insulation Principles,* Earth Sheltered Structures Fact Sheet 05, Univ. of Minnesota for U.S. DOE, May 1981, 8 pp. (AUA publication 102).

Underground Space Center (Univ. of Minnesota), 1982. *Earth Sheltered Residential Design Manual,* Van Nostrand Reinhold Co., New York.

Valko, P., 1972. "The Effect of Shape and Orientation on the Radiation Impact on Buildings," *Teaching the Teachers on Building Climatology,* CIB Proceedings No. 25, Document D20:1973, Svensk Byggtjanst.

Wiener, I.S., 1955. "Solar Orientation: Application of Local Wind Factors," *Progressive Architecture,* vol. 36. no. 2, February 1955, pp. 112-118.

Building Plan

Add-On Greenhouse Space Heating System (with mass) retrofit manual, ERDA Technology Transfer Director, 2 Rockefeller Plaza, Albany, NY 12223.

Add-On Sunspace Space Heating System (without mass) retrofit manual, ERDA Technology Transfer Director, 2 Rockefeller Plaza, Albany, NY 12223.

Bier, J., 1978. "Vertical Solar Louvers: A System for Tempering and Storing Solar Energy," *2nd NPSC,* AS/ISES, vol. 1, pp. 209-213.

Bier, J., 1979. "Performance of a Low-Cost Owner-Built Home using Vertical Solar Louvers," *3rd NPSC,* AS/ISES, pp. 643-647.

Bilgen, E., and M. Chaaban, 1982. "Solar Heating-Ventilating System using a Solar Chimney," *Solar Energy* (Pergamon), vol. 28, no. 3, pp. 227-233.

Block, I., 1980. "Attic Ventilators and Energy Conservation," *ASHRAE Journal,* vol. 22, no. 2, February 1980, pp. 46-49.

Brown, E.J., and W.H. Kapple, D.H. Percival, 1958-1959. *Construction for Attic Ventilation,* Technical Note No. 9 (collected papers), SHC/BRC, 20 pp.

Dunlavy, H.A., ed., 1977. *The Handbook of Moving Air,* American Ventilation Association, P.O. Box 7464, Houston, TX 77008, 54 pp.

Dutt, G.S., 1979. "Condensation in Attics: Are Vapor Barriers Really the Answer?", *Energy and Buildings* (Elsevier Sequoia), vol. 2, no. 4, December 1979, pp. 251-258.

Dutt, G.S. and J. Beyea, 1977. *Attic Thermal Performance: A Study of Townhouses at Twin Rivers,* CES Report No. 53, Princeton Univ., Princeton, NJ, 46 pp+ App.

Dutt, G.S. and J. Beyea, 1979. *Hidden Heat Losses in Attics: Their Detection and Reduction,* CES Report No. 77, Princeton Univ., Princeton, NJ, 54 pp.

Dutt, G.S. and D.T. Harrje, 1982. "Improved Thermal Designs for Light Frame Structures," *Design and Performance of Light Frame Structures: Wall and Floor Systems,* conference proceedings, Forest Products Research Society, Madison, WI.

Energy Conservation and Solar Heating for Greenhouses, Northeast Regional Agricultural Engineering Service, May 1978, 48 pp. (available from Agricultural Extension Office of participating Northeastern states and from Northeast Regional Agricultural Engineering Service, Riley Robb Hall, Cornell Univ., Ithaca, NY 14853).

Evans, M., 1980. "Thermal Properties of Internal Walls and Floors," Chapter 10, *Housing, Climate and Comfort,* Halsted Press/John Wiley and Sons, Inc., New York, pp. 101-103.

Faiman, D., 1980. "Rotating Prism Wall as a Passive Heating Element," *Solar Energy* (Pergamon), vol. 25, no. 6, pp. 563-564.

Fineblum, S.S., 1982. "Cooling Small, Internally Heated Buildings by Natural Convection and Solar Chimney-Augmented Ventilation," *1982 Annual Meeting,* ASES, pp. 643-648.

Jones, R.W., 1980. *Passive Solar Heating of Buildings with Attached Greenhouse, Progress Report,* Univ. of South Dakota, 28 pp. (NTIS Report No. DOE/CS/30242-3).

Kusuda, T. and J.W. Bean, 1981. *Savings in Electric Cooling Energy By The Use of a Whole-House Fan,* NBS Technical Note 1138, 39 pp. (U.S. GPO SN 003-003-02317-5).

Labs, K. and D. Watson, 1982. "Passive Solar Design for Light-Frame Construction," *Design and Performance of Light Frame Structures: Wall and Floor Systems* conference proceedings, Forest Products Research Society, Madison, WI.

Lee, B.E. and M. Hussain, B. Soliman, 1980. "Predicting Natural Ventilation Forces Upon Low-Rise Buildings," *ASHRAE Journal,* vol. 22, no. 2, February 1980, p. 35+.

McFarland, R.D. and R.W. Jones, 1980. "Performance Estimates for Attached Sunspace Passive Solar Heated Buildings (with Tables)", Los Alamos Scientific Laboratory, P.O. Box 1663, Los Alamos, NM 87545, 56 pp.

Mears, D.R., ed., 1979. *Solar Energy for Heating of Greenhouses and Greenhouse-Residence Combinations,* 4th Annual Conference Proceedings, U.S. DOE and Rutgers Univ., 214 pp. (Dept. of Biological and Agricultural Eng., Cook College/Rutgers Univ., P.O. Box 231, New Brunswick, NJ 08903).

Morrison, G.L., 1980. "Passive Solar Energy Storage in Greenhouses," *Solar Energy* (Pergamon), vol. 25, no. 4, pp. 365-372.

Nottage, H.B. and J.G. Slaby, W.P. Gojsza, 1952. "A Smoke-Filament Technique for Experimental Research in Room Air Distribution," Paper 1461, *ASHVE Transactions,* vol. 58, pp. 399-404.

Reppert, M.H., ed., 1979. *Summer Attic and Whole-House Ventilation,* NBS Special Publication 548, 153 pp. (U.S. GPO SN 003-003-02089-3).

Shick, W.L. and R.A. Jones, 1976. *Illinois Lo-Cal House,* SHC/BRC Council Notes, vol. 1, no. 4, Spring 1976, 8 pp.

Simon, M.J., 1947. *Your Solar House,* Simon and Schuster, New York, 125 pp.

Solar Energy for Heating of Greenhouses and Greenhouse-Residence Combinations, 3rd Annual Conference Proceedings, U.S. DOE and Colorado State Univ., 1978, 117 pp. (Solar Energy Applications Laboratory, Colorado State Univ., Fort Collins, CO 80523).

Swet, C.J., 1980. "Phase Change Storage in Passive Solar Architecture," *5th NPSC,* AS/ISES, pp. 282-286.

The Thermal Mass Pattern Book: Guidelines for Sizing Heat Storage in Passive Solar Homes, Total Environmental Action, Inc., 1980, 11 pp. (TEA, Church Hill, Harrisville, NH 03450).

van der Mersch, P.L. and P.J. Burns, C.B. Winn, 1981. "A Simplified Method for the Design of Cylinder Water Walls for Passive Solar Heating," *1981 Annual Meeting,* AS/ISES, pp. 872-876.

Wolfert, C.K. and H.S. Hinrichs, *Fundamentals of Residential Attic Ventilation,* H.C. Products Co., P.O. Box 68, Princeville, IL 61559, 1974.

Wray, W.O. and J.D. Balcomb, 1979. "Sensitivity of Direct Gain Space Heating Performance to Fundamental Parameter Variations," *Solar Energy* (Pergamon), vol. 23, pp. 421-425.

Building Envelope

Adams, L., 1976. "Thermal Conductance of Air Spaces," ASHRAE *Journal,* vol. 18, no. 3, March 1976, pp. 37-38.

Anderson, L.O. and G.E. Sherwood, 1974. *Condensation Problems in Your House: Prevention and Solution,* Agriculture Information Bulletin No. 373, USDA Forest Products Laboratory, 39 pp. (U.S. GPO SN 0100-03318).

Bassett, M.R., 1981. *Sensitivity of Residential Heating Energy to Building Envelope Thermal Conductance,* Building Research Note No. 177, DBR/NRCC, Ottawa, 17 pp.

Berthier, J., 1973. *Weak Thermal Points or Thermal Bridges,* NBS Technical Note 710-7 (U.S. GPO No. C13.46:710-7).

Blick, E.F., 1980. "A Simple Method for Determining Heat Flow through Earth Covered Roofs," L.L. Boyer, ed., *Earth Sheltered Building Design Innovations* conference proceedings, Oklahoma State Univ. Arch. Extension, Stillwater, OK 74078, April 1980, pp. III.19-III.23.

Blount, S.M., 1958. *Sprayed Roof Cooling Systems,* Industrial Experimental Program, School of Engineering, Bulletin No. 9, North Carolina State Univ., Raleigh, 20 pp. (reprinted by Spray Cool Systems, Ltd., Sunspace Corp., Atlanta, GA).

Bomberg, M. and K.R. Solvason, 1980a. *How to Ensure Good Thermal Performance of Cellulose Fibre Insulation, Part I: Horizontal Applications,* Building Research Note No. 157, DBR/NRCC, Ottawa, 26 pp.

Bomberg, M. and K. Solvason, 1980b. *How to Ensure Good Thermal Performance of Cellulose Fibre Insulation, Part II: Exterior Walls,* Building Research Note No. 158, DBR/NRCC, Ottawa, 18 pp.

Bomberg, M. and K. Solvason, 1980c. *How to Ensure Good Thermal Performance of Blown Mineral Fibre Insulation in Horizontal and Vertical Installations,* Building Research Note No. 167, DBR/NRCC, Ottawa, 28 pp.

Bowen, R.P., and C.J. Shirtliffe, G.A. Chown, 1981. *Urea-formaldehyde Foam Insulation: Problem Identification and Remedial Measures for Wood Frame Construction,* Building Practice Note 23, DBR/NRCC, Ottawa, 59 pp.

Burt, Hill, Kosar, Rittelmann Associates, 1977 (edited by K. Williams). *Planning and Building the Minimum Energy Dwelling,* Craftsman Book Co., Solana Beach, CA.

Carlsson, B. and A. Elmroth, P.A. Engvall, 1980. *Airtightness and Thermal Insulation: Building Design Solutions,* D37:1980, Svensk Byggtjanst, Stockholm, 144 pp.

Conrad, G.R. and G.T. Pytlinski, T.C. McConnell, 1981. "Assessment of Contemporary Residential Roof Surfaces as Nocturnal Radiators and Solar Collectors," *Passive Cooling '81,* AS/ISES, pp. 251-255.

Construction Details for Air Tightness, 1980. Proceedings No. 3, NRCC 18291, DBR/NRCC, Ottawa, 58 pp.

Crowther, K., and B. Melzer, 1979. "The Thermosiphoning Cool Pool: A Natural Cooling System," *3rd NPSC,* AS/ISES, pp. 448-451.

Dexter, M.E., 1981. "Energy Conservation Guidelines for including Mass and Insulation in Building Walls," *Thermal Performance of the Exterior Envelope of Buildings,* Special Publication No. 28, ASHRAE, pp. 52-57.

Duffin, R.J. and G. Knowles, 1981. "Temperature Control of Buildings by Adobe Wall Design," *Solar Energy* (Pergamon), vol. 27, no 3, pp. 241-249.

Ebele, C., 1980. *Experimental Testing of Cooling by the Roof Radiation Trap.* research report, School of Architecture and Urban Planning, Univ. of California/Los Angeles, April 1980.

Elmroth, A., 1978. *Well-Insulated Airtight Buildings: Design and Construction,* D10:1978, Svensk Byggtjanst, Stockholm, 33 pp.

Energy Saving Homes: The Arkansas Story, Owens-Corning Fiberglass Corp., Insulation Operating Division, Fiberglass Tower, Toledo, OH 43659, 1980, 50 pp. (free).

Evans, M., 1980. "Required Thermal Performance for Walls and Roofs," Chapter 9, *Housing, Climate and Comfort,* Halsted Press/John Wiley and Sons, Inc., New York, pp. 88-100.

Evans, M., 1980. "Thermal Properties of Roofs and Walls," Chapter 8, *Housing, Climate and Comfort,* Halsted Press/John Wiley and Sons, Inc., New York, pp. 76-87.

Givoni, B., 1977. "Solar Heating and Night Radiation Cooling by a Roof Radiation Trap," *Energy and Buildings* (Elsevier Sequoia), vol. 1, 1977, pp. 141-145.

Givoni, B., 1981b. "Experimental Studies on Radiant and Evaporative Cooling of Roofs," *Passive Cooling '81,* AS/ISES, pp. 279-283.

Godfrey, R.D. and K.E. Wilkes, A.G. Lavine, 1981. "A Technical Review of the "M" Factor Concept," *Thermal Performance of the Exterior Envelope of Buildings,* Special Publication No. 28, ASHRAE, pp. 472-485.

Haggard, K.L., 1977. "The Architecture of a Passive System of Diurnal Radiation Heating and Cooling," *Solar Energy* (Pergamon), vol. 19, pp. 403-406.

Hay, H.R. and J.I. Yellott, 1969a. "Natural Air Conditioning with Roof Ponds and Movable Insulation," *ASHRAE Transactions,* vol. 75, part I, pp. 165-177.

Hay, H.R. and J.I. Yellott, 1969b. "International Aspects of Air Conditioning with Movable Insulation," *Solar Energy* (Pergamon), vol. 12, no. 4, pp. 427-438.

Homma, H. and R.W. Guy, 1981. "Ventilation of Back Space of Building Enclosure Siding for Solar Heat Gain Reduction," *Thermal Performance of the Exterior Envelope of Buildings,* Special Publication No. 28, ASHRAE, pp. 856-874.

Hopman, R., 1979. "Night Sky Cooling: Two Passive Strategies," *3rd NPSC,* AS/ISES, pp. 466-470.

Houghten, F.C., and C. Gutberlet, H.T. Olson, 1940. "Summer Cooling Load as Affected by Heat Gain through Dry, Sprinkled, and Water-covered Roofs," Paper 1157, ASHVE *Transactions,* vol. 46, pp. 231-246 (the work upon which subsequent ASHRAE *Guide* recommendations were based).

Jones, R.A., 1975. *Moisture Condensation,* SHC/BRC Council Notes, vol. 1, no. 1, 8 pp.

Latta, J.K., 1973. *Walls, Windows and Roofs for the Canadian Climate,* Special Technical Publication No. 1, NRCC 13487, DBR/NRCC, Ottawa, 87 pp+ App.

Lowinski, J.F., 1979. "Thermal Performance of Wood Windows and Doors," PH-79-6, No. 2, ASHRAE *Transactions,* vol. 85, part I, pp. 548-564.

Misselhorn, D.J., 1979. "Some Problems with Insulation on Suspended Ceilings," ASHRAE *Journal,* vol. 21, no. 3, March 1979, pp. 46-49.

Mitalas, G.P., 1978. *Relation between Thermal Resistance and Heat Storage in Building Enclosures,* Building Research Note No. 126, DBR/NRCC, Ottawa, 12 pp.

"Moisture in Building Construction," Chapter 21, *1981 Fundamentals Handbook,* ASHRAE, pp. 21.1-21.22.

Morgan, C.F., 1979. "Thermal Performance of Ventilated Skin Roofs," *4th NPSC,* AS/ISES, pp. 719-722.

Niles, P.W.B., 1976. "Thermal Evaluation of a House using a Movable Insulation Heating and Cooling System," *Solar Energy* (Pergamon), vol. 18, no. 5, pp. 413-419.

Niles, P.W. and H.R. Hay, K.L. Haggard, 1976. "Nocturnal Cooling and Solar Heating with Water Ponds and Movable Insulation," *ASHRAE Transactions,* vol. 82.

Olgyay, V., 1963. "Solar Control," Chapter 7, *Design with Climate,* Princeton Univ. Press, Princeton, NJ, pp. 63-83.

Palmiter, L., 1981. "Optimum Conservation for Northwest Homes," *6th NPSC,* AS/ISES, pp. 223-227.

Quirouette, R.L. and E.C. Scheuneman, 1981. *Estimating Energy Savings from Reinsulating Houses,* Building Practice Note No. 20, DBR/NRCC, Ottawa, 38 pp.

Reagan, J.A. and D.M. Acklam, 1979. "Solar Reflectivity of Common Building Materials and its Influences on the Roof Heat Gain of Typical Southwestern U.S.A. Residences," *Energy and Buildings* (Elsevier Sequoia), vol. 2, no. 3, pp. 237-248.

Robinson, D.A., 1979. "The Art of the Possible in Home Insulation," *Solar Age,* vol. 4, no. 10, October 1979, pp. 24-32.

Rogers, T.S., 1951. *Design of Insulated Buildings for Various Climates,* Architectural Record Books/McGraw-Hill Book Co., New York, 119 pp.

Rudoy, W. and R.S. Dougal, 1979. "Effects of the Thermal Mass on Heating and Cooling in Residences," PH-79-11, No. 3, ASHRAE *Transactions,* vol. 85, part I, pp. 903-916.

Shaaban, A.K., 1981. *Double Shell Systems for Passive Cooling,* Report No. CAED-R81.1, College of Architecture and Environmental Design Research Center, Texas A & M University, College Station, TX 77843, 57 pp.

Shurcliff, W.A., 1981. *Superinsulated Houses and Double-Envelope Houses,* Brick House Publishing Co., Andover, MA, 182 pp.

Smith, I.E. and S.D. Provert, 1979. "The Effectiveness of Reflective Foil as Insulation," *Appl. Energy,* vol. 5:85.

Speltz, J.J. and G.D. Meixel, 1981. "A Computer Simulation of the Thermal Performance of Earth Covered Roofs," T.L. Holthusen, ed., *The Potential of Earth Sheltered and Underground Space* (AUA 1981 Annual Meeting), Pergamon Press, New York, pp. 91-108.

Stephenson, D.G., 1976. *Determining the Optimum Thermal Resistance for Walls and Roofs,* Building Research Note No. 105, DBR/NRCC, Ottaw, 13 pp.

Tavana, M. and R. Kammerud, H. Akbari, T. Borgers, 1980. "A Simulation Model for the Performance Analysis of Roof Pond Systems for Heating and Cooling," *1980 Annual Meeting,* AS/ISES, pp. 816-820.

"Thermal Insulation and Water Vapor Retarders," Chapter 20, *1981 Fundamentals Handbook,* ASHRAE, pp. 20.1-20.32.

Tsongas, G.A. and R.S. Carr, G.D. Katz, 1980. "Optimum Insulation R-Value in South-Facing Walls," *5th NPSC,* AS/ISES, pp. 321-325.

Tsongas, G.A. and F.G. Odell, J.C. Thompson, 1981. "A Field Study of Moisture Damage in Walls Insulated without a Vapor Barrier," *Thermal Performance of the Exterior Envelope of Buildings,* Special Publication No. 28, ASHRAE, pp. 801-815.

Tye, R.P. and A.O. Desjarlais, J.G. Bourne, S.C. Spinney, 1981. "The Effective Thermal Resistance of an Insulated Standard Stud Wall Containing Air Gaps," *Thermal Performance of the Exterior Envelope of Buildings,* Special Publication No. 28, ASHRAE, pp. 956-977.

Van Straaten, J.F., 1967. "Thermal Insulation," Chapter 9, *Thermal Performance of Buildings,* American Elsevier Publishing Co., Inc., New York, pp. 133-172.

Van Straaten, J.F., 1967. "Heat Transfer through Opaque Elements Under Periodically Fluctuating Conditions," Chapter 7, *Thermal Performance of Buildings,* American Elsevier Publishing Co., New York, pp. 81-103.

Vild, D.J., and M.I. Erickson, G.V. Parmelee, A.N. Cerny, 1955. "Periodic Heat Flow through Flat Roofs," *Heating, Piping and Air Conditioning,* vol. 27, no. 7, July 1955, pp. 145-151.

Wang, F.S., 1981. "Comparative Studies of Vapor Condensation Potentials in Wood Framed Walls," *Thermal Performance of the Exterior Envelope of Buildings,* Special Publication No. 28, ASHRAE, pp. 836-846.

Ward, D.B., 1977. *Building Design for Radiation Shielding and Thermal Efficiency,* No. TR-85, Defense Civil Preparedness Agency, 44 pp. (U.S. GPO 1978: 0-251-364).

"Water Cooled Roofs," *Architectural Forum,* vol. 84, no. 6, June 1946, pp. 165-168 (architectural application of research by Houghten *et al.*).

Weidt, J.L., and R.J. Saxler, W.J. Rossiter, Jr., 1980. *Field Investigation of the Performance of Residential Retrofit Insulation,* NBS Technical Note 1131, 67 pp. (U.S. GPO).

Wilson, A.G., 1961. *Air Leakage in Buildings,* Canadian Building Digest 23, DBR/NRCC, 4 pp.

Yellott, J.I., 1976. "Early Tests of the 'Skytherm' System," *Passive Solar Heating and Cooling* conference proceedings, Los Alamos Scientific Laboratory, pp. 54-62 (NTIS No. LA 6637-C).

Yellott, J.I. and Hay, H.R., 1969. "Thermal Analysis of a Building with Natural Air Conditioning," ASHRAE *Transactions,* vol. 75, part I, pp. 178-188.

Building Openings

A Comparison of Products for Reducing Heat Loss through Windows, U.S. DOE, Energy Efficient Windows Program, Lawrence Berkeley Laboratory, Bldg. 90, Room 3111, Univ. of California, Berkeley, CA 94720, 1982, 8 pp.

Aitken, D.W., 1981. "The Use of Air Flow Windows and Blinds for Building Thermal Control and for Solar-Assisted Heating, Cooling, and Lighting," *6th NPSC,* AS/ISES, pp. 611-615.

Barnes, P.R., 1979. "Roof Overhang Design for Solar Control," *4th NPSC,* AS/ISES, pp. 153-157.

Bowley, W.W. and P.W. McFadden, 1978. "Energy Loss by Infiltration through Windows," *Technology for Energy Conservation* conference proceedings, Albuquerque, pp. 544-555 (Information Transfer).

Brambley, M.R. and M. Godec, 1982. "Effectiveness of Low-Emissivity Films for Reducing Energy Consumption in Greenhouses," *1982 Annual Meeting,* ASES, pp. 25-30.

Croiset, M. and H. Bizebard, 1973. *Ventilation Air Inlets for Buildings,* NBS Technical Note 710-6, 62 pp (U.S. GPO No. C13.46:710-6). '

Cumali, Z.O. and R. Sullivan, 1980. "Determination of the Maximum Beneficial Window Size for Single Family Residences," *5th NPSC,* AS/ISES pp. 1007-1011.

Dietz, A.G.H., 1979. "Materials for Solar Energy Collectors," A.A.M. Sayigh, ed., *Solar Energy Applications in Buildings,* Academic Press, New York, pp. 17-39.

Ducar, G.J. and G. Engholm, 1965. "Natural Ventilation of Underground Fallout Shelters," Article No. 1916, ASHRAE *Transactions,* vol. 71, part I, pp. 88-100 (includes comparative tests of different ventilator caps).

Evans, B.H., 1976. "Design of Skylights," *Solar Radiation Considerations in Building, Planning and Design* conference proceedings, BRAB, pp. 146-155.

Evans, M., 1980. "Design of Openings for Air Movement and Ventilation," Chapter 13, *Housing, Climate and Comfort,* Halsted Press/John Wiley and Sons, Inc., New York, pp. 125-131.

Evans, M., 1980. "Design of Openings for Sun and Light," Chapter 12, *Housing, Climate and Comfort,* Halsted Press/John Wiley and Sons, Inc., New York, pp. 109-124.

Fairey, P.W. and W. Bettencourt, 1981. " 'La Sucka': A Wind-Driven Ventilation Augmentation and Control Device," *Passive Cooling '81,* AS/ISES, pp. 196-200.

Fuchs, R. and J.F. McClelland, 1979. "Passive Solar Heating of Buildings Using a Transwall Structure," *Solar Energy* (Pergamon), vol. 23, no. 2, pp. 123-128.

Gery, M.E.C. "Plans for the Solar TAP Panel," *New Roots* Reprints, P.O. Box 548, Greenfield, MA 01302, 5 pp.

Hagan, D.A. and B. Wadsworth, L. Palmiter, 1980. "Preliminary Results of Thermosiphon Air Panels Retrofit," *5th NPSC,* AS/ISES, pp. 1046-1050.

Harrison, D., 1975. "Beadwalls," *Solar Energy* (Pergamon), vol. 17, no. 5, pp. 317-319.

Hastings, S.R. and R.W. Crenshaw, 1977. *Window Design Strategies to Conserve Energy,* NBS Building Science Series 104, 209 pp. (U.S. GPO SD No. C13.29/2:104).

Hoglund, I. and B. Wanggren, 1980. *Studies of the Performance of Weatherstrips for Windows and Doors,* D4:1980, Svensk Byggtjanst, Stockholm, 59 pp.

Holleman, T.R., 1951. *Air Flow through Conventional Window Openings,* Research Report No. 33, Texas Engineering Experiment Station, Texas A & M College, College Station.

Jaros, A.L., 1976. "Selection of Glass and Solar Shading to Reduce Cooling Demand," *Solar Radiation Considerations in Building Planning and Design* conference proceedings, BRAB, pp. 156-175.

Jones, R.E., 1980. "Effects of Overhang Shading of Windows Having Arbitrary Azimuth," *Solar Energy* (Pergamon), vol. 24, no. 3, pp. 305-312.

Jones, R.E. and R.F. Yanda, 1981. "Finite Width Overhang Shading of South Windows," *1981 Annual Meeting,* AS/ISES, pp. 867-871.

Kessler, H.J. and J. Peck, 1980. "Architectural Implications of Passive and Hybrid Wall-Mounted Transparent Solar Collectors for Production Homes," *1980 Annual Meeting,* AS/ISES, pp. 734-738

Kohler, J., 1981. "TAPFLOW," *Solar Age,* vol. 6, no. 3, March 1981, pp. 48-51.

Lampert, C.M., 1982. "Advanced Heat-Mirror Films for Energy-Efficient Windows," *1982 Annual Meeting,* ASES, pp. 733-738.

Langa, F. and B. Flower, D. Sellers, 1982. "Solar Glazings: A Product Review," *Rodale's New Shelter,* vol. 3, no. 1, January 1982, pp. 58-69.

Langdon, W.K., 1980. *Movable Insulation,* Rodale Press, Emmaus, PA, 379 pp.

Lewis, D. and W. Fuller, 1979. "Restraint in Sizing Direct Gain Systems," *Solar Age,* vol. 4, no. 12, December 1979, pp. 28-32.

Mercer, R. and J. McClelland, L. Hodges, R. Szydlowski, 1981. "Recent Developments in the Transwall System," *6th NPSC,* AS/ISES, pp. 178-182.

Olgyay, A. and V. Olgyay, 1957. *Solar Control and Shading Devices,* Princeton Univ. Press, Princeton, NJ, 201 pp.

Peck, J. and H.J. Kessler, C.N. Hodges, 1979. "The ClearView Solar Collector System," *Technology for Energy Conservation* conference proceedings, Tucson, pp. 208-212 (Information Transfer).

Peck, J. and T.L. Thompson, H.J. Kessler, 1980. "Windows for Accepting or Rejecting Solar Heat Gain," *5th NPSC,* AS/ISES, pp. 985-989.

Peck, J. and T.L. Thompson, H.J. Kessler, C.N. Hodges, 1979. "Recent Design and Performance Data for the Hybrid ClearView Solar Collector System," *3rd NPSC,* AS/ISES, pp. 597-602.

Prowler, D., ed., 1979. *Solar Glazing,* Topical Conference of the Mid-Atlantic Solar Energy Association, 50 papers (available through ASES).

Rubin, M. and R. Creswick, S. Selkowitz, 1980. "Transparent Heat Mirrors for Windows: Thermal Performance," *5th NPSC,* AS/ISES, pp. 990-994.

Rubin, M. and S. Selkowitz, 1981. "Thermal Performance of Windows Having High Solar Transmittance," *6th NPSC,* AS/ISES, pp. 141-145.

Sasaki, J.R. and A.G. Wilson, 1962. *Window Air Leakage,* Canadian Building Digest 25, DBR/NRCC, 4 pp.

Schubert, R.P. and B. Kennedy, 1981. "The Testing of Full Scale Ventilator Cap Types to Determine Their Effect on Natural Ventilation," *Passive Cooling '81,* AS/ISES, pp. 201-205.

Schubert, R.P. and B. Kennedy, 1981. "The Testing of Full Scale Ventilator Cap Types to Determine their Effect on Natural Ventilation," *1981 Annual Meeting,* AS/ISES, pp. 901-905.

Shaviv, E., 1980. "Sun-Shades as a Passive Cooling Element," *5th NPSC,* AS/ISES, pp. 762-766.

Shurcliff, W.A., 1980. *Thermal Shutters and Shades,* Brick House Publishing Co., Andover, MA.

Sobin, H.J., 1981. "Window Design for Passive Ventilative Cooling: An Experimental Model-Scale Study," *Passive Cooling '81,* AS/ISES, pp. 191-195.

Symposium on Window Management as it Affects Energy Conservation in Buildings, ASHRAE (MF)DE-79-5, Detroit Meeting, June 1979 (microfiche edition).

Symposium on Energy Conservation through Improved Fenestration Treatment, ASHRAE (MF)DA-76-10, 1976 (microfiche edition).

Symposium on Energy-Related Performance of Windows, ASHRAE (MF)PH-79-6, Philadelphia Meeting, January 1979 (microfiche edition).

Theilacker, J.C. and S.A. Klein, W.A. Beckman, 1981. "Solar Radiation Utilizability for South-Facing Vertical Surfaces Shaded by an Overhang," *1981 Annual Meeting,* AS/ISES, pp. 853-856.

Thermosyphoning Air Panel Space Heating System retrofit manual, ERDA Technology Transfer Director, 2 Rockefeller Plaza, Albany, NY 12223.

Utzinger, M., 1980. "A Simple Method for Sizing Overhangs," *Solar Age,* vol. 5, no. 7, July 1980. pp. 37-39.

Van Straaten, J.F., 1967. "Heat Transfer through Non-opaque Elements," Chapter 8, *Thermal Performance of Buildings,* American Elsevier Publishing Co., Inc., New York, pp. 104-132.

Weidt, J.L. and J. Weidt, S. Selkowitz, 1981. "Field Air Leakage of Newly Installed Residential Windows," *Thermal Performance of the Exterior Envelope of Buildings,* Special Publication No. 28, ASHRAE, pp. 149-159.

Wilson, A.T. and B.L. Stickney, 1980. "An Experimental Comparison of Retrofit Vertical Air Collectors," *5th NPSC,* AS/ISES, pp. 1042-1045.

"Window Insulation," feature series, *Solar Age,* vol. 7, no. 1, January 1982, pp. 42-58.

Windows for Energy Efficient Buildings, occasional publication of the Lawrence Berkeley Laboratory, c/o Stephen Selkowitz, Room 3111, Bldg. 90, 1 Cyclotron Road, Berkeley, CA 94720, 1979 and 1980 issues.

Yellott, J.I., 1979a. "Fenestration and Heat Flow through Windows," D. Watson, ed., *Energy Conservation through Building Design,* Architectural Record Books/McGraw-Hill Book Co., New York, pp. 114-140.

Yellott, J.I., 1979b. "Glass: An Essential Component in Passive Heating Systems," *3rd NPSC,* AS/ISES, pp. 95-99.

Yellott, J.I. and W.B. Ewing, 1976. "Energy Conservation—Exterior Shading of Fenestration Techniques," *ASHRAE Journal,* vol. 18, no. 7, July 1976, pp. 23-30.

Information Sources

ASES (formerly AS/ISES). American Solar Energy Society; Membership services: ASES, 1230 Grandview Avenue, Boulder, CO 80302; Nonmember publications sales, 27 Harrison Street, Bridgeport, CT 06604

ASHRAE. American Society of Heating, Refrigerating and Air-Conditioning Engineers, Inc.; Publications Office: 1791 Tullie Circle N.E., Atlanta, GA 30329.

AS/ISES. American Section, International Solar Energy Society. Name changed in 1982 to American Solar Energy Society (ASES). See ASES.

AUA. American Underground-Space Association; 221 Church St. SE, Dept. of Civil & Mineral Eng., Univ. of Minnesota, Minneapolis, MN 55455

BRAB. Building Research Advisory Board, National Academy of Sciences; Printing and Publishing Office: 2101 Constitution Ave. N.W., Washington, DC 20418.

BRE. Building Research Establishment; Distribution Unit, Garston, Watford, Great Britain WD2 7JR

CES. Center for Environmental Studies; Princeton Univ., Princeton, NJ 08540.

CMHC. Canadian Mortgage and Housing Corp.; 1500 Merivale Road, Ottawa, Canada.

Conservation and Renewable Energy Inquiry and Referral Service (formerly National Solar Heating and Cooling Information Center), P.O. Box 8900, Silver Spring, MD 20907. Toll-free: 800/523-2929; 800/462-4983 (Pennsylvania); 800/523-4700 (Alaska and Hawaii).

DBR/NRCC. Division of Building Research, National Research Council of Canada; Ottawa, Canada K1A OR6.

HUD. Department of Housing and Urban Development; Division of Building Technology and Standards, 451 7th Street S.W., Washington, DC 20410

Information Transfer, Inc., 1160 Rockville Pike, Rockville, MD 20852.

NAHB/RF. National Association of Home Builders Research Foundation, Inc.; P.O. Box 1627, 627 Southlawn Lane, Rockville, MD 20850.

National Climatic Center. Environmental Data Service, Federal Building, Asheville, NC 28801.

NBRI. National Building Research Institute, Council for Scientific and Industrial Research (CSIR), Box 395, 001 Pretoria, South Africa.

NTIS. National Technical Information Service; 5285 Port Royal Road, Springfield, VA 22161.

SHC/BRC. Small Homes Council/Building Research Council; Univ. of Illinois at Urbana/Champaign, One East Saint Mary's Road, Champaign, IL 61820.

Svensk Byggtjanst (The Swedish Council for Building Research),Box 7853, S-103 99, Stockholm, Sweden.

Underground Space Center, 28 Appleby Hall, 128 Pleasant St. S.E., Univ. of Minnesota, Minneapolis, MN 5545.

UNIPUB. Box 433, Murray Hill Station, New York, NY 10016.

U.S. GPO. U.S. Government Printing Office; Superintendent of Documents, Washington, DC 20402

Index